Strategies for Promoting Sustainable Hospitality and Tourism Services

Maximiliano Emanuel Korstanje
University of Palermo, Argentina

Babu George
Fort Hays State University, USA

Alexandru-Mircea Nedelea
Stefan cel Mare University of Suceava, Romania

IGI Global
PUBLISHER of TIMELY KNOWLEDGE

A volume in the Advances in Hospitality, Tourism, and the Services Industry (AHTSI) Book Series

Published in the United States of America by
IGI Global
Business Science Reference (an imprint of IGI Global)
701 E. Chocolate Avenue
Hershey PA, USA 17033
Tel: 717-533-8845
Fax: 717-533-8661
E-mail: cust@igi-global.com
Web site: http://www.igi-global.com

Library of Congress Cataloging-in-Publication Data

Names: Korstanje, Maximiliano E., editor. | George, Babu P., editor. | Nedelea, Alexandru, 1971- editor.
Title: Strategies for promoting sustainable hospitality and tourism services / Maximiliano Emanuel Korstanje, Babu P. George, Alexandru-Mircea Nedelea, editors.
Description: Hershey, PA : Business Science Reference, 2020. | Includes bibliographical references and index. | Summary: "This book discusses interdisciplinary perspectives in the areas of global tourism and highlights cultural boundaries of strategic knowledge management through case studies"-- Provided by publisher.
Identifiers: LCCN 2020001187 (print) | LCCN 2020001188 (ebook) | ISBN 9781799843306 (hardcover) | ISBN 9781799852803 (paperback) | ISBN 9781799843313 (ebook)
Subjects: LCSH: Sustainable tourism--Management. | Hospitality industry--Environmental aspects. | Hospitality industry--Management.
Classification: LCC G156.5.S87 S78 2020 (print) | LCC G156.5.S87 (ebook) | DDC 338.4/791--dc23
LC record available at https://lccn.loc.gov/2020001187
LC ebook record available at https://lccn.loc.gov/2020001188

This book is published in the IGI Global book series Advances in Hospitality, Tourism, and the Services Industry (AHTSI) (ISSN: 2475-6547; eISSN: 2475-6555)

British Cataloguing in Publication Data
A Cataloguing in Publication record for this book is available from the British Library.

For electronic access to this publication, please contact: eresources@igi-global.com.

Advances in Hospitality, Tourism, and the Services Industry (AHTSI) Book Series

Maximiliano Korstanje
University of Palermo, Argentina

ISSN:2475-6547
EISSN:2475-6555

Mission

Globally, the hospitality, travel, tourism, and services industries generate a significant percentage of revenue and represent a large portion of the business world. Even in tough economic times, these industries thrive as individuals continue to spend on leisure and recreation activities as well as services.

The Advances in Hospitality, Tourism, and the Services Industry (AHTSI) book series offers diverse publications relating to the management, promotion, and profitability of the leisure, recreation, and services industries. Highlighting current research pertaining to various topics within the realm of hospitality, travel, tourism, and services management, the titles found within the AHTSI book series are pertinent to the research and professional needs of managers, business practitioners, researchers, and upper-level students studying in the field.

Coverage

- Casino Management
- Health and Wellness Tourism
- Hotel Management
- Customer Service Issues
- Food and Beverage Management
- Travel Agency Management
- Cruise Marketing and Sales
- Service Design
- Tourism and the Environment
- Service Training

IGI Global is currently accepting manuscripts for publication within this series. To submit a proposal for a volume in this series, please contact our Acquisition Editors at Acquisitions@igi-global.com or visit: http://www.igi-global.com/publish/.

The Advances in Hospitality, Tourism, and the Services Industry (AHTSI) Book Series (ISSN 2475-6547) is published by IGI Global, 701 E. Chocolate Avenue, Hershey, PA 17033-1240, USA, www.igi-global.com. This series is composed of titles available for purchase individually; each title is edited to be contextually exclusive from any other title within the series. For pricing and ordering information please visit http://www.igi-global.com/book-series/advances-hospitality-tourism-services-industry/121014. Postmaster: Send all address changes to above address.

Titles in this Series

For a list of additional titles in this series, please visit:
http://www.igi-global.com/book-series/advances-hospitality-tourism-services-industry/121014

Accelerating Knowledge Sharing, Creativity, and Innovation Through Business Tourism
Muhammad Waseem Bari (Government College University, Faisalabad, Pakistan) Sadia Shaheen (Government College University, Faisalabad, Pakistan) and Meng Fanchen (Beijing Institute of Technology, hina)
Business Science Reference • © 2020 • 300pp • H/C (ISBN: 9781799831426) • US $225.00

Contemporary Management Approaches to the Global Hospitality and Tourism Industry
Abraham Pius (Arden University, UK) Husam H. Alharahsheh (University of Wales Trinity Saint David, UK) and Adenike A. Adesanmi (University of East London, UK)
Business Science Reference • © 2020 • 290pp • H/C (ISBN: 9781799822042) • US $225.00

Global Entrepreneurial Trends in the Tourism and Hospitality Industry
Debasish Batabyal (Amity University, Kolkata, India)
Business Science Reference • © 2020 • 300pp • H/C (ISBN: 9781799826033) • US $225.00

Handbook of Research on the Impacts, Challenges, and Policy Responses to Overtourism
Cláudia Ribeiro de Almeida (University of Algarve, Portugal) Alfred Quintano (Independent Researcher, Malta) Moisés Simancas (University of La Laguna, Spain) Raquel Huete (University of Alicante, Spain) and Zélia Breda (University of Aveiro, Portugal)
Business Science Reference • © 2020 • 447pp • H/C (ISBN: 9781799822240) • US $285.00

Economic, Educational, and Touristic Development in Asia
Bryan Christiansen (Global Training Group, Ltd, UK) and Hakan Sezerel (Anadolu University, Turkey)
Business Science Reference • © 2020 • 241pp • H/C (ISBN: 9781799822394) • US $215.00

Industrial and Managerial Solutions for Tourism Enterprises
Atilla Akbaba (Izmir Katip Çelebi University, Turkey) and Volkan Altıntaş (Izmir Katip Celebi University, Turkey)
Business Science Reference • © 2020 • 336pp • H/C (ISBN: 9781799830306) • US $225.00

Handbook of Research on Smart Technology Applications in the Tourism Industry
Evrim Çeltek (Gaziosmanpasa University, Turkey)
Business Science Reference • © 2020 • 569pp • H/C (ISBN: 9781799819899) • US $295.00

Global Opportunities and Challenges for Rural and Mountain Tourism
Devkant Kala (University of Petroleum and Energy Studies, India) and Satish Chandra Bagri (Hemvati Nandan Bahuguna Garhwal University, India)
Business Science Reference • © 2020 • 354pp • H/C (ISBN: 9781799813026) • US $225.00

IGI Global
PUBLISHER of TIMELY KNOWLEDGE

701 East Chocolate Avenue, Hershey, PA 17033, USA
Tel: 717-533-8845 x100 • Fax: 717-533-8661
E-Mail: cust@igi-global.com • www.igi-global.com

Editorial Advisory Board

Table of Contents

Detailed Table of Contents

This chapter not only explores the dilemmas of tourism security and safety in these difficult days but reflects the author's own experience in the field. In 2015, when he first wrote this chapter, the number one issue in tourism security was how to handle terrorism's multiple forms. Terrorism had gone through several mutations since the 1960s. The terrorism of 2015 was very different from that of September 11th, 2001 and in reality even more different from the acts of terrorism against the tourism industry that had begun in the 1960s. Tourism continued to suffer from acts of terrorism that had occurred throughout the world. Although each act of terrorism hurt the industry, massive damage on the macroscale had not occurred in the nine years from 2001-2015, or in the years prior to 2001. In 2020, Covid-19 has destroyed thousands of lives, and in tsunami fashion, it has economically swept away everything that came before it. In Europe the virus wiped out tourism in such major tourism nations as France, Italy, Spain, and the United Kingdom.

Airbnb is viewed as an advocate of small micro medium enterprises (SMMEs) in order to boost their revenue. Furthermore, in 2016, Airbnb released a report stating that its community generated R2.4 billion in economic activity in SA, which is the estimated sum of guest spending and host income. It is important to elucidate that SA's tourism fraternity has been amongst the best performing sectors amid the challenges it is faced with. The sharing economy concept requires all tourism stakeholders to work together (public-private partnership) in order to see its fruition. The sharing economy represents the power of the collaborative consumption and become a means to create an additional value chain for the tourism industry by decreasing the barriers of entry. The Airbnb Africa Academy is a pro-poor skills development and support initiative that was piloted in South Africa, and it has increased access and the success of hosts on the Airbnb platform.

Buenos Aires city in the days post convertibility crisis appealed to the adoption of tourism as a fresh economic alternative to struggle against poverty and the financial crises. The currency exchange, as well as the abandonment of the convertibility system, favoured Argentina in many ways. For example, Buenos Aires was selected by neighbouring countries as a main tourist destination. This chapter reviews the effects of tourism in the urban landscape, such as real estate speculation, gentrification process, only to name few. The authors stress the role played by urban heritage in the process of touristification—as some voices adhere—which is conducive to the transformation of old industrial neighbourhoods.

The chapter is an attempt towards measurement of tourist satisfaction towards different selected destination attributes from responses of four hundred randomly collected sample data through a structured questionnaire. The collected data were analyzed with the help of suitable statistical tools. The result of the chapter identified different attributes of the destination and classified them into satisfying, neutral, and dissatisfying attributes. The chapter also provides an overview of overall satisfaction of the tourists as well as discovers the important factors of destination attributes. The findings conclude some recommendations and suggestions to develop strategies to increase tourist arrival to ensure repeat visit.

Over years, management and marketing have been influential and useful to organize the working condition in modern corporations in tourism fields. The classical literature suggests that motivated workers enhance their commitment to the goal of the corporation. In this process, the assistance of managers is vital to achieve the success. In doing so, the program of incentive (monetary awards) gives to workers a reason to internalize management decisions. This not only is not truth, but also, we have found in this case-study opposite evidence. The capital somehow disorganizes the human relations and harmony in service organizations. Of course outcomes of this review should not be extrapolated to other unit of analysis; this is valid only for this organization.

Culture is playing an increasingly important role in tourism, and food is one of the key elements of culture. Tourists enjoy indigenous food, particularly items of local or ethnic nature. Furthermore, knowledge of the local, regional, and national cuisine has become an interest for tourists. The concept of consuming local food or drink is considered first-hand cultural experience, and it is on top of the tourist attraction list. In India, the promotion of food as a component of its destination attractiveness is in the budding phase. The context of this contribution is to underpin such linkages between tourism and food that can add to the cultural value of the destinations. The present empirical research is aimed to explore the role of food, culture, and tourism in sustaining the tourism of India. The current study will also attempt to address the role of regional food in promoting the cultural tourism of a particular destination. Further, the study will explore how the local food helps the tourist to recall the cultural heritage of a tourist's destination.

Chapter 7
Revenue Management in Portugal: A Strategy to Compete 97
Marine Ferreira, Polytechnic Institute of Porto, Portugal
Teresa Dieguez, Polytechnic Institute of Cávado and Ave, Portugal

The world is constantly changing, and the hotel industry is one of the sectors where we can feel it the most. Nowadays, the hotel market is dynamic, diverse, difficult, and dangerous, which creates new challenges for hotel managers, particularly in terms of revenue, creation, and optimization. In order to overcome that, revenue management emerges as a crucial price management tool to face competitiveness and business growth. This study aims to analyze the practices and advantages of revenue management, as well as understand if its implementation in Portugal influenced the revenue and growth of the hotel industry. It also intends to analyze whether there is adequate revenue training or if further training should be required. A quantitative methodology was used, and 284 answers were collected. From those answers, 115 were validated, analyzed, and discussed. Conclusions were made, and finally some limitations were presented as well as suggestion for future research.

Chapter 8
The Engineering of Territorial Tourism Study Case of Algeria 122
Mohammed El Amine Abdelli, University of Zaragoza, Spain
Nadia Mansour, University of Sousse, Tunisia
Smaili Nabila, University of Tizi-Ouzou, Algeria

Tourism is a vital sector that presents one of the ways of economic diversification of a country. It has become a major economic activity that plays an important role in bringing people, countries, and regions together. In addition, it is an area promoted as a strategic sector since it provides a high percentage of foreign exchange earnings and represents a major share of GDP. Also, it provides a large number of jobs, especially for the most vulnerable groups in the society, including women, young people, and the low-skilled. Boosting this sector would result in remedying the inadequacy of the budgets allocated to the sector to allow maximum investments and to improve the business climate.

Chapter 9
A Futuristic Approach to Sustainable Tourism Development: Lessons From Kandovan Village...... 140
Hamed Rezapouraghdam, Bahçeşehir Cyprus University, Cyprus
Habib Alipour, Eastern Mediterranean University, Turkey
Arash Akhshik, Eastern Mediterranean University, Turkey

In spite of the enormous efforts and implementation of the masterplans for the development of rural areas in Iran, the majority face economic, social, and ecological challenges that jeopardize their future, particularly where tourism has been prescribed as a panacea. Among those, Kandovan, known as the only inhabited rocky village in the world that attracts half a million visitors, annually entails excessive consideration and as a case embraces imperative contributions to the sustainability-future research domain. This study aims to contribute to an enhanced understanding of the nexus between "sustainable development" and the "future" in tourist destinations. Through the analysis of the secondary data, this study takes a scenario-planning approach and discusses the current social, environmental, and economic themes: the foundation of four potential future scenarios. The underpinned framework in the context of Kandovan offers noteworthy implications both in theory and practice of tourism by proposing an area-based planning model according to the emerged alternative scenario.

Chapter 10
Monginis: A Unique Innovation in the Franchise Model 158
Christo Fernandes, Independent Researcher, India
Babu George, Fort Hays State University, USA
Ajit Mishra, Independent Researcher, India

Monginis was started in 1970 at Chembur, Mumbai, India, with just one store. The franchise model that it pioneered in the industry embraced some real-time practices of human development that were carefully captured through the recruitment strategy, job engagement, training, positive work environment, along with strategic practice of embracing local taste in product that has summoned success. How will Monginis reach out to every Indian? While the franchise model has led to expansion and growth, the company does not wish to sit idly on past laurels, but how should it move ahead? This is an updated case study that touches upon two additional dimensions: 1) getting children to bake a cake program and 2) training intervention and buddy support to the franchisee team.

Chapter 11
Modeling Destination Competitiveness: The Unfamiliar Shift for Destination Rebranding, Restructuring, and Repositioning With DMOs.. 171
Bindi Varghese, Christ University, India
Shazin Aboobacker, Christ University, India

Tourism is a tactical economic practice across the globe, but the urban and provincial transformations in the industry are strongly contemplated in the light of an unfamiliar shift in tourism business. This chapter discusses an integrated concept with a framework relating systematic approach of managing the destination and its competitiveness. An investigation on the impact on tourism and the recent narrative of national, regional, and local planning approach directs towards efficient destination management organizations (DMO) in practice for future development. This has proceeded by the formation of a competitive approach, emphasizing on the DMO roles and responsibilities helpful for a destination management during an unfamiliar business trend. Modeling destination competitiveness demands an absolute mechanism through destination rebranding, restructuring, and repositioning with DMOs for enabling competency.

Foreword

THE RELATION OF JOHNSON'S RASSELAS TO VOYAGE, EXPERIENCE, SUSTAINABILITY, AND HAPPINESS TO UNDERSTAND MOBILITY IN A COVID-19 CONTEXT

Introduction

For many countries, tourism is considered as a godsend because of its many benefits such as its ability to: protect and maintain built and natural heritage (Timothy & Boyd, 2006); improve the negative image of some destinations (Avraham, 2015); alleviate poverty (Holden, 2013); promote more sustainable forms of exploitation; create jobs (Mshenga & Owuor, 2009); etc. All these benefits of the tourism industry could be said to have the capacity to contribute to improve local residents' quality of life (Croes, Rivera, Semrad & Khalizadeh, 2017), and particularly families with low-income (Mcabe, Joldersmna & Li, 2010). That said, the positive image and assets of some destinations supported with an efficient marketing strategy have resulted in those destinations to be over visited by tourists, subsequently leading to the disruption of the life of local residents, and the destruction of their heritage, etc. (Seraphin, Sheeran & Pilato, 2018). This Janus-faced character of tourism (Sanchez & Adams, 2008), could be argued to make mankind less unhappy (but not happy), as every positive impact of the industry generates a negative one (Sanchez & Adams, 2008). In *The history of Rasselas, prince of Abyssinia*, written by Samuel Johnson in 1759, it is also argued that mankind is not meant to be happy but less unhappy.

All these points which remain open in tourism-related literature lead scholars to re-think the nature of sustainability and sustainable tourism in the days of COVID-19. The importance of a book of the calibre of *Strategies for promoting Sustainable Hospitality and Tourism Services* is threefold. At a first glimpse, the fourteen chapters forming this volume provide readers with empirical-based models to understand the state of emergency. Secondly, the chapters, which have multidisciplinary perspectives as well as a multicultural view, focus on the role of culture as a leading force that helps the promotion of more sustainable destinations. Third, the book combines a critical character of the current practices of sustainable tourism with good practices to overcome the ecological and now economic crisis just after COVID-19. The present foreword re-discusses critically the role of happiness and ethics in view of a world that has changed forever.

As the previous backdrop, this text is focusing on the nature of mankind through an analysis of *The history of Rasselas, prince of Abyssinia* (reprinted version of 1985), to investigate the phenomenon of (over) mobility which is one of the causes of overtourism (Seraphin & Ivanov, 2020), alongside its total opposite, undertourism, generated by the COVID-19 context which has put an end to the tourism industry and cognate sectors (Jamal & Budke, 2020). The concept of 'Happiness' (or 'quality of life') is going to be central in this analysis, as we are arguing that people unhappiness is leading to (over) mobility, even in lockdown period caused by the break out of COVID-19 (Jamal & Budke, 2020). This argument is by and large supported by *The history of Rasselas, prince of Abyssinia.* Indeed, Johnson's (1985) main argument is that by nature mankind is unsatisfied and as a result is constantly looking for new experiences, and it is this constant research of experiences that make mankind less unhappy. Voyage, therefore mobility is presented in *The history of Rasselas, prince of Abyssinia* as a way to reach this happiness. This argument is also supported with the fact that in France, the outbreak of COVID-19 has not stopped individuals to move in and out cities (INSEE, 2020). Instead, the outbreak of the virus has led to an exodus from major cities to countryside or coastal areas (INSEE, 2020). As of the 17th March 2020, Paris had lost 11% of its residents (INSEE, 2020).

The objective of this foreword is to establish parallels between Rasselas' experience and the phenomenon of over and under mobility. If sufficient parallels are established, we will be able to talk about the 'Rasselas Syndrome'. From a methodological point of view, the topic of over and under mobility will be discussed through a philosophical lens for many reasons: First, because tourism is an international and intercultural industry, and issues related to the industry needs to be analysed using a tool with the same characteristics (Clement & Demonque, 1995; Rosch, 2013). Second, O' Gorman (2007) adopted a philosophical approach to analyse the hospitality sector. As for *The history of Rasselas, prince of Abyssinia* (Johnson, 1985) it was chosen because: 'Rasselas is a moral tale' (Johnson, 1985: 9); travel writing as tale of a life experience have a witness value and thus is trustworthy (De Ascaniis & Grecco-Marasso, 2011); travel writings have the potential to influence travellers (Mansfield, 2017), therefore, studying *The history of Rasselas, prince of Abyssinia* could help us to understand what motivates people to travel. And finally, because Johnson (1985: 103) is calling for pragmatism when he claimed that: 'The ruins of their architecture are the schools of modern builders'.

Samuel Johnson's Rasselas: Background

The Quest for Happiness

The question of happiness is central in *The history of Rasselas, prince of Abyssinia*. What is happiness? How can happiness be attained? The overall conclusions are: First, happiness is a state of mind; second, there is no recipe for happiness (Smith, 1996); and finally, total happiness is unattainable (Zadeh & Pirnajmuddin, 2013). Experiences such as traveling make us less unhappy but not happy (Johnson, 1985). To show that total happiness is not possible, Johnson (1985: 23) argues that: 'Marriage has many pains, but celibacy has no pleasure'. More related to the topic of this foreword, is the case of Imlac, a well-travelled character in *The history of Rasselas, prince of Abyssinia,* who is not presented as a happy person but as less unhappy than other people: 'I am less unhappy than the rest, because I have a mind replete with images, which I can vary and combine at pleasure' (Johnson, 1985: 68). To reinforce the fact that happiness is not possible for mankind and that there is no recipe for happiness, the travel writting ends with chapter 49, entitled: 'A conclusion in which nothing is concluded' (Johnson,1985: 149). The main reason why mankind can't be happy (but less unhappy) is based on his inability to appreciate what he has, and also because of the nature of his desires that are constantly renewed (Clement & Demonque, 1995). Indeed, between chapter 1 and chapter 14, Rasselas is introduced to the reader as one of the princes of the Happy Valley, an allegory of the Garden of Eden (a paradise), and yet, he voiced his dissatisfaction with life.

Johnson enjoyed traveling to distant countries and this is how he developed his knowledge of the world (Zadeh & Pirnajmuddin, 2013). Johnson's Rasselas opposes two main school of thoughts regarding happiness and mobility. One side that supports the fact that we should be happy with what we have, and the other one that supports that trying to get more is all about vanity and won't lead to anything good (Aden, 1961; Zadeh & Pirnajmuddin, 2013).

Voyage and more generally experience through mobility is presented rather negatively in Johnson's Rasselas. Because of his will to travel the world to gain experience and happiness, Rasselas is depicted in the travel writing as a fallen angel (Smith, 1996). The happy valley, symbol of home is presented very positively from chapter 1 to chapter 11: 'All the diversities of the world were brought together, the blessings of nature were collected, and its evils extracted and excluded' (Johnson,1985: 40). Once Rasselas voiced his will to leave the happy Valley, it was suddenly depicted as a prison: 'Teach me the way to break my prison (…) We may escape by piercing the mountain' (Johnson, 1985: 70). As for the external world, it is depicted very negatively: 'You may soon repent your curiosity. The world, which you figure to yourself smooth and quiet as the lake in the valley, you will find a sea foaming with tempests, and boiling wit whirlpools: You will be sometimes overwhelmed by the waves of violence, and sometimes dashed against rocks of treachery' (Johnson,1985: 69). In the COVID-19 context, Paris is presented as prison for their inhabitants who are leaving the city to countryside and coastal areas where they have their second home (INSEE, 2020; Letelegramme [Online]; Liberation [Online]).

Voyage and experience through mobility is also presented as human needs to be fulfilled. To be less unhappy mankind must constantly have something to pursue and desire. Mobility (traveling) is presented as an option. Indeed, quite early in the travel writing we are reminded that men have always been keen to travel and that some places have always been very popular with travellers. The pyramids in Egypt are one of them (Johnson, 1985). The exodus of Parisians to the countryside and coastal areas because of the breakout of COVID-19 (INSEE, 2020; Letelegramme [Online]; Liberation [Online]), are reminders of one the very first forms of tourism, which were health related (Lickorish & Jenkins, 1997), and nowadays known as wellness tourism (Stainton, 2020 [Online]). Interestingly, in Rasselas, crowded areas are presented as opportunities to learn from others: 'This said Imlac to the prince is the place where travellers and merchants assemble from all the corners of the earth. You will find men of every character, and every occupation' (Johnson, 1985:73). Today, crowded areas are presented negatively, under the umbrella of overtourism (Panayiotopoulos & Pisano, 2019).

Equally important, ethical issues related to mobility were already highlighted in Rasselas. Among these, 'overtourism', assimilated to the invasion and violation of space due to mankind unwise use of technology and more broadly, unwise use of knowledge to serve self-interests instead of serving the public welfare: 'If men were all virtuous, returned the artist, I should with great alacrity teach them all to fly. But what would be the security of the good, if the bad could at pleasure invade them from the sky? (…) Even this valley, the retreat of princes, the abode of happiness, might be violated by the sudden descent of some of the naked nations that swarm on the coast if the southern sea' (Johnson,1985: 52). The world would be a better place if mankind was more virtuous (Clement & Demonque, 1995; Smith, 1996). The differences between voyage/travel and tourism are epitomising the difference between a virtuous and non-virtuous use of knowledge. In the COVID-19 context, mobility is associated with death, as Parisians, when moving to countryside, and coastal areas, they are contributing to spread the virus, what subsequently triggers distress, anger, resentment (tourismphobia) from locals living in those areas (Letelegramme [Online]; Nouvelobs [Online]).

Travelers (and Special Interest Tourism) vs Tourists (and Mass Tourism)

Campistron (2018) opposes tourism and voyage. Indeed, he argues that voyage is a philosophical quest, an adventure that benefits the traveller and the wider community. This view can be supported with the achievements and contributions of some well-known travellers. Among these are, Herodotus, known as the first travel writer. In the 5th century he toured many countries including Greece. His traveling contributing to a better understanding of the cultures of the countries surrounding ancient Greece (Edgell, 2017). As for Marco Polo, the worldliest travellers of his era, he 'gave the world a rare glimpse of the culture and geography of China (…) change the future of travel' (Edgell, 2017: 54-55). Charles Darwin is another example of traveller whom research benefitted the general public. He visited the Galapagos Islands where he studied birds and in particular Finches, which led him to his theory of evolution and the book entitled 'On the origin of species' (Edgell, 2017). Equally important, Campistron (2018) argued that the tourism industry can't contribute to the above because it is merely an entertainment industry.

As opposed to voyage, tourism starts when there is no more adventure. Tourism is also a secluded, self-centred activity and as such self-contained individuals. The objective of tourists is not to discover the world but to ignore it (Campistron, 2018). Equally important, people involved in tourism are looking for their own benefits such as maximising their free time (Campistron, 2018), or their safety. In the COVID-19 context, the second major reason for this exodus (triggered by the pandemic) of Parisians, is due to the fact that many Parisians have a second home (Letelegramme [Online]). This form of tourism (second home tourism), which represents a significant part of domestic tourism (Hiltunen, 2008), since Jaakson (1986), is associated with terms such as surety; time and distance.

Special Interest Tourism (SIT) occurs 'when the traveller's motivation and decision-making are primarily determined by a particular special interest with a focus either on activity/ies and/or destinations and settings' (Hall & Weiler, 1992 cited in Trauer, 2006: 186). SIT appeared to accommodate the varied and specialised needs and tastes of tourists and is to be opposed to mass consumption and non-commercialised individual travel (Trauer, 2006). This form of tourism emerged in the 1980s (Trauer, 2006) and was stimulated by a need for cultural and environmental holidays (Yeoman, Robertson, Ali-Knight, Drummond & McMahon-Beattie, 2009). SIT contributes to enhance the image of a destination; to enrich tourists' experiences and is profitable to a wider range of providers (Jin & Sparks, 2017). Among the other benefits of this form of tourism, can be mentioned: minimise negative economic, environmental and social impacts; generates greater economic benefits for local people and enhances the well-being of host communities; improves working conditions and access to the industry; involves local people in decisions that affect their lives and life chances; make positive contributions to the conservation of natural and cultural heritage, embracing diversity; provides more enjoyable experiences for tourism though move meaningful connections with local people, and a greater understanding of local cultural, social and environmental issues; provides access for physically challenged people; and is culturally sensitive, encourages respect between tourists and hosts and build local pride and confidence (Edgell, 2017). Second home tourism, which is a SIT (Hiltunen, 2008; Jaakson, 1986) in the pandemic context of COVID-19 is associated with death and tourismphobia.

Based on our findings so far, it could be argued that mass tourism generates tourists, whereas Special Interest Tourism (SIT) is more likely to generate (unwanted) travellers/voyagers. This is to some extent supported by Edgell (2017) who argued that Herodotus (at the origin of cultural tourism); Marco Polo (frontrunner for cultural and geotourism) and Charles Darwin (for ecotourism) were responsible travellers. The Parisians in the current pandemic context are considered as irresponsible travellers (Letelegramme [Online]; Liberation [Online]).

Happiness and the Mobility of Travellers/Voyagers and Tourists

Rasselas decided to leave the Happy Valley (home) to travel the world because he was unhappy and was looking for happiness (Aden, 1961; Leyburn, 1955; Smith, 1996; Zadeh & Pirnajmuddin, 2013). Travelling therefore plays a role in mankind quest for happiness. On that basis, SIT can make us less unhappy as it serves the general public rather than an individual interest. However, in crisis context, SIT serves individual interest, and sometimes at the expense of the general public. The fact that some destinations are overcrowded resulting in some people being unhappy (INSEE, 2020; Milano, 2017; Seraphin et al, 2018) could be explained by a non-virtuous management of destination and also non-virtuous use of knowledge by mankind. This trend is unlikely to change. Since the 1980s, air traffic has doubled every 15 years—a trend that is expected to continue. In 2016, nearly 4 billion people travelled by plane, a number expected to reach 7.2 billion by 2035.4 (The Travel and Tourism Competitiveness Report, 2017). Even in an under tourism context, such as COVID-19 (Jamal & Budke, 2020), destination management is pointed out (Letelegramme [Online]; Liberation [Online]); Nouvelobs [Online]).

Descartes (1637), Epicure (IV B.C) and Shopenhauer (1818) all cited in Clement and Demonque (1995), argued that desire is a natural and constantly renewed feeling, and that not all our desires can be met, hence the reason that mankind can only be less unhappy (and not happy). Equally important, they argued that mankind only desires what it cognitively acknowledges as achievable and affordable. Nowadays, traveling is more and more acknowledged as an affordable desire due to its low cost (Seraphin & Ivanov, 2020). As for the constant desire to travel it is exacerbated by effective online or offline marketing strategies (Morgan, Pritchard, & Pride, 2013).

In our day and age, this desire to travel is easily exacerbated thanks to the growing trend in developed countries to change life and start a new one for self-fulfilment (Chartier, 2017). For instance, in France, 70% of the citizens want to start a new life. 15% actually jump the hoop (Chartier, 2017). Many factors have triggered this trend: (a) As life expectation is increasing, people have an opportunity to have more than just one life (b) working environments are more and more stressful and leading to burn-out (c) people are getting more and more individualistic and self-centred (d) A need for self-fulfilment (e) a will to make our life meaningful (f) and a return to nature, that has been amplified by the breakout of COVID-19 (Chartier, 2017; INSEE, 2020).

Based on the above, the phenomenon of (over) mobility could be summarised by the formula below:

$$\left(Over\right) mobility\ in\ a\ non\text{-}crisis\ context = \left(Cognitive\ desire \times low\ cost\ of\ traveling\right) + \left(Cognitive\ desire \times Marketing\ strategies\right) + \left(\frac{Cognitive\ desire \times Growing\ trend\ to\ change\ life}{self\text{-}fulfilment}\right)$$

$$\left(Over\right) mobility\ in\ a\ crisis\ context = \left(Cognitive\ desire \times Safety\ desire\right) + \left(Cognitive\ desire \times Second\ home\right) + \left(\frac{Cognitive\ desire \times Growing\ trend\ to\ change\ life}{self\text{-}fulfilment}\right).$$

Conclusion

The history of Rasselas, prince of Abyssinia reveals that travelling plays a role in mankind quest for happiness, but can only makes mankind less unhappy because happiness is unattainable. SIT therefore voyage in a non-crisis context, has revealed to serves the general public interests rather than individual one; whereas in crisis context it is the other way round.

Additionally, the study of *The history of Rasselas, prince of Abyssinia* alongside the study of recent trends in traveling and mankind behaviour have revealed that (Over) mobility is a consequence of an exacerbation of mankind cognitive desire to travel by more and more effective marketing strategies; the low of traveling; the need for mankind self-fulfilment.; the survival instinct in crisis context. Many parallels have been established between Rasselas (enlightenment) and the present day (contemporary world) when it comes to traveling behaviour of mankind. The 'Rasselas syndrome' could be defined as a permanent and compulsive desire of mankind to travel in order to reach an unattainable happiness. From a practical point of view, the results of this research provide evidence that a good understanding of mankind nature is important in order to tackle issues related to mobility. From a conceptual point of view, this foreword has conceptualised the issue of (over) mobility (even in lockdown context).

Hughes Seraphin
University of Winchester, UK

REFERENCES

Aden, J. M. (1961). Rasselas and the vanity of human wishes. *Criticism, 3*(4), 295–303.

Amey, K. (2015). *10 places you can forget about visiting*. Available at: http://www.dailymail.co.uk/travel/travel_news/article-3147092/A-royal-reprieve-London-Italy-s-secret-religious-archive-island-inhabited-tribe-kill-sight-10-places-forget-visiting.html

Avraham, E. (2015). Destination image repair during crisis: Attracting tourism during the Arab Spring. *Tourism Management, 47*, 224–232. doi:10.1016/j.tourman.2014.10.003

Buckley, J. (2017). *Florence launches campaign telling tourists how to behave*. Available at: http://www.cntraveler.com/story/florence-launches-campagin-tellingtourists-how-to-behave

Campistron, M. (2018). *Tourisme: On part pour oublier le monde plutot que pour le decouvrir*. Available at: https://www.nouvelobs.com/societe/20180718.OBS9887/tourisme-on-part-pour-oublier-le-monde-plutot-que-pour-le-decouvrir.html

Chartier, C. (2017). Changer de vie? Chiche. *L'Express, 3441*, 24–44.

Clement, E., & Demonque, C. (1995). *Philosophie*. Hatier.

Coldwell, W. (2017). *First Venice and Barcelona: Now anti-tourism marches spread across Europe*. Available from: http://www.theguardian.com

Croes, R., Rivera, M. A., Semrad, K., & Khalizadeh, J. (2017). *Happiness and tourism: Evidence from Aruba*. Institute for Tourism Studies. doi:10.13140/RG.2.2.29257.85602

De Ascaniis, S., & Grecco-Morasso, S. (2011). When tourist give their reasons on the web: The argumentative significance of tourism related UGC. *Information and Communication Technology in Tourism*, 125-137.

Edgell, D. L. (2017). Sustainable tourism development: Yesterday, today and tomorrow. *Journal of Hospitality & Management, 15*(1), 45–59.

Hiltunen, M. J. (2008). Environmental impacts of rural second home tourism. Case Lake District in Finland. *Scandinavian Journal of Hospitality and Tourism, 7*(3), 243–265. doi:10.1080/15022250701312335

Holden, A. (2013). *Tourism, poverty and development*. Routledge. doi:10.4324/9780203861547

INSEE. (2020). *Population présente sur le territoire avant et après le début du confinement – Premiers résultats*. Available at: https://www.insee.fr/fr/information/4477356

Jaakson, R. (1986). Second-home domestic tourism. *Annals of Tourism Research, 13*(3), 367–391. doi:10.1016/0160-7383(86)90026-5

Jamal, T., & Budke, C. (2020). Tourism in a world with pandemics: Local-global responsibility and action. *Journal of Tourism Futures*. doi: org/ doi:10.1108/JTF-02-2020-0014

Jin, X., & Sparks, B. (2017). Barriers to offering special interest tour products to the Chinese outbound group market. *Tourism Management, 59*, 205–215. doi:10.1016/j.tourman.2016.08.007

Johnson, S. (1985). *The history of Rasselas, Prince of Abissinia*. Penguin Classics.

Le Nouvel Observateur (L'Obs). (2020). *Les Parisiens se réfugient à Belle-Ile-en-Mer: Personne n'a pensé qu'ils seraient aussi cons*. Available at: https://www.nouvelobs.com/coronavirus-de wuhan/20200318.OBS26261/les-parisiens-se-refugient-a-belle-ile-en-mer-personne-n-a-pense-qu-ils-seraient-aussi-cons.html?utm_term=Autofeed&utm_medium=Social&utm_source=Twitter#Echobox=1584558864

Le Télégramme. (2020). *Coronavirus: tension sur Belle-Ile-en-Mer avec les résidents secondaires*. Available at: https://www.letelegramme.fr/dossiers/confrontes-au-coronavirus-des-bretons-temoignent/coronavirus-tension-sur-belle-ile-en-mer-avec-les-residents-secondaires-17-03-2020-12528216.php?utm_term=Autofeed&utm_medium=Social&utm_source=Twitter#Echobox=1584472161

Leadbeater, C. (2017). *Anti-tourism protesters in Barcelona slash tyres on sightseeing buses and rental bikes*. Available at: http://www.telegraph.co.uk

Leyburn, E. D. (1955). No romantic absurdities or incredible fictions: The relation of Johnson's Rasselas to Lobo's voyage to Abyssinia. *Modern Language Association, 70*(5), 1059–1067. doi:10.2307/459886

Libération. (2020). *En quittant Paris, on accélère la propagation du coronavirus*. Available at:https://www.liberation.fr/france/2020/03/16/en-quittant-paris-on-accelere-la-propagation-du-coronavirus_1781946

Lickorish, L. J., & Jenkins, C. L. (1997). *An introduction to tourism*. Butterworth Heinemann.

Mansfield, C. (2017). Travel writing in place branding. A case study on Nantes. *Journal of Tourism, Heritage & Services Marketing*, *3*(2), 1–7.

McCabe, S., Joldersma, T., & Li, C. (2010). Understanding the benefits of social tourism: Linking participation to subjective well-being and quality of life. *International Journal of Tourism Research*, *12*(6), 761–773. doi:10.1002/jtr.791

Milano, C. (2017). *Overtourism y Turismofobia. Tendencias globales y contextos locales*. Ostelea School of Tourism & Hospitality.

Morgan, N., Pritchard, A., & Pride, R. (2013). *Destination brands. Managing place reputation* (3rd ed.). Butterworth-Heinemann.

Mshenga, M., & Owuor, G. (2009). Opportunities for micro and small scale business in the tourism sector: The case of the Kenya coast. *Journal of Business and Management*, *2*(2), 52–68.

O'Gorman, K. D. (2007). The hospitality phenomenon: Philosophical enlightenment? *International Journal of Culture, Tourism and Hospitality Research*, *1*(3), 189–202. doi:10.1108/17506180710817729

Panayiotopoulos, A., & Pisano, C. (2019). Overtourism dystopias and socialist utopias: Towards an urban armature for Dubrovnik. *Tourism Planning & Development*, *16*(4), 393–410. Advance online publication. doi:10.1080/21568316.2019.1569123

Paris, N. (2017). *Tourists have turned Oxford into 'hell' local claim*. Available at: http://www.telegraph.co.uk/travel/news/oxford-tourist-hell-overcrowding-residents-locals-complain/

Park, H. (2014). *Heritage tourism*. Routledge.

Rosch, F. (2013). Hooray! Hooray! The end of the world has been postponed! Politics of peace in the adventures of Tintin? *Politics*, *34*(3), 225–236. doi:10.1111/1467-9256.12024

Routledge, P. (2001). 'Selling the rain', resisting the sale: Resistant identities and the conflict over tourism in Goa! *Social & Cultural Geography*, *2*(2), 221–240. doi:10.1080/14649360120047823

Sanchez, P. M., & Adams, K. M. (2008). The Janus-faced character of tourism in Cuba. *Annals of Tourism Research*, *35*(1), 27–46. doi:10.1016/j.annals.2007.06.004

Seraphin, H., & Ivanov, S. (2020). Overtourism: A revenue management perspective. *Journal of Revenue and Pricing Management*. Advance online publication. doi:10.105741272-020-00241-7

Séraphin, H., Sheeran, P., & Pilato, M. (2018). Over-tourism and the fall of Venice as a destination. *Journal of Destination Marketing & Management*, *9*, 374–376. doi:10.1016/j.jdmm.2018.01.011

Smith, D. H. (1996). Patterns in Samuel Johnson's Rasselas. *Studies in English Literature*, *36*(3), 623–639. doi:10.2307/450802

Stainton, H. (2020). *Types of tourism: A glossary.* Available at: https://tourismteacher.com/types-of-tourism-glossary/

Tapper, J. (2017). *As touting for punt trips becomes a crime, is tourism overwhelming Britain's cities.* Available at: http://www.theguardian.com

Timothy, D. J., & Wall, G. (1997). Selling to tourists: Indonesian street vendors. *Annals of Tourism Research*, *24*(2), 322–340. doi:10.1016/S0160-7383(97)80004-7

Trauer, B. (2006). Conceptualizing special interest tourism—frameworks for analysis. *Tourism Management*, *2*(7), 183–200. doi:10.1016/j.tourman.2004.10.004

Yeoman, I., Robertson, M., Ali-Knight, J., Drummond, S., & McMahon-Beattie, U. (Eds.). (2004). *Festival and events management : An international arts and culture perspective*. Elsevier Butterworth-Heinemann. doi:10.1016/B978-0-7506-5872-0.50010-7

Zadeh, M. R. M., & Pirnajmuddin, H. (2013). The orbit of pursuit in Johnson's Rasselas. *Journal of Language Teaching and Research*, *4*(2), 401–405. doi:10.4304/jltr.4.2.401-405

Preface

INTRODUCTION

Thinking the problem of sustainability in times of COVID-19 is not an easy task. Coronavirus disease not only tittered the tourism industry on the brink of collapse but also led mankind to re-think critically its place in this world. The tourism industry, doubtless, will face major risks and challenges in the years to come, in a post-COVID-19 climate. As Tim Ingold puts it, one of the aspects of the current contaminations as well as the impossibilities of nation-states to reverse the climate change is not given by the lack of interests in sustainable issues but also in the cultural constraints left by the capitalist system. He coins the term *dwelling and relational perspectives* to denote the territorial and cultural organizations of hunter-gatherers and modern citizens. While the former signals to relational bondage with nature, laying the foundations towards communion with the earth, the latter refers to an economy of stocks and markets which subordinated the nature to the human rationality (Ingold 2000; 2005). In this respect, as Ingold adheres, it is impossible to reverse the impacts of consumer´s culture on the planet without changing the mainstream cultural values of capitalism. Though much attention was paradoxically paid to the climate change and the effects of greenhouse gas emissions to the global economy, less (efficient) efforts have been coordinated by the side of authorities to reduce the gas emissions (Korstanje & George, 2012). In a seminal book, Chaturvedi & Boyle (2015) bemoan that the mass-media elaborates, packages and disseminates terror-based messages in order to create a climate of fear as never before. The spectatorship feels to be captivated by what experts dubbed as “the spectacle of terror” revolving around climate change. This does not mean that climate change does not exist, nor its impacts are minimized, but the spectacularisation of climate change ushers lay-people in a paralysis where there is no action. People believe that no matter they can do, climate change cannot be reversed. When this happens, a paradoxical situation emerges. Senior sociologist Anthony Giddens called this *the Paradox of Giddens.* He argues convincingly that whenever members of society do not see the impacts of their decisions they move as nothing happens. People often make evaluations according to the game of costs and gains. When no costs are perceived, there is no intention to change attitudes or established customs. It is unfortunate that nations systematically failed to cooperate towards a global agreement for the reduction of gas emission nor the adoption of shared politics and policies.

In the tourism fields, the problem of pollution and climate change has historically occupied a central position. Several studies have focused on the economic and social impacts of climate change to the tourism industry (Nicholls, 2006; Becken & Hay, 2007; Weaver 2011; Scott, Hall, Gossling 2012). These studies originally initiated by scholars, who live in countries dependent on tourism and ecology as Australia and New Zeeland, focused on the concerns to adapt to the tourism industry to the negative effects of climate change. In their book *Tourism and Climate change: impacts, adaptation and mitigation,* Scott, Hall & Gossling (2012) alert that scholars should distinguish between sustainable tourism and sustainable tourism development. Climate change poses a serious risk not only for the tourism industry but for the entire planet. Western thinkers do not understand the real essence of climate change as well as greenhouse gases. Technology would play a leading role in monitoring the steps and policies to mitigate the effects of climate change. Authors start from the premise that the impossibilities of nations to reverse climate change puts experts in the urgency to elaborate mitigation-related programs oriented to protect the tourism industry. Since its inception, the tourism industry was culturally programmed to expand to the four corners of the world. The tourism industry not only paved the ways for the rise of globalization but also cemented the technological breakthroughs that affirmed the authority of capitalism as a hegemonic cultural and economic project. Today´s tourism is affected by climate change but at the same time, it is unable to slow its unstoppable advance. This begs the question to what extent may we change our behavior without changing our cultural values?

One of the pioneering voices in addressing a question like this was undoubtedly Jost Krippendorf. For him, tourism seems not to be good or bad. All depends on the steps and the ways sustainability is instrumentally used. Krippendorf toys with the belief that those critical scholars (coming from Marxist traditions) who allude to tourism as an ossifying mechanism misunderstand its nature. Even industrialism, the precondition towards modern capitalism, brought some benefits such as poverty alleviation, the application of technologies to mobilities without mentioning the adoption of progress as the main guideline to follow. However, these benefits, which originally disposed of our world a safer place, engendered a biased notion revolving around the notion of progress. Needless to say, ethics plays a leading role in the configuration of sustainability in the tourist system. Modernity is mainly based on the belief that local resources are unlimited and growth is inevitably entwined to consumption. More consumption more progress, exegetes of modernity recognize. Of course, as Krippendorf acknowledges, further economic progress equates with fairer wealth distributions, but this does not always occur. To accelerate the reversibility of climate change, this ideal should be changed. This suggests that our current (unsustainable) practices are legitimated in our cultural values.

As the previous backdrop, the present book discusses critically the role played of culture to a reconfiguration of new emerging sustainable practices and politics. This seems to be a point which remains unexplored in the tourism fields. Culture is often venerated or applauded as the remedy to all maladies, but in fact, under some conditions, the obsession for cultural consumption leads gradually to a climate of alienation and commoditization (Korstanje 2012; Staiff, Bushell & Watson, 2013). In this respect, Comaroff & Comaroff have studied how heritage and tourism, far from resolving long-dormant inter-ethnic disputes, aggravate old rivalries. Although tourism originally engages with substantial advances to reduce poverty in ethnicities which were historically relegated from the economic growth, no less true is that nation-states impose higher taxes over these communities, opening the doors to the rise of conflict and violence (Comaroff & Comaroff 2009). Neither the top-down nor the bottom-up planning suffices to explain why sustainability fails to avoid conflict. Here it is precisely the needs of editing a book about sustainability in a moment of history when tourism goes through an inevitable crisis. Is the culture part of the problem or the solution?

The case of creative tourism would be a self-explanatory description of how the paradox of culture works. Greg Richards (2011) sheds light on the evolution—if not expansion—of creative tourism as a new cultural trend oriented to placate mass-consumption. Creative tourism should be understood as a new form of tourism—probably born in the turn of the century—that gives guests the possibility to elaborate a craft or another product with hosts. This type of tourism suggests active participation by the side of tourists. As Richards observed, over the recent years creative tourism has led scholars to a confusing position anyway. At a closer look, creative tourism has been promoted as an authentic form of consumption while initially, the activity involved a few persons. It is important to mention here that it provided some aboriginal tribes fresh resources to struggle against poverty. But in the time, the activity tended to be commoditized and more people demanded to travel to these sites. The paradox lies in the fact the sense of attractiveness which often associates to tourism ushers the destination –unless controlled or regulated- to the over-crowding. This moot-point was recently studied by scholars who are captivated by the over-tourism or tourist-phobia in different cities of Europe and Asia (Seraphin, Sheeran & Pilato 2018; Koens, Postma & Papp 2018; Routledge 2010).

It is safe to say that even though each chapter can be separately read, each one deals with the problem of sustainability in contexts highly-dependent on tourism and hospitality industries. To some extent, it is important not to lose the sight of the fact that capitalism and tourism often commoditize culture as a unilaterally-packaged product offered to first-world tourists. Rich tourists are psychologically motivated to visit exotic places and remote destinations to feel a type of archetypical return to nature. This essentialized vision of the "*Otherness*" echoes the old visions of the noble savage that alimented the curiosity of European readerships during the colonial rule (Korstanje 2012). Having said this, the host-guest encounter seems to be fraught of political discordances and tensions. When locals -geographically in the global South- internalize the stereotypes, narratives and discourse created in the Global North, their autonomy to introduce radical shifts is undermined. The center-periphery dependency is centered on a symbolic tie which continues the legacy of colonialism. This means that ethics plays a crucial role in the consolidation of more sustainable tourism. This asks for the following point: to what extent may we make a more sustainable culture when basically we commoditize climate change as a spectacle (of visual consumption)?

The point is brilliantly addressed by Dr Peter Tarlow in the first chapter which is entirely limited to the discussion of ethics, terrorism and marketing. He eloquently observes how the war against terrorism, which inspired the American policies just 9/11, is continued in a war against COVID-19. Today, communities have an opportunity to make a more sustainable world in the post-COVID-19 context. We passed from the struggle of locals against *over-tourism* to a new (radicalized) world where mobilities have a marginal position. Paradoxically, tourism accelerated the dissemination of the virus and is this virus nowadays one of its major threats.

The second chapter, which is authored by Unathi Sonwabile Henama & Lwazi Apleni, describes the advantages and disadvantages of *Airbnb* to promote sustainable destinations. With a focus in South Africa, the chapter proffers a coherent diagnosis that explains how this digital platform champions for the promotion of small businesses. They theorize on the benefits of a shared economy as a new surfacing concept that helps tourism to deal with complex scenarios. Shared economy, as well as Airbnb, notably contributes to tourism supply change management and local economy. Furthermore digital technologies—like Airbnb—allow the disposition of post-disaster accommodation in contexts of emergencies.

The third chapter (written by Maximiliano Korstanje) analyzes the problem of heritage consumption in the post-2001 context in Buenos Aires Argentina. The end of convertibility, which accelerated the renounce of former president Fernando de La Rua ignited a new climate where instability and violence emerged. In consequence, Argentina faced one of the most severe crises in its history. Tourism and heritage consumption not only accelerated the recovery of Buenos Aires as an international destination but rechanneled new geographical transformation in the city, in what geographers know as *gentrifica-*

tion process. Tourism boosted the economy while engendering serious material asymmetries between old and new neighbors.

The fourth chapter, in charge of Lakhvinider Singh, continues the discussion left by Chakraboty, revolving around the opportunities of heritage tourism to boost India. In this instance, the involving author takes the study case of Jaipur (India) describing in detail the interplay between pre and post-purchase behavior in tourists who consume heritage. As Professor Singh brilliantly notes, there is a gap between consumption and expectations which the present study intends to unravel.

The fifth chapter (M. Korstanje) explores the problem of monetary rewards in a rent-a-car company in Argentina. The gathered information reveals two important assumptions. On one hand, monetary rewards increase exponentially the levels of sales in the organization but on another, it creates the conditions for competence, greed and conflict. Top-ranked managers enter in serious conflicts with desk staff because of the implementation of monetary rewards. The final remarks suggest that monetary rewards should be at least complemented with non-monetary rewarding-programs.

In the sixth chapter, Priya Singh & Ashaq Husain Najar offer an interesting investigation on the contribution of culture and above all food tourism to the authentic representation of Indian culture. Based on the example of food tourism, authors argue that food is a quintessential feature of culture which integrates countless elements such as behavior, tradition, dress, cuisine and moral values into an all-encompassing landscape. It is vital to note here that the chapter centers on branding destination theory and its ramification to sustainability.

The seventh chapter, in the hands of authoritative authors as Marine Ferreira and Teresa Dieguez, reviews the Revenue Management model and its potential applications to Portugal. Discovered originally in the 60s decade in the US, as authors agree, RM poses as a valid and efficient instrument to promote more sustainable destinations. RM not only optimized the gains and capital of investors but also it facilitates rapid sustainable growth. Though poorly developed in Portugal, RM still offers a fertile ground for the incorporation in the tourism and hospitality industries.

Mohammed El Amine Adbelli, Nadia Mansour and Smaili Nabila, in the eighth chapter, interrogate furtherly on the economic problems left by colonialism. With a basis in Algeria, they hold the thesis that tourism not only helps in alleviating poverty creating jobs for under-skilled people but also attracts foreign investors to diversify the economy. As authors alert, nation-states should not be dependent of the tourism industry. When this happens any crisis may wreak havoc in the local economy. Rather, tourism revitalizes the necessary tendons to decentralize the economy towards more sustainable paths.

The ninth chapter, Hamed Rezapouraghdam, Habib Alipour and Arash Akhshik allude to the decentralization of the economy as a solution to some of the problems we often face in under-developed economies. The dependency of the tourism sector is, in fact, one of the vicious of underdeveloped nations. These researchers call attention to the urgency to adopt futuristic approaches in rural communities. To put the same in other terms the challenges of policymakers to design more sustainable destinations rest not only in the fact that the industry dynamism is unpredictable, chaotic and very complex but today the industry is facing major socio-economic, cultural and ecological challenges which should be addressed and resolved. The futurist approach is, in this part, an essential toolkit of tourism development and of course needs further discussion, as authors conclude.

In the tenth chapter, Christo Fernandez, Babu P. George and Ajit Mishra explore the railway bakery chain known as Monginis Food ltd situated CST railways station (Mumbai) which receive monthly more than 4 millions of peoples. Its growth was undoubtedly accompanied by the expansion of a franchisee system that led to becoming a national brand. Today the brand has a strong presence in other countries as Egypt and is aimed at the UK prestigious market. Authors enumerate the policies and steps that incorporated Monginis to be a successful and sustainable option for la India. Among the challenges posed of this company, authors mention the rise of new competitors which place it in jeopardy, the proliferation of online options, the misuse of branding name, and the escalating of prices or price wars without mentioning some hygiene issues which discredits the firm in the recent decades.

The eleventh chapter, which is presented by Bindi Varghese and Shazin Aboobacker, reminds the importance to analyze competition as one of the touchstones of the tourism system. The rise of competition among destinations, as well as the needs of adopting sustainable practices, leads towards a difficult position for policy-makers. Authors explain that the system requires a conflictive dialogue of multiple stakeholders each one negotiating per their own interests. DMO model facilitates proper coordination towards more efficient ways of tourism exploitation. The chapter holds that the different sectors forming tourism and hospitality ushers these industries to an inevitable fragmentation. To emerge as a successful option, tourism should be supported by a strong interplay of laws, policies, and decision-making processes articulated towards an all-encompassing model.

Ultimately, the last chapter (B. Nair & *Sathiyabamavathy)* explores the conditions for high competitive environments for promoting pro-heritage consumption. Authors argue convincingly that local heritage, when dully regulated and controlled, may help to boost local economies as well as collective self-esteem in the community, preserving cultural values otherwise would be forgotten or placed under the dust of oblivion. Authors magisterially review several works and documents in sustainable tourism which are helpful for policymakers and professionals. The chapter gives a clear distinction between amenities expectation and amenities perception.

Last but not least, even if the future of tourism remains uncertain in the years to come, it is important to remind its essential nature. Tourism is more than a resilient activity; rather it is part of the resiliency of society. The lessons of 9/11 teach us that sooner or later tourism re-appears –probably not in the same form- to help the society to understand and deal with the trauma. Like a dream that revitalizes our psychological frustrations, tourism re-organizes the social bondage in order for the society keeps united.

Maximiliano E. Korstanje
University of Palermo, Argentina

Babu George
Fort Hays State University, USA

Alexandru Mircea Nedelea
Stefan cel Mare University of Suceava, Romania

REFERENCES

Becken, S., & Hay, J. E. (2007). *Tourism and climate change: Risks and opportunities* (Vol. 1). Multilingual Matters.

Chaturvedi, S., & Doyle, T. (2015). *Climate terror: A critical geopolitics of climate change*. Palgrave Macmillan. doi:10.1057/9781137318954

Comaroff, J. L., & Comaroff, J. (2009). *Ethnicity, Inc.* University of Chicago Press. doi:10.7208/chicago/9780226114736.001.0001

Giddens, A. (2011). The Politics of Climate change. Polity Press.

Ingold, T. (2000). *The perception of the environment: essays on livelihood, dwelling and skill*. Psychology Press.

Ingold, T. (2005). Epilogue: Towards a politics of dwelling. *Conservation & Society*, 501–508.

Koens, K., Postma, A., & Papp, B. (2018). Is overtourism overused? Understanding the impact of tourism in a city context. *Sustainability*, *10*(12), 4384. doi:10.3390u10124384

Korstanje, M. (2012). Reconsidering cultural tourism: An anthropologist's perspective. *Journal of Heritage Tourism*, *7*(2), 179–184. doi:10.1080/1743873X.2011.639883

Korstanje, M. E., & George, B. (2012). Global warming and tourism: Chronicles of apocalypse? *Worldwide Hospitality and Tourism Themes*, *4*(4), 332–355. doi:10.1108/17554211211255684

Nicholls, S. (2006). Climate change, tourism and outdoor recreation in Europe. *Managing Leisure*, *11*(3), 151–163. doi:10.1080/13606710600715226

Richards, G. (2011). Creativity and tourism: The state of the art. *Annals of Tourism Research*, *38*(4), 1225–1253. doi:10.1016/j.annals.2011.07.008

Routledge, P. (2001). 'Selling the rain', resisting the sale: Resistant identities and the conflict over tourism in Goa. *Social & Cultural Geography*, *2*(2), 221–240. doi:10.1080/14649360120047823

Scott, D., Hall, C. M., & Gossling, S. (2012). *Tourism and climate change: Impacts, adaptation and mitigation*. Routledge. doi:10.4324/9780203127490

Seraphin, H., Sheeran, P., & Pilato, M. (2018). Over-tourism and the fall of Venice as a destination. *Journal of Destination Marketing & Management*, *9*, 374–376. doi:10.1016/j.jdmm.2018.01.011

Staiff, R., Bushell, R., & Watson, S. (Eds.). (2013). *Heritage and tourism: Place, encounter, engagement*. Routledge. doi:10.4324/9780203074619

Weaver, D. (2011). Can sustainable tourism survive climate change? *Journal of Sustainable Tourism*, *19*(1), 5–15. doi:10.1080/09669582.2010.536242

Acknowledgment

We, the editors want to recognize the support of our families who donated part of the time spend in this collection as well as the kindness and patience of a leading publisher like IGI global and its supportive staff in answering all our questions and doubts. Hope this collection is welcomed by the Academia in view of the difficult days to come, and the efforts and time dedicated to gather all these high-quality chapters. Lastly, we wish to thank reviewers and authors by their time and hope counting with them in a near project soon.

Chapter 1
Tourism, Terrorism, Morality, and Marketing:
A Study of the Role of Reciprocity in Tourism Marketing

Peter E. Tarlow
Texas A&M University, USA

ABSTRACT

This chapter not only explores the dilemmas of tourism security and safety in these difficult days but reflects the author's own experience in the field. In 2015, when he first wrote this chapter, the number one issue in tourism security was how to handle terrorism's multiple forms. Terrorism had gone through several mutations since the 1960s. The terrorism of 2015 was very different from that of September 11th, 2001 and in reality even more different from the acts of terrorism against the tourism industry that had begun in the 1960s. Tourism continued to suffer from acts of terrorism that had occurred throughout the world. Although each act of terrorism hurt the industry, massive damage on the macroscale had not occurred in the nine years from 2001-2015, or in the years prior to 2001. In 2020, Covid-19 has destroyed thousands of lives, and in tsunami fashion, it has economically swept away everything that came before it. In Europe the virus wiped out tourism in such major tourism nations as France, Italy, Spain, and the United Kingdom.

INTRODUCTION

In 2015 when I first wrote this chapter the number one issue in tourism security was how to handle terrorism's multiple forms. Terrorism had gone through several mutations since the 1960s. The terrorism of 2015 was very different from that of September 11th 2001, and in reality even more different from the acts of terrorism against the tourism industry that had begun in the 1960s. Despite the fact that the face of terrorism had changed and in 2015 was often intertwined with the illegal narcotics trade there was a certain consistency about it. Tourism continued to suffer from acts of terrorism that had occurred

DOI: 10.4018/978-1-7998-4330-6.ch001

throughout the world. By 2015 the tourism industry was learning (and in some ways had learned) to deal with terrorism's challenges. Although each act of terrorism hurt the industry, massive damage on the macro scale had not occurred in the nine years from 2001- 2015, or in the years prior to 2001. In many locations despite crimes and acts of terrorism much of the industry thrived and even as late as the forth quarter of 2019 a major issue facing the tourism industry was not a lack of tourists due to acts of terror but rather what came to be known as "overtourism". Cities such as Venice, Italy and Barcelona, Spain struggled to balance tourism with their city's quality of life. They struggled to maintain their cultural integrity, to allow a quality of life for their residents, and at the same time to permit tourism to flourish. The most common issue, prior to the Covid -19 crisis was not tourism's survival but its sustainability in places filled with an excess of visitors.

In early 2020, everything changed. Tourism went from one of the world's strongest and most successful industries to one of the world's weakest industries. Not only did tourism face a crisis but also multiple industries that were dependent on tourism or connected to the tourism industry had to fight for their survival. For example, a possible unforeseen consequence of the fall of tourism was the collapse of the oil market. The April 20th, 2020 headlines from the British Broadcasting Company (BBC) "US oil prices turn negative as demand dries up" (BBC, 2020). would have seemed ludicrous just a few months prior. An industry that worried about overtourism in 2019 now just a few months later worried that it might not survive. The "culprit" for this freefall and near collapse was the coronavirus, later called Covid-19. The virus, soon to become a worldwide pandemic, seemed to emerge almost from nowhere. Covid-19 has destroyed thousands of lives and in tsunami fashion it has economically swept away everything that came before it. In Europe the virus wiped out tourism in such major tourism nations as: France, Italy, Spain, and the United Kingdom. The Covid-19 medical and economic infection "migrated" to the Americas and by late March of 2020 most travel and tourism as we had known it just a few months prior had ceased to exist. If the terrorism attacks of 2001 were a major wave that knocked the tourism industry to its knees, the Covid-19 virus was an economic tsunami that will take many months until we know the full extent of virus' economic impact. Even a superficial perusal of world's media provides insights into the depth of this problem. For example the March 17th edition of *Forbes* notes that: "International travel has since taken a plunge, amid the Coronavirus pandemic. Flights have been grounded. Borders have been closed. The spread of COVID-19 has dramatically derailed businesses, communities, and livelihoods across the globe. Now the questions seems to be: What will travel and tourism look like, and how will it influence the global economy, once the dust settles and being mobile is safe again" (Forbes, 2020).

Such a major health and economic crisis was bound to make people wonder: Did the virus occur out of nothing? Was its occurance accidental or purposeful? Can or is it likely that such an occurance might happen again? Part of the issue surround the Covid-19 outbreak is its cause. Currently, (April 2020), there are three major theories surrounding the cause for the Covid-19's outbreak: These are:

1. that the consuming of a sick animal purchased at the Wuhan wet market infected someone and then the virus "jumped" from one human being to another,
2. The virus accidently leaked at the Wuhan Biological Lab. The lab was studying corona viruses at the time and is located in close proximity to the "wet market"
3. the virus was intentionally leaked as an experiment in biological warfare.

Currently no one has proven any of these three theories to be false, although most experts do not believe that China intentionally created and leaked the virus. Many scientists believe the first theory. For example, the journal Science News stated that: "The analysis of public genome sequence data from SARS-CoV-2 and related viruses found no evidence that the virus was made in a laboratory or otherwise engineered" (Science Daily, 2020).

Others disagree and believe that it may have been due to lax adherences to biosafety protocols. For example, the journal Business Insider noted in its April 16th editions: "The COVID-19 outbreak started in the city of Wuhan, home to some of China's top research laboratories. One of them — the Wuhan Institute of Virology — has done research on coronaviruses that originate in bats, as is thought to be the case with the novel virus. The Chinese government denied that the lab was to blame early on, instead saying the outbreak started at a wet market in the city. But there were holes in that theory. The market didn't sell bats, and the first patient — as well as multiple other early cases — didn't have any connection to the market, as Yahoo News pointed out" (Business Insider, 2020).

Currently we do not know which theory is correct If however for purposes of argument we assume that the third theory were to be correct, the world would then have to grapple with multiple issues. These issues would entail such questions as: what would be the proper response? Does the theory of proportionality work under such circumstances and what would be a proportional response? Might a proportional response lead others to calculate that such a biological attack might be worth the cost?

These are questions of proportionality and many of these same questions are exactly the same questions that the world pondered regarding acts of terrorism in 2015.

Just a few weeks after the Al Qaeda attacks against a Paris magazine and a kosher supermarket, the world read of the terrorist attacks against a major hotel in Libya: (BBC, 2015).

Once again, terrorism had touched the world or tourism and once again, the world was reminded that tourism presents terrorists with not only soft targets, but it also provides terrorist with a great deal of publicity. From the terrorists' perspective we may call this form of publicity, "purposeful negative marketing" (PNM). A few weeks later, Copenhagen, Denmark saw almost a repeat performance. This time terrorist(s) attacked a café in the presence of the French Ambassador; a shooting at a local synagogue soon followed the attack. The headlines on the CNN website perhaps best summarize the situation. "We have tasted the ugly taste of fear" (Danish) Prime Minister says" (CNN, 2015).

A short time after the attacks in Paris, the city was quiet, perhaps too quiet. Associated press reporters noted that tourists have simply disappeared. A mid –January news article by Thomas Adamson perhaps summed up the situation best when it stated: "Among the tourists who were still braving visits, many took comfort in the extra security presences. With 10,000 troops deployed across the country including 6,000 in the Paris region alone, the security operation put in motion after the attacks is the most extensive ion French soil in recent history. The (Bryan Texas) *Eagle*, page A-3, January 19, 2015). The dearth of tourists however was short lived, as the French were able to assure the world that they had taken full control of the situation, employed some ten thousand troops to sensitive locales, and have given the impression that the terrorist attacks were an anomaly.

The terrorism attacks in many parts of Europe remind us that terrorism is as much about "purposeful negative marketing" as it is about death and destruction. It is of note that when Norbert Vanhove writes about detriments of tourism demand he mentions the following as noted by Middelton:

1. Economic factors
2. Comparative factors
3. Demographic factors
4. Geographic factors
5. Socio-cultural factors
6. Mobility
7. Government/regulation
8. Media communication
9. Information and communication technology (Vanhove, pp.50-51)

Vanhove does not list safety and security as a negative determent nor does he mention either of them in his work, Despite the fact that tourism security was an issue in 2005, Vanhove takes the typical marketing approach that safety and security have nothing to do with tourism marketing.

Terrorism goes beyond violence. Terrorism produces fear and through fear seeks the destruction of economies and reputations, and as such, it is a major threat to the tourism industry. Perhaps it is for this reason that we cannot separate terrorism from tourism and tourism from marketing. To understand terrorism is to see it as a negative form of marketing. Instead of encouraging people to visit a place, terrorism seeks to discourage visitors. It seeks to empty hotel rooms and to transform the vibrant into the decadent. It creates xenophobia and a sense of mistrust and fear both of the foreign visitor and even the local citizen. Lewis and Chambers remind us that: "For the customer, perception is reality. The point is so critical that it is worth repeating: *Perception is reality.* Perhaps one of the greatest mistakes that we make as marketers is thinking that what we perceive is also what the customer perceives." (Lewis, Chambers, p.139)

To see terrorism as merely attacks against the innocent is to misunderstand terrorism. Tourism scientists and professionals must see terrorism in its fullness, the use of murder and destruction as a purposeful negative marketing. It is the attempt to destroy not only lives but also economies. Perhaps Milan Kunders said it best when he states that: "Business has only two functions - marketing and innovation." (Brainy Quote, nd.).

Because tourism is meant to be joyful we often forget that tourism is business; it is a major business. Depending on how we define tourism it is perhaps even the world's largest business. Those outside of the profession tend to forget that marketing is the lifeblood of tourism. Without making a profit hotels die, restaurants close, and attractions cease to be. To know tourism, to work in tourism is to know the truth of Kunders' statement in relationship with the world's largest industry: tourism. Thus, without marketing tourism ceases or comes close to ceasing. Richard Gartell notes that: "From the perspective of the consumer, destinations are perceived as those geographic areas that have attributes, features, attractions, and services that appeal to the perspective user" (Gartell: 4)

Although it may be an overstatement to say that tourism is nothing more than marketing, it is not incorrect to argue that marketing is a major part of what tourism professionals do. Tourism is the selling of an intangible product, the providing people with a way to spend their extra income. Tourism is a system that connects hundreds of millions of people not only with other human beings but also with their hopes and dreams. The tourism marketer's job is to create a world in which dreams can be turned into reality, fantasies become new and ever changing experiences which in turn become memories that last a lifetime.

To accomplish this goal is not an easy process. To attract its clientele, to transform the unreal into the real, and to bring people who work all year into the realm of the unreal, the tourism industry must create the proper ambiance. A part of that ambiance is the supposition that the visitor does not have to worry about his security, that s/he will be able to visit the locale and return to his or her home without fear of bodily harm. Thus, with a few exceptions, such as what we might call: "bellicose dark tourism", the tourism industry relies on a state of peace or at least a state of perceived safety and security.

This chapter uses the tourism industry as a metaphor for the world of marketing. It also understands that there is a close relationship between the positive marketing of the tourism industry and the negative marketing promulgated by terrorism. The chapter views both industries: tourism and terrorism as symbolic of the interaction between marketing and its potential clientele. From this perspective to be involved in the world of tourism is to be involved in marketing. Unfortunately, as Islamic State has taught the world the same is also true of the world of terror.

As such tourism is much more than mere travel. Tourism is the totality of future hopes blending with past experiences. Just as in marketing, tourism is the reversal of grammatical tenses. It is the place in time where we dream of a future so as to remember it once the event has passed. As such tourism serves to remind us that marketing is the act of making the future precede the past, and the transformation of a past into a future event. Terrorism, from a sociological viewpoint, has a great deal of similarities with tourism. The difference is that while tourism is both positive and seeks to build, terrorism reverses the roles. Terrorism seeks to turn the beautiful into the horrific, to transform peace into war and human understanding into interpersonal fear and distrust. Ironically tourism and terrorism, in the world of marketing, are mirror images of each other. Terrorism is the reversal of the values of tourism.

If we see tourism as a universal metaphor for marketing, and the tourist as a metaphor for consumer, we can begin to understand the importance of the client's physical wellbeing and the realization of the dream that is tourism. Simply put, most visitors tend to avoid areas of conflict and tourism industry professionals go out of the way to claim that their destination is one that is safe and secure. Terrorist marketers do the exact opposite of tourism marketers. For example, even a superficial review of tourism conferences and literature will reveal a preference for the term "safe" or "safety" over the term "secure" and "security". Although the terms technically have different meanings, tourism professionals tend to use safety and avoid the term security not for academic clarity but because the marketing image of safety is easier to sell then that of security. In this chapter safety, security and surety are used as interchangeable terms. Many tourism professionals and marketers fear that even a hint of insecurity will have a negative impact on the tourism industry.

Despite the best efforts to deemphasize security threats, both tourism and government officials, and the general public know that tourism does not exist in a perfectly safe and secure world. Were they to forget this fact, terrorism marketers do their best to remind the potential tourism client of this fact. Terrorism marketers promote fear, and this lack of security has been a major tourism marketing challenge. Since the attacks of September 11, 2001 tourism officials, even in western nations, have had to confront not only issues of crime and cyber crime, but also issues of terrorism, cyber terrorism and the potential for bio-terrorism. This interlocking of tourism, and safety with the fear of terrorism and war presents a number of both practical and ethical dilemmas for tourism scholars and for tourism marketers, often called in the tourism profession: practitioners.

For example, if we return to the case of the terrorism attacks in Paris, we find an example of not only the impact of security on tourism but also the interrelationship between marketing and security. The French at first reported that there has not been a drop in visitations to Paris due to the well-publicized attacks against the Magazine *Charlie Hebdo* and the kosher supermarket. A week later these initial statements had to be changed. It would be a mistake to believe that one incident will change the way people perceive a location, but if the locale is hit by multiple attacks, or the media consistently remind the public of these attacks, or the locale suffers from a consistent stream of negative publicity then what might have been a single tragic incident becomes a part of that locale's image and can create a great deal of negative publicity. Tarlow has noted that a city's negative safety and security image acts as a break on its economy. For example he has written: …the perception of crime is enough to keep people away (from the locale) thereby transforming a negative perception into a negative economy. (Tarlow 2014 p. 25)

The question then of what to do about terrorism, however, is more than a mere military or political question it is also a question of perceptions and perceptions form the essence of marketing. If terrorism attacks the economic heart of a locale, then that locale's lifeblood is threatened. If terrorism is understood to be purposeful negative marketing, then there is a clear relationship between a political stance and a marketing position. If tourism is all about publicity and terrorism seeks purposeful negative publicity, terrorism becomes a form of negative tourism marketing. Tarlow has written: "Attacks on tourism provide a great deal of publicity and terrorism seeks publicity. Tourism centers where there is a great deal of media coverage. An attack against a tourism center creates an instant impact on the locale, which can lead to long-term perception consequences." (Tarlow, 2014, p.99)

Theoretical Issues in Tourism Security

In 2009 Tarlow published in the *Sage Handbook of Tourism Studies* (edited by Jamal and Robinson) a highly detailed theoretical perspective of tourism security, safety and surety.

Tarlow noted that tourism surety (used here to signify both tourism security and safety) is about risk and recognizes that despite both the public's and the industry's desire for total safety and security, reality dictates that there will always be some security problems. We may then argue that to be alive is to learn to live with an element of risk. From this perspective both the consumer (the visitor/tourist) and the provider (the tourism entity) must accept that fact that risk is ever present and that tourism security is an ongoing and changing concept. Tourism surety is about managing risk all forms of risk, be that risk, physical, psychological, economic, reputational, or in 2020 bio-security. The great Covid-19 pandemic underlined the fact that although risk is ubiquitous, different locales will have different types of risk. The world learned once again during the worldwide Covid-19 pandemic, that the presence of risk is universal but the type of risk and the risk's intensity will vary from locale to locale and will change with time. We may say that marketing, in this context, is the presentation of risk in a way that the risk produces the least amount of economic or reputational harm possible. Marketing in tourism is the attempt to control both positive and negative perceptions of risk while at the same time reinforcing the dream-like aspects of tourism.

Some of the Early Attempts to Develop a Theory of Tourism Safety/Security

Human beings, either consciously or unconsciously, tend to develop paradigms that become theories and from these theories they develop policies. A literature review of the last four or five decades makes this principle clear. When we review the literature we see that tourism scholars of the 1980s tended to blame the visitor for tourism crime. Many of these scholars wrote from a Marxist perspective and assumed that a disparity in the distribution of wealth would produce a tendency toward crime and/or violence.

For example, the works of scholars such as Meda Chesney-Liind and Ian Lind asked questions either directly or indirectly such as: do tourists produce crime or is crime a by-product of conspicuous wealth? The assumption was that tourists incite crime due to the fact that locals may become jealous or resentful of the tourists' wealth and leisure. From these scholars' perspective this jealousy or resentment turns to frustration and frustrations then turn into crime. This Marxist perspective is still very much alive. For example, when asked what the US could do to confront ISIS, the US State Departments' spokesperson, Marie Harf, indicated that one way to lessen terrorism was to provide more job opportunities for young men in the Middle East. Some of the US media immediately saw a connection with Marxism and Ms. Harf's statement produced headlines such as: "Thursday, 19 February 2015 Item title *Obama Anti-terrorism "Jobs for Jihadists" Concept Is a Marxist Idea.* (Duke, 2015).

Returning to tourism, the basic theoretical underpinning is that tourists have expendable income, and therefore they provoke issues of social disparity that causes locals to turn resentment into crime. This theoretical perspective sees tourism as symbolic of an "unfair" distribution of wealth and crime (or terrorism) as a product of the hopelessness. Thus, it is not the criminal or terrorist who is to be blamed but rather the social conditions that transform the person into being a terrorist. We may call this perspective the "Robin Hood Tourism Security Approach." Just as in Marxism, there are the "good" and the "evil" (the bourgeois and the proletariat) so too does tourism act as a conductor of symbolic wealth. This theoretical perspective does not accept the notion that the thief might be richer than the victim or that the thief (criminal) might enjoy his or her "work". Thus, despite the fact that many terrorists come from wealth and not poverty the Marxist tends to side with the perceived proletariat. Those in the upper class (tourists or visitors) are evil by nature. Thorstein Veblen's *The Theory of the Leisure Class* is the example par excellence of this form of thinking. Veblen defines leisure as "the non-productive consumption of time (1) from a sense of the unworthiness of productive work, and (2) as an evidence of the pecuniary ability to afford a life of idleness" (Veblen, 1899, p. 46). In this theoretical paradigm there is no middle class and the fact that the proletariat might benefit from tourism does not exist. To some extent this position continues to exist. For example, former two-time U.S. presidential candidate Bernie Sanders criticized the skiing industry seeing it as a form of domination of the poor by the wealthy. The newspaper, The Washington Examiner in February of 2020 quoted Sanders as saying: "What we are talking about is the kind of developments that will provide decent jobs for our people," he said. "Very often, and this is a concern within our own state, is that industries such as the skiing industry and other tourist-attracting industries in Vermont ... they provide very low-paying, menial jobs." (Lin, 2020).

The Functionalist Approach to Tourism Security

In the 1990s some scholars took an approach quite different from that of the Marxist school. For example, Tarlow has written: "A counter theoretical perspective to the Marxist approach would be a Durkheim way of viewing tourism crime. The theories of David Emile Durkheim form the basis for Functionalism. Functionalist theory predicts that society is a stable living system and that a change in any one aspect of the system produces unexpected consequences in other parts of the system. Thus, from a Functionalist perspective crime is not a result of a wealth-disparity but rather the result of the introduction of new social groupings into society and thereby taking it from a stable system to an unstable system. In such a paradigm, crimes exist when instability enters into the system." (Tarlow: Sage: p.467). The pandemic of 2020 has caused such an interruption. With over 2,000,000 (two million) confirmed cases and at least 150,000 dead due to the disease Covid has become the most destabilizing event in the history of tourism. The World Tourism Organization (UNWTO) in late March of 2020 stated that: "Based on the latest developments (quarantine measures, travel bans & border closures in most of Europe, which represents 50% of international tourism, and in many countries of the Americas, Africa and the Middle East), the evolutions in Asia and the Pacific and the patterns of previous crises (2003 SARS and 2009 global economic crisis). UNWTO estimates international tourist arrivals could decline by 20% to 30% in 2020. This would translate into a loss of 300 to 450 US$ billion in international tourism receipts (exports) – almost one third of the US$ 1.5 trillion generated globally in the worst-case scenario." (UNWTO, 2020)

These numbers represent the direct impact. They do not necessarily indicate the indirect impacts, the fear that people might have to travel, to stay in hotels or to eat at buffet-style restaurants. The World Travel Council takes even a dimmer view stating: "With airlines and TSA reporting a 96% drop in air travel, nearly 80% of hotel rooms laying empty and the CDC renewing the No Sail Order for cruise ships, Coronavirus has all but stopped the travel industry in its tracks. The World Travel & Tourism councils estimates up to 75 million jobs are at immediate risk and in 2020 alone, the global economy could stand to lose $2.1 trillion." (Alpha sense, 2020). Functionalists would go beyond the mere numbers. They would want to know how these losses impact everything from families to lacl of eduction, what the long-range impact of such a social disrubtion would be on society and who might benefit and who might suffer due to these disruptions

These scholars tended to see tourists not as the aggressors but rather as the victims, and in the case of the current pandemic they would see the industry also as a victim. Therefore they were worthy of protection. For example in 1997 Pizam, Tarlow and Bloom published an article entitled: "Making Tourists Feel Safe: Whose Responsibility is It?[1] These authors assumed that:

1. Tourists and the tourism industry are not crime producers but rather wealth producers and as such deserve protection from the perspective of personal safety and industrial safety and security. Thus there was to be a symbiotic relationship between the tourism industry, the security industry and marketers.
2. The question now was who was responsible for the protection of the industry and its customers. Thus the concept that the tourism industry and the security industries (both public and private) need to work together was born. It was based on this theoretical perspective that in 2004 the US Department of Justice asked Ronald W. Glensor and Kenneth J. Peak to published a pamphlet in its series "problem oriented guides for police, #26 entitled "Crimes Against Tourists"
3. As tourism security was now a legitimate issue for discussion conferences that were both academic and applied began to grow. For example the Las Vegas International Tourism Safety and Security Conference is now in its 22nd year.

Even within the functionalist approach there are differences. For example, scholars such as Yoel Mansfeld and Abe Piza, tend to emphasize such questions as:

- Why do incidents of security, such as crime, terrorism, wars, riots, and civil unrest exist at tourism destinations?
- What are the motivators of the perpetrators/offenders"

We may call these scholars part of the "Israeli school of thought". Often, but not always, crime and terrorism are seen as two sides of the same coin with at times common results.

Some of the non-Israelis take a different view. For example, Tarlow works under the assumption that acts of crime and terrorism are fundamentally different in nature, although he admits that the results to the industry may overlap or at times be similar. Thus, he has written: "Terrorism is often confused with criminal behavior. In the world of tourism, however, terrorism and crime are very different social ailments. Criminals, especially those who are business-people rather than criminals of passion, seek a parasitic relationship with tourism. Indeed, it may be stated that tourism criminals, be they freelancers of part of an organized group, need the tourism industry to succeed in order for them to be successful. Terrorists, on the other hand, seek to destroy nations (or governments) often through random deaths that lead to an economic collapse." (Tarlow, as found in Wilks et al, p.79, 2006)

The September 11, 2001 attacks impacted the world of tourism and changed not only the way its leaders understood violence but brought home to them the interlocking of tourism and security. A similar change took place in 2020. The Covid-19 virus brought home to tourism's leaders the interaction between health and tourism. Just as the September 11, 2001 attacks ended tourism's naivety and forced the industry to face the fact that terrorism could destroy the tourism industry, so too the Covid 19 virus forced tourism's leaders to understand that issues of health were a part of tourism security. Tourism scholars now understood that without tourism security tourism could not survive. Thus, in 2006 two major works were published. *Tourism in Turbulent Times*, edited by Jeff Wilks, Donna Pendergast and Peter Leggat and A second major work, Tourism Security & Safety, also published by Elsevier in 2006 and edited by Yoel Mansfeld and Abraham Pizam. In both cases these scholarly books saw tourism security from the perspective of Talcott Parsons functionalist approach in which tourism was a living system and everything from issues of health safety to food illnesses to acts of publicized crime to terrorism would have a part to play in the industry's ability to survive.

Proportionality in an Age of Pandemics

In the first decade of the twentieth century a new use of the word proportionality or proportional response crept into the political vocabulary. The word proportionality has multiple meanings and nuances and when using the term, clarity must be sought. A simple definition has been that a response to an act of aggression should be met with a response that is neither weaker nor stronger than the damage done by the aggressor. The question is, is there an interrelationship between political proportionality, marketing and tourism?

These questions become even more difficult when we apply them to the massive destruction of people's livelihoods due to the outbreak of Covid-19. Additionally, is there a proportional response to what many people believe was the mismanagement of the Coronavirus (Corvid-19) on the part of China? How ought the world to react to a nation that due to an act of nature, by accident or intentionally destroyed much of the world's tourism industry? Assuming that that the virus occurred naturally or by accident does not justify the fact that China failed to inform the world that this was a highly contagious virus until it was too late. To make matters still more complicated there is an academic and political dispute as to what China's role was in the spread of Covid-19, how much of the data were intentionally held back and if the world should hold China responsible for its lack of forthrightness. Were China's actions merely a cover-up for poor bio-safety policies, or were this actions either intentional or unintentional acts of aggression?

In comparison with early acts of intentional terrorism the challenges posed by Covid-19 seem daunting. Europe, and to some extend the United States, has argued for what Europeans consider to be a proportional response to an act of terrorism. Despite the fact that this is considered a politically correct term, there is no clear definition of what a proportional response is, especially when it comes to tourism. To make matters even more difficult, many Europeans and academics rarely distinguish between acts of crime and acts of terrorism. For example the Doha declaration of the United Nations Office of Drugs and Crime (UNODC) has stated: "The role of the criminal justice system in countering terrorism is a challenging one. Indeed, the primary objective of counter-terrorism strategies must be to prevent terrorist incidents from taking place, and in some cases law enforcement agencies are able to prevent terrorist attacks from occurring. However, some existing criminal justice practices are less effective when it comes to preventing terrorist conspiracies from achieving their aim or a terrorist threat is too extensive for available resources to cope with. A forward-looking, preventive and well-funded criminal justice strategy against terrorist violence requires a comprehensive system of substantive offences, investigative powers and techniques, evidentiary rules and international cooperation." (Kkienerm, 2020). The declartion then goes on to demonstrate the confluence of crime and terrorism when it states: "in terms of where these international crimes are prosecuted, with the exception of the Special Tribunal for Lebanon, there are no international criminal courts or tribunals with dedicated jurisdiction over terrorist crimes. Some of the most heinous and high profile cases may be heard by an international criminal court or tribunal, such as the International Criminal Court, the jurisdiction of which is sourced in the Rome Statute of the International Criminal Court 1998 (Rome Statute)." ((Kkienerm, 2020). If terrorism is not separate from crime but rather a subcategory of the general perspective then the question must be asked if European concepts of proportionality must be assigned to both acts of crime and of terrorism. Although, at the writing of this chapter, there is no evidence of intentional bio-terrorism or crime, the fact that Covid-19 has demonstrated how vulnerable the world economy is might inspire a future biological attack, were that to occur the question would then become how does reciprocity work in a world economically frozen by such an attack?

If proportionality is understood as an economic term, then the answer becomes some form of sanctions. The logic being that X nation or group has hurt Y's economy (read: tourism) and thus, in a tit-for-tat response Y hurts X's economy. If proportionality is measured in military terms, then does it mean that in the case of a terrorism attack, the victimized nation should seek out an equal number of citizens of the other side or members of the said terrorism group to destroy? If proportionality is to be understood in biological terms then does proportionality allow for a counter bio or bio-chemical attack?

What then is proportionality and how does it morph from a military term into one that deals with tourism and marketing? In its simplest format proportionality means: when A attacks B, how should B respond? The European Union defines it as: "Similarly to the principle of subsidiary, the principle of proportionality regulates the exercise of powers by the European Union. It seeks to set actions taken by the institutions of the Union within specified bounds. Under this rule, the involvement of the institutions must be limited to what is necessary to achieve the objectives of the Treaties. In other words, the content and form of the action must be in keeping with the aim pursued." (EU Union Law, 2020).

The ICRC defines proportionality as: "Launching an attack which may be expected to cause incidental loss of civilian life, injury to civilians, damage to civilian objects, or a combination thereof, which would be excessive in relation to the concrete and direct military advantage anticipated, is prohibited" (ICRC, 2020).

The International Committee for the Red Cross (ICRC) admits that this definition is somewhat questionable and that a number of nation-states have refused to accept this definition. The Galestone Institute offers the following definition: "Proportionality in international law is not about equality of death or civilian suffering, or even about [equality of] firepower. Proportionality weighs the necessity of a military action against suffering that the action might cause to enemy civilians in the vicinity. "Under international humanitarian law and the Rome Statute, the death of civilians during an armed conflict, no matter how grave and regrettable does not constitute a war crime.... even when it is known that some civilian deaths or injuries will occur. A crime occurs if there is an intentional attack directed against civilians (principle of distinction) or an attack is launched on a military objective in the knowledge that the incidental civilian injuries would be clearly excessive in relation to the anticipated military advantage (principle of proportionality)."—Luis Moreno-Ocampo, Chief Prosecutor, International Criminal Court.

"The greater the military advantage anticipated, the larger the amount of collateral damage -- often civilian casualties -- which will be "justified" and "necessary." — Dr. Françoise Hampton, University of Essex, UK." (Bryen, 2014).

Bryen goes onto state: "Proportionality in international law is not about equality of death or civilian suffering, or even about firepower returned being equal in sophistication or lethality to firepower received. Proportionality weighs the military necessity of an action against the suffering that the action might cause to enemy civilians in the vicinity" 2015).

The question of proportionality may at first seem to be of minor importance to tourism marketers. However, if tourism requires some form of peaceful coexistence then the way a government reacts to a terrorism threat impacts not only its politics, but also the capability of a tourism community to sell itself on the world stage. This leads to several basic questions, especially in a world of pandemics: These are:

1. Is proportionality an acceptable principle in the war on terrorism?
2. How would the concept apply to a pandemic provoked either through carelessness or on purpose?
3. Does proportionality tie the hand of tourism marketers?
4. Is the concept of proportionality moral or does the concept do more harm than good?
5. Does this concept go against human nature?
6. Does the concept of proportionality lead to an Orwellian state of no-war no-peace or even to continual war?

The above questions make tourism professionals ask if the principle of proportionality is an acceptable principle in the war on terrorism especially in this post-Covid 19 age and what would be its impact on tourism and tourism marketing? One side of the argument proposes that a government's response should be "proportional", meaning that the response should be fit the crime. The other side of the debate would take the position that security is a life and death matter for tourism (marketing) and as such there cannot be proportionality, but rather a terrorist attack on innocent civilians at a tourism center deserves an all-out response that destroys the enemy. Both positions have both applied and ethical problems.

THE THEORETICAL MARKETING ISSUE

In the first decades of the twenty-first century the various outbreaks of hostilities in the Middle East that have pitted several Middle Eastern nations against various terrorists group have reintroduced the concept of "proportionality." One side argues that it does not matter who began a war or how many human shields are used, a state may only do proportional damage to its terrorist enemy. To do more is disproportional and therefore illegitimate.

Thus, in its Gaza wars, Israel was faced with a "marketing" challenge in Europe. Many Europeans stated that although Israel had a right to defend itself its response was not proportional. They condemned Israel for what they considered to be a non-proportional response to the firing of missiles at its civilian centers. This leads to the following questions: "What is a proportional response?" and is the demand for proportionality in international affairs both realizable and moral? How should Israel have handled the marketing side of the wars and how did international criticism of lack of proportionality impact Israeli tourism? Furthermore, was the principle of proportionality applied universally or was it aimed solely at Israel as a form of passive anti-Semitism on the part of Europeans and academicians? If missiles had been aimed at European civilian centers would Europe have acted differently?

From a marketing perspective, questions such as those below become essential:

- Is there such a thing as a proportional response?
- Is war as much an issue of marketing as it is of fighting?
- Who defines proportionality and by what measurement scale?
- Is proportionality a legal principle, a moral principle, or a marketing principle used by one side against the other
- In the case of bio-terrorism or war against a tourism industry is there such a thing as proportionality and how would it be measured vis-à-vis not only damage to life but long-term economic damage?

To judge this principle we need to consider exactly what is the meaning of the idea of proportionality in foreign affairs. The (US) Council of Foreign Affairs provides historical background on this principle stating: "The doctrine originated with the 1907 Hague Conventions, which govern the laws of war, and was later codified in Article 49 of the International Law Commission's 1980 Draft Articles on State Responsibility (PDF). The doctrine is also referred to indirectly in the 1977 Additional Protocols of the Geneva Conventions. Regardless of whether states are party to the treaties above, experts say the principle is part of what is known as customary international law. According to the doctrine, a state is legally allowed to unilaterally defend itself and right a wrong provided the response is proportional to the injury suffered"

Does proportionality then mean a mere tit-for tat response? Even if this tit-for-tat doctrine is adopted, its meaning is unclear. If, for example, one side targets a school filled with children with a rocket attack then is the other side expected to target its enemy's school? Taken to the extreme, one wonders if the US and its allies used proportional responses in WWII. Should Japan's attack on Pearl Harbor have been met with nothing more than a US attack against a Japanese naval base? To confuse this principle still further, it is not clear if the principle applies or not to terrorist attacks or to attacks of a biological nature. Does this principle only apply to attacks of a military nature or attacks aimed at economic destruction?

If we take the idea of proportionality literarily then as Bret Stephens has proposed: "For every single rocket that falls randomly on Israeli soil, an Israeli missile will hit a carefully selected target in Gaza. Focusing the minds of Hamas on this type of "proportionality" is just the endgame that Israel needs" Of course, one may ask if this is not merely a formula for an Orwellian 1984 scenario in which wars never end.

Is the philosophical argument over proportionality also an argument dealing with issues of marketing? Is the term one that propagandists use to attack their opposition or is it a term that describes a moral code ensconced into international law?

Those who favor proportional responses have yet to create a proportionality measurement scale. When it comes to issue of crime, specific guidelines (often called laws) can be established. For example, the penalty for theft can be set at X number of years in prison or a fine costing the perpetrator X amount in a fine. This tit-for-tat system can even be employed, although some would argue unfairly, to issues of murder. Thus, there are societies that argue that the taking of a life should result in the death penalty. Others argue against the death penalty instead they seek to impose harsh jail punishments and note that a proportional response is, from their perspective, immoral.

Is there a relationship between the political term proportionality and the marketing term? Proportionality, like any form of marketing is often more subjective than it is objective, it is an inexact science. There are no clear guidelines for proportional responses and what may be considered a legitimate proportional response in one society may not be considered a proportional response in another society. Furthermore there are no definitions of what proportionality means. For example, if a terrorist kills 10 people then do we find ten terrorists to kill? Is proportionality then a marketing tool, based on the idea that whoever gets his/her message out faster and to a wider audience wins? In the case of an attack against the health of a population does the injured party then attack its enemy in a like manner? Do we measure proportionality by economic hurt, taking of an equal number of lives, reputational destruction; Mutual destruction of landmarks etc?

Of course no one knows how one creates a proportional reputational response. Proportionality can also lead to situations ad absurdum. In WWII should the US have lost more soldiers to Hitler's armies in order to maintain proportional responses? Should the allies have murdered 6 million Germans to create a proportional response to the murder of six million Jews? If the answer is yes, then the next question is when? Should these six million Germans have been murdered during the hostilities or as a punishment after the hostilities had ended? This leads to the deeper philosophical question if there is a proportional response to evil? When we base proportionality against these questions then it becomes clear that proportionality does not mean eye for an eye, but rather an action that is less damaging to the perpetrator than it was to the victim. Thus, Paul M. Bischke has written:

An eye for an eye, and a tooth for a tooth." Sounds very hard-nose doesn't it? It seems like an ancient recipe for harshness that modern society has long ago outgrown. Not so. Few passages in the Bible are as badly misunderstood as this one. The "eye for an eye" maxim is not about harshness; it's about proportional retribution. (2015).

Finally, the biggest problem that proportionality proponents have is that it has had no historical success. In the area of crime, governments have attempted multiple combinations of proportionality and so far none has succeed in stopping crime.

On the other side of the ledger is the idea of a non-proportional response. This argument goes something like: if X hurts Y then Y will do so much damage to X that X will not want to continue to fight. "Non-proportionalists" argue that this is no such thing as a fair war; that to fight is to risk all and that peace comes through strength. Their leading example is the Roman Empire, Ancient Rome had a policy of pure conquest and understood that the party that does not win, has lost. A more modern example is the multiple and non-proportional Jordanian attacks on the Islamic State after the horrific murder by burning of the Jordanian pilot.

There are, however, several problems with the total victory scenario. Perhaps the greatest problem is that non-proportional responses leave no room for moral growth and the innocent suffer along with the guilty. People who oppose non-proportional responses point to the US having dropped an atomic bomb on Hiroshima as non-proportional, noting that thousands of innocent lives were taken by this one act.

Because proportionality is often framed in an ethical or moral framework and much of the Western world uses the Bible as the basis of moral guides, we now turn to issues of proportionality in the Biblical text. Although the modern world does not accept the Biblical view as its only framework for morality, we can argue that the Biblical text serves as a starting point to respond to the question: In a world of terrorism in which whole economies are put at risk alongside the lives of hundreds if not thousands of people, is proportionality a moral response to terrorism or does it create negative marketing and turns a bad situation into a worse one?

PROPORTIONALITY IN A BIBLICAL FRAMEWORK

The Hebrew Bible

The Bible, although universally read is very much a Middle Eastern work. Although it was not intended to be a marketing guide we also can read the text as a textbook about marketing. It is a book that reflects the harshness of war. The Hebrew Bible does not attempt to sugarcoat reality. Alongside its ethical and moral teachings the text also presents us with: pain and death, human suffering and ethical dilemmas. In one book we read much of humanity's loftiest thoughts and at the same time some of its greatest tragedies.

A superficial examination of Hebrew Scripture (and also the Christian Testament, called by Christians: New Testament) would seem to argue on the side of proportionality. For example, Hebrew Scripture's principle of "Ayin Tachat Ayin, Shen tachat Shen" (Eye for an eye/tooth for a tooth) as seen in Exodus 21:23-27 and in Deuteronomy 19: 16-21 would seem to support a tit-for-tat concept of proportionality.

Western readers of the text often confuse this philosophical statement with vengeance. The text follows the guidelines for proportionality. Thus, with only one exception, that of the "Arei-Miklat" or "Cities of Refuge" found in Numbers 35: 9-30, it is the courts that are to carry out these acts of proportionality, (often called "the laws of equivalency"). Yet for many proponents of proportionality, it is to be practiced more on the macro than on the micro level or it is to be spoken of but not used. Liberal thought for example traditionally has opposed the death penalty even though the death penalty is a perfect proportional response to murder. In a like manner, liberal thought would condemn the idea that if an army killed X number of enemy soldiers than the opposing army would have the right to kill an equal number of soldiers' lives: no more and no less. It may be noted that God does not employ proportionality when it comes to Cain's murder of Able. Instead, Cain is marked but does not lose his life.

The guiding principle of the law of equivalency seems to be that the use of proportionality would limit conflicts and thus, save innocent lives. In other words there would be an equal about of pain (justice) given to the person who caused the pain.

Another aspect of proportionality as found in the Bible may be called 'statistical proportionality." For example, Biblical law provides proportional (statistical measures) for criminal acts. Yet here the text is careful to provide disproportional responses (you pay more in the fine that the value of what you steal) as the Biblical text understands that an exact tit-for-tat response in a world in which not all crimes are caught would make crime profitable. Concerning "statistical proportionality"] Horwitz notes that: "In all cases where Scripture required a wrong-doer to pay more than restitution or actual damages, the excess payment went to the injured party and not to the government or the community." (Horowitz: 163).

The following table helps to clarify the different formats of proportionality and to whom these laws applied. It should be noted that the when referring to proportionality the text is clearly referring to

Table 1. Overview of proportionality

Type of Proportionality	Against Whom	To Be Used on Macro or Micro Level?
Equivalent proportionality	An Individual found guilty of harming another person	Micro
Statistical proportionality	Someone who has done material damage	Micro
Macro/Marketing Proportionality	Enemy states	Macro

MICRO PROPORTIONALITY VERSUS MICRO PROPORTIONALITY IN HEBREW SCRIPTURE

There are no direct texts referring to the principle of proportionality on the macro level, but there are a number of Biblical vignettes that would establish a different principle on the macro level from that which we have discussed on the micro level.

The following texts will serve to illustrate the point that while the text views the need for proportional responses on the micro level as not only necessary but also prudent, it takes a very different approach to macro level proportionality. Not only would it be a marketing nightmare to explain a Biblical proportional response to a public forum such as the United Nations, but it would go against not only the spirit but also the letter of Biblical law.

If we view the concept of conflict as an occurrence not between individuals but rather between groups then it is the Biblical view that war is a part of life. Violence begins almost with creation. Lemech's mass murders (Genesis 4:23-24) may have tipped the scales in God's non-proportional decision in the Noah story to wipe life off the face of the earth. Throughout Genesis and Exodus disproportional responses seem to be the norm rather than the exception. Thus, in the case of Sodom and Gomorrah (Genesis 1.8), there can be no doubt that God's desire to destroy these cities, as symbols of evil are a disproportional use of force. The same pattern is seen in the Book of Exodus. What are the Ten Plagues if a non-proportional response to slavery? (Exodus 7 et al). On the other hand, we can see the tenth plague, the killing of Egypt's firstborn sons, as a direct proportional response to Pharaoh's order to kill all of Israel's male babies. A war crimes investigation of the Tenth Plague would clearly put God on trial? The culmination of this "war" between Pharaoh and God is the destruction by drowning of Egypt's entire army at the Red Sea (Exodus 14:26-31). In the case of Korach's rebellion against Moses (Numbers 16), the Earth literally swallows up those who participated in the failed coup d'état.

Taken together the Biblical tales relating to the macro level indicate a series of collective principles. Included in these are:

There is a major difference between micro level jurisprudence and international law as interpreted on the macro level. Although the Bible appears to promote proportionality on the micro scale, all out war does not seem to demand proportional responses, but rather disproportional responses that lead to one side's claiming victory.

On the micro level the laws of proportionality apply so as to move hostilities away from the personal and instead develop a methodology that permits a rational approach to retribution

The Biblical text then posits that the micro levels and the macro levels of reciprocity have nothing in common. In fact, it is a methodological error to confuse the two. The eye for an eye theory may work in a court of law but it is simply not a part of warfare and may only lead to greater warfare. On the macro level, micro crimes coalesce into collective evil. This means that

- In the face of evil, there can be no proportional response. Instead evil must be confronted and destroyed
- To choose not to destroy evil is in and of itself evil
- Once a war is declared it is a mistake not to fight to win the war. To fight a half war is to perpetuate war and thus to do more damage than good.

In wars there is collateral damage. There is, however, a major difference between choosing to bring about the destruction of innocent civilians and the accidental death of innocent civilians. While in both cases unjust death does occur, the two should not be confused

When we examine the Hebrew Bible's principles concerning proportional versus non-proportional responses it becomes clear from the Biblical perspective that those calling for macro level proportional responses may at best be naive and at worst guilty of seeking a perpetual state of war. In fact, the Bible seems to present its readers with the notion that in times of war, the use of proportionality may be more than foolish, it may be nothing more than an alliance with evil and thus to be condemned as evil.

Proportionality in the New Testament and Christian Thought

The Hebrew Bible's same concepts also apply in the Christian world. For example, we read in the Gospel of Matthew 5:38-42 the following: "Woe to you, teachers of the law and Pharisees, you hypocrites! You give a tenth of your spices: mint, dill and cumin. But you have neglected the more important matters of the law– justice, mercy and faithfulness. You should have practiced the latter, without neglecting the former. You blind guides! You strain out a gnat but swallow a camel"

Even if we extract the anti-Semitism found in this passage (the text provides a non-proportional analysis of the Pharisees), two concepts become clear: (1) the text is not referring to political proportionality but rather to issues of proportional taxes and (2) as in so much of Hebrew Scripture it is referencing the micro rather than the macro. Another possible concept of proportionality is also found in Mathew when we read "You have heard that it was said, an eye for an eye and a tooth for a tooth.' But I say to you, do not resist an evil person; but whoever slaps you on your right cheek, turn the other to him also." The problem here is that we are given a non-proportional response in the negative. In other words, the text is indicating that in the face of evil, the response is nothing. Such a response is clearly not practical in the political world and the verse has always been interpreted on the micro rather than on the macro level.

Proportionality: Tourism and Marketing

When tourism professionals use the term proportionality they often mean it in the mathematical sense of the term rather than in the term's political sense. Proportionality, in this sense, tends to fall into two categories: Direct proportionality where as "X" increases then "Y" increases in a proportional level. For example, if a man is paid a set wage of X dollars an hour, then we can argue his or her wage will also increase at a proportional rate. Tourism marketers often take the position that more advertising will lead to greater numbers of visitors and that a greater number of visitors twill then translate into more revenue.

Inverse proportionality argues that as "X" increases then "Y' decreases. Thus, from the perspective of science as we move away from a sound, it appears to us that the sound diminishes. Translating this term into tourism we may argue that as violence increases into a specific location then the number of visitors to that locale will decrease.

From the perspective of inverse proportionality we can see the relationship between political proportionality, purposeful negative marketing and political proportionality. Tourism leaders during the Covid-19 crisis were faced with new issues of proportionality. With the halting of most transportation, there was no amount of marketing that would solve the current crisis. In a like manner how does one maintain social distancing in high human contact locations such as sporting events, theatrical performances or casinos? Additionally, can tourism employees argue that their being laid-off is a non-proportional reaction to a situation for which they bear no responsibility?

CONCLUSION

Tourism leaders are often faced with the question of: are they better off if their government takes a non-proportional stance that brings hostilities to a rapid conclusion, even if this stance may produce unwanted civilian casualties or should tourism learn to tolerate a stance that ends in a stalemate? In a pandemic situation, the tourism industry rapidly collapsed, thousands were thrown out of work, and the industry became almost totally dependent on the good graces of governments. For example, the Wall Street Journal reports that: "Reeling from the coronavirus crisis, U.S. airlines are seeking over $50 billion in financial assistance from the government, more than three times the size of the industry's bailout after the Sept. 11 attacks." https://www.wsj.com/articles/airlines-seek-up-to-50-billion-in-government-aid-amid-coronavirus-crisis-11584378242 <April 19, 2020>

As seen above tourism professionals use direct and indirect concepts of proportionality in their preparation of budgets, in policy development and in marketing. Often overlooked, however, is an understanding of political proportionality and how it impacts tourism marketing becomes vital for tourism leaders. There are thus a number of reasons that tourism-marketing professionals need to understand proportionality, not just in the mathematical or business sense but also in the political and military sense and the impact of these latter meanings on tourism marketing. Among these reasons are:

1. Military and political concepts of proportionality impact the industry's bottom line. A perfect example of the overlap between proportionality in the political-military sense and tourism is in the current debate in the United States over the "visa waiver' program. The visa waver program permits citizens of specific countries to enter the United States without a passport. Some have argued that as the potential for terrorism increases it is essential to roll back or cut this program out entirely.

Others have argued that as it becomes more difficult for visitors to enter the United States, visitors will shy away and that the nation will lose a vital source of income. Thus, we see an interfacing between tourism policy and issues of political and military proportionality.

2. Political proportionality determines how a nation presents itself to the world. In an industry that thrives on peaceful coexistence a continual state of no-war-no peace causes major reassessments in the way that tourism marketing is done.

This interfacing between marketing and political proportionality also impacts tourism in a number of ways. Among these are: (1) Proportionality can become a means for "negative marketing" or a way for one side to discredit the other side. Thus, claiming a non-proportional response is a means to undercut opposition and to argue that the other side is not playing fair. Proportionality sounds fair, and makes the party clamoring for a proportional response seem responsible but in reality is a form of negative marketing akin to a political campaign. (2) International criticism of a locale impacts that locale's ability to promote itself, the locale is turned into a pariah and visitors are stigmatized, and (3) tourism needs peace and proportional responses have traditionally lead to states of stalemates. We can then argue that if proportional responses toward terrorism are taken, then we have a political status quo in which tourism leaders are clearly placed on the defensive. If on the other hand, terrorists fear attacking a particular location, this may mean that tourism leaders can focus on a more direct (positive form) of proportionality such as increased marketing. In an age when we know that bio-terrorism or warfare is capable of total economic destruction along with the deaths of tens of thousands of people, how we react might also determine the extent of future hostilities. Tourism in such a world can simply not exist. Thus the decisions made today will determine the strength of the industry for the foreseeable future. Western nations often claim proportional responses as a way to take the high moral ground. Clearly no one desires that civilians suffer. Yet, we have to ask ourselves. Is it better to be in a state of perpetual war? Can the tourism industry exist in a perpetual war? Do proportional responses not only lengthen conflicts but also create negative marketing situations in which not only civilians face never-ending conflicts but also thousands are hurt both economically and morally? If tourism is a means to create better understanding between peoples, then can its destruction due to mismanaged ideas of proportionality benefit or harm the world?

If tourism is about the pursuit of understanding between peoples then do proportional responses produce conflict resolutions or merely moral standoffs? George Orwell published his famous book, *1984* in 1949. The novel then set in the future depicts a world in continual conflict. His is a world where Ministries of War are called Ministries of Peace, Marketing becomes a "Ministry of Truth" in which anything but the truth is stated. Were Orwell alive today would he see the use of "proportional responses" as a tool by which governments chose to take the "high moral ground" so as to keep us in a state of war? How would Orwell have understood our interconnected world where pandemics infect not only cities or provinces, but nations and continents?

REFERENCES

Chesney-Lind, M., & Lind, I. Y. (1986). Visitors as Victims: Crimes against Tourists in Hawaii. *Annals of Tourism Research, 13*(2), 167–191. doi:10.1016/0160-7383(86)90036-8

Gartell, R. B. (1988). *Destination Marketing*. Kent Hunt Publisher.

Lewis, R. (1988). *Chambers Richard*. Marketing Leadership in Hospitality.

Pizam, A. (1997). Article. *Journal of Travel Research, 36*(1), 23–28. doi:10.1177/004728759703600104

Tarlow, P. (2014). *Tourism Security*. Elsevier.

Tourism Safety and Security. (2009). In *The Sage Handbook of Tourism Studies*. SAGE.

Vanhove, N. (2005). *The Economics of Tourism Destinations*. Elsevier.

Veblen, T. (1899). *The Theory of the Leisure Class*. Mentor Boos.

ADDITIONAL READING

Altschuler, W. (2020, March 17). How To Help The Travel And Tourism Industry During The Coronavirus Crisis. Retrieved May 15, 2020, from https://www.forbes.com/sites/wendyaltschuler/2020/03/17/how-to-help-the-travel-and-tourism-industry-during-the-coronavirus-crisis/

Bischke, U. M. (2015, January 30). An Eye for an Eye. Retrieved May 15, 2020, from http://www.november.org/razorwire/rzold/04/0407.html

Bowen, C., Fidgeon, P., & Page, S. J. (2014). Maritime tourism and terrorism: Customer perceptions of the potential terrorist threat to cruise shipping. *Current Issues in Tourism, 17*(7), 610–639. doi:10.1080/13683500.2012.743973

Bryen, S. (2014, July 20). The Doctrine of Proportionality. Retrieved May 15, 2020, from https://www.gatestoneinstitute.org/4462/proportionality-doctrine

Bryen, S. (2014, July 20). The Doctrine of Proportionality. Retrieved May 15, 2020, from https://www.gatestoneinstitute.org/4462/proportionality-doctrine

Collman, A. (2020, April 16). The US is investigating the theory that the COVID-19 virus was accidentally released from a Chinese lab. Retrieved May 14, 2020, from https://www.businessinsider.com/us-investigating-chinese-lab-covid-19-theory-2020-4

COVID-19's Impact On Travel Creates Widespread Uncertainty. (n.d.). Retrieved May 15, 2020, from https://www.alpha-sense.com/insights/covid-19-travel-uncertainty

Duke, S. (2015, February). Obama Anti-terrorism "Jobs for Jihadists" Concept Is a Marxist Idea. Retrieved May 15, 2020, from https://www.thenewamerican.com/usnews/item/20150-obama-anti-terrorism-jobs-for-jihadists-concept-is-a-marxist-idea

Enders, W., & Sandler, T. (1991). Causality between transnational terrorism and tourism: The case of Spain. *Studies in Conflict and Terrorism, 14*(1), 49–58.

Fletcher, J., & Morakabati, Y. (2008). Tourism activity, terrorism and political instability within the Commonwealth: The cases of Fiji and Kenya. *International Journal of Tourism Research, 10*(6), 537–556. doi:10.1002/jtr.699

Henderson, J. C., Shufen, C., Huifen, L., & Xiang, L. L. (2010). Tourism and terrorism: A hotel industry perspective. *Journal of Tourism. Hospitality & Culinary Arts, 2*(1), 1–14.

Higgins-Desbiolles, F. (2003). *Hostile meeting grounds: Encounters between the wretched of the earth and the tourist through tourism and terrorism in the 21st century* (Doctoral dissertation). Elsevier.

Kkienerm. (n.d.). Counter-Terrorism Module 4 Key Issues: Criminal Justice Responses. Retrieved May 15, 2020, from https://www.unodc.org/e4j/en/terrorism/module-4/key-issues/criminal-justice-responses.html

Korstanje, M. E. (2016). *The rise of thana-capitalism and tourism*. Taylor & Francis. doi:10.4324/9781315457482

Korstanje, M. E. (2017). *Terrorism, Tourism and the End of Hospitality in the'West*. Springer.

Lex Access to European Union law. (n.d.). Retrieved May 15, 2020, from http://europa.eu/legislation_summaries/glossary/proportionality_en.htm

Libya hotel attack: Five foreigners among nine killed. (2015, January 25). Retrieved from https://www.bbc.com/news/world-africa-31001094

Libya hotel attack: Five foreigners among nine killed. (2015, January 28). Retrieved May 15, 2020, from https://www.bbc.com/news/world-africa-31001094

Lim, N. (2020, February 13). Bernie Sanders once slammed tourism industry for providing 'low-paying, menial jobs'. Retrieved May 15, 2020, from https://www.washingtonexaminer.com/news/bernie-sanders-once-slammed-tourism-industry-for-providing-low-paying-menial-jobs

Pizam, A., & Smith, G. (2000). Tourism and terrorism: A quantitative analysis of major terrorist acts and their impact on tourism destinations. *Tourism Economics, 6*(2), 123–138. doi:10.5367/000000000101297523

Quotes, M. (n.d.). Retrieved May 15, 2020, from https://www.brainyquote.com/quotes/keywords/marketing.html

Richter, L. K., & Waugh, W. L. Jr. (1986). Terrorism and tourism as logical companions. *Tourism Management, 7*(4), 230–238. doi:10.1016/0261-5177(86)90033-6

Rule 14. Proportionality in Attack. (n.d.). Retrieved May 15, 2020, from https://www.icrc.org/customary-ihl/eng/docs/v1_cha_chapter4_rule14

Science Daily. (n.d.). Retrieved March 17, 2020, from https://www.sciencedaily.com/releases/2020/03/200317175442.htm

Sönmez, S. F. (1998). Tourism, terrorism, and political instability. *Annals of Tourism Research, 25*(2), 416–456. doi:10.1016/S0160-7383(97)00093-5

Sönmez, S. F., Apostolopoulos, Y., & Tarlow, P. (1999). Tourism in crisis: Managing the effects of terrorism. *Journal of Travel Research*, *38*(1), 13–18. doi:10.1177/004728759903800104

Stephens, B. (2009, January 07). An Endgame for Israel. Retrieved May 15, 2020, from https://www.wsj.com/articles/SB123120028301755589

Tarlow, P. (2011). Tourism disaster management in an age of terrorism. *International Journal of Tourism Anthropology*, *1*(3-4), 254–272. doi:10.1504/IJTA.2011.043709

Tarlow, P. (2011). Tourism risk management in an age of terrorism. *Economía Autónoma*, *4*(7), 18–30.

Tarlow, P. (2014). *Tourism security: strategies for effectively managing travel risk and safety*. Elsevier.

Tarlow, P. E. (2006). Terrorism and tourism. In *Tourism in turbulent times* (pp. 103–116). Routledge. doi:10.1016/B978-0-08-044666-0.50014-2

US oil prices turn negative as demand dries up - BBC News. (n.d.). Retrieved May 15, 2020, from https://www.bbc.com/news/business-52350082

World Tourism Organization. (n.d.). Retrieved May 15, 2020, from https://www.unwto.org/impact-assessment-of-the-covid-19-outbreak-on-international-tourism

Yan, H. (2015, February 16). Danish police kill man suspected in attacks. Retrieved May 15, 2020, from https://www.cnn.com/2015/02/15/europe/denmark-shooting/index.html

KEY TERMS AND DEFINITIONS

COVID19: It is known as well SARS-co2, and surfaces recently in the end of 2019 in the Chinese city of Wuhan. This virus outbreak not only shocked the world but also stopped a great part of commercial activities in the world.

Pandemic: It is an epidemic disease spread across a large region or reaching the four continents.

SARS: This signals to Severe Acute Respiratory Syndrome, a viral respiratory disease emerged in the 2000s.

Terrorism: This is the use of illegal violence for political or religious goals.

Tourism Security: An emerging discipline oriented to study how to strength security and safety in the tourist system.

Chapter 2
Airbnb as a Tourism Supply Chain Champion for Small Businesses:
The Case of South Africa

Unathi Sonwabile Henama
https://orcid.org/0000-0002-1111-0729
Department of Tourism Management, Tshwane University of Technology, South Africa

Lwazi Apleni
https://orcid.org/0000-0001-8521-6174
Department of Recreation and Tourism, University of Zululand, South Africa

ABSTRACT

Airbnb is viewed as an advocate of small micro medium enterprises (SMMEs) in order to boost their revenue. Furthermore, in 2016, Airbnb released a report stating that its community generated R2.4 billion in economic activity in SA, which is the estimated sum of guest spending and host income. It is important to elucidate that SA's tourism fraternity has been amongst the best performing sectors amid the challenges it is faced with. The sharing economy concept requires all tourism stakeholders to work together (public-private partnership) in order to see its fruition. The sharing economy represents the power of the collaborative consumption and become a means to create an additional value chain for the tourism industry by decreasing the barriers of entry. The Airbnb Africa Academy is a pro-poor skills development and support initiative that was piloted in South Africa, and it has increased access and the success of hosts on the Airbnb platform.

DOI: 10.4018/978-1-7998-4330-6.ch002

INTRODUCTION

"The tourism industry has established itself as one of the leading industries in the world economy, being promoted for its economic benefits across all countries. The economic benefits include improving the balance of payments, attracting foreign exchange, in addition to increasing state coffers through the taxation of non-residents. Tourism is an export industry, where the tourism product offering can be consumed at the destination area, which means that the bulk of the value adding happens at the destination area" Henama (2017a: 72). Tauoatsoala *et al.* (2015) noted that tourism activitties are associated with the hosting of the tourism consumer at the local area, because tourism consumption happens at the destination area. This is due to the simultaneous consumption and production of the tourism product offering. "Tourism has greater impact economic impact on an economy, as the injection of tourism revenue has a direct, indirect and induced impact on the economy" Sifolo & Henama (2019: 198). "Following the first racially inclusive democratic elections in 1994, the South African Government efforts to eliminate poverty have been frustrated by the continued shedding of jobs from the formal economy" Henama (2014b: 1).

"Tourism can catalyze other industries such as manufacturing and agriculture when the tourism industry is planned to be pro-poor and benefit the destination area" Strydom *et al.* (2019: 9). In the case of South Africa, tourism has been recorded as the "new gold" as the growth of tourism is similar to the mining exploits of South Africa in the 1970s. The tourism industry has grown in leaps and bounds in South Africa (SA) since the first democratic elections in 1994 according to Henama (2018). It has grown in its economic contribution and is now called the "new gold", and is a major employer within the economy. In South Africa, the National Development Plan (NDP) recognizes tourism as one of the key drivers of employment and economic growth. The NDP envisages the promotion of SA as a tourist and business events destination. Tourism is projected to grow the tourism industry, and is expected to attract 20 million tourists by the year 2030. The direct economic impact of the tourism industry has significant indirect and induced impacts. The Table below depicts the contribution of the sector in 2018.

Table 1. Socio-economic contribution of tourism in 2018

Economic Indicator	Contribution
International arrivals	10.5 million
Direct jobs	720 000 direct jobs
Total jobs	1.5 million direct and indirect jobs accounting for 4.5 percent total employment in South Africa
GDP – Direct contribution	R130 billion, accounting for 2.8 percent contribution to the total GDP
GDP – Total contribution	R425.5 billion, accounting for 8.6 percent total contribution to the economy
Total Tourism Foreign Direct Spend (TTFDS)	R58.57 billion
Domestic Trips	13 million
Total Direct Domestic Spend (TDDS)	R18.29 billion

Source: Parliamentary Monitoring Group, 2019

As depicted in Table 1 above, the contribution of tourism to the economy is measured against economic indicators such as direct contribution to the GDP; total contribution to the GDP; direct contribution to employment; total contribution to employment; international tourist arrivals; domestic holiday trips; visitor exports; and travel and tourism investments. The international arrivals and domestic trips have a direct impact on other economic indicators such as contributing to the GDP and job creation, and the Tourism Committee will be conducting close oversight on these two indicators in the 2019/20 Financial Year. In order for tourism to be developmental, it as to improve the Quality-of-Life and standard of living of the locals, which would in turn, make them tourism ambassadors of their destination and being tourism friendly. "When locals benefit from the tourism industry, they become better hosts for tourists, due to the value domain" Henama and Sifolo (2017: 48). This can be done by developing infrastructure that would be used by both tourists and locals, being a tourism legacy that would improve the Quality-of-Life of locals.

Theoretical Framework: Tourism Supply Chain Management

The supply chain includes a series of activities that are connected in the processes of transforming raw materials into finished and final products and about the connecting of processes from production to customer delivery. Therefore, supply chain can be conceived to be enterprises cooperating in different areas in the network of supplying products and services. "Customers' demands will be met through the supply chain product offerings" Badenhorst-Weiss and Nel (2011: 3). The tourism industry is dominated by a few large principals that dominate the more capital-intensive sectors such as aviation and tour operating, and most businesses in tourism are SMMEs. According to Jaremen and Nawrocka (2015), the largest tourism companies usually represent corporations integrated both horizontally and vertically. Tourism Supply Chain (TSC) is unique as the tourism industry is highly fragmented and heterogenous, consists of many different components including accommodation, transportation, dinning, shopping and sight-seeing. The tourism industry is an amalgam of various suppliers and providers, which means that the purchase of a tourism product offering includes companies that are located in other industries and sectors to deliver the tourism product offering. The diversity of the tourism required greater coordination and collaboration to deliver the tourism experience.

The tourism product offering does not transfer ownership to the customer, and simultaneous production and consumption means that the consumption only happens at the destination area. The tourism customer has to undertake an element of travel to consume at the destination area, which makes tourism supply chain transport intensive. The variability in delivery makes controlling quality and standards. Ghaderi *et al.* (2018) defined Tourism Supply Chain (TSC) as a network of tourism organizations engaged in different activities, ranging from the supply of different components of tourism products or services such as flights and accommodation to the distribution and marketing of the final tourism product at a specific tourism destination and involves a wide range of participants in both the private and public sectors. Dordevic (2010) noted that TSC partly includes players from hospitality as well as travel. In the context of South Africa, the tourism industry is private sector led, and the public service is supposed to create an enabling environment to allow for the private sector to develop tourism product offerings to meet client's needs. The emergence of ICT has had a revolutionary impact on the supply chains, especially in tourism. According to Jaremen and Nawrocka (2015), ICT has also strengthened tourist's independence in tourism product construction and thus upgraded its importance among the components of the tourism supply chain.

Font *et al.* (2008) noted that the sustainability in tourism depends strongly on the developmemt of better linkages between supply and demand. "TSC framework description also considers power relationships and business links among partners" Jaremen and Nawrocka (2015: 105). Sifolo (2015) viewed that that tourism supply chain management has a potential panacea to the looming tourism sector malaise. The key challenge for tourism enterprises is to balance between competition for customers and market share, and sharing among stakeholders. Molefe *et al.* (2018: 14) state that "Tourism enterprises must invest in understanding the do's and don't's of tourism supply chain management to ensure that they continue to mainatin supply and demand". The sharing economy has changed the supply chain systems. The sharing economy has provided an avenue to transact on underutilized assets, being a new way for inventory management. "Sharing economy business models are hosted through digital platforms that enable a more precise, real-time measuremet of spare capacity and the ability to dynamically connect that capacity with those who need it" PWC (2015: 15).

Sharing Economy

In the case of the sharing economy, it can be associated with Customer-to-Customer (C2C) transactions, and this has been the driving force of the sharing economy, which includes companies such as Airbnb and Uber. The sharing economy is about Peer-to-Peer (P2P) or person-to-person transactions, facilitated by digital platforms for collaborative consumption. "Peer review systems operate as a control mechanism in the sharing economy. The peer review system has both positive and negative features. It promotes trust in the sharing economy" Henderson (2016: 28). The peer review systems and ratings can be abused, and are undertaken to prevent market failure. Henderson (2016) notes that online reviews are a significant driver of consumer behaviour, providing a way for customers to discover, evaluate and compare products and services on the web. Customers are more inclined to believe the reviews of those that have experienced a stay at an Airbnb property. Guest reviews of their stay are compulsory as part of their differentiate feature, and as part of quality assurance.

Oskam, and Boswijk (2016) noted that the sharing economy can be distinguished from other economic forms:

- Sharing is about consumer-to-consumer platforms and not about renting or leasing a good from a company (business-to-business). In the latter case, we would speak of the product-service economy, where a consumer gains access to a product whilst the service provider retains ownership.
- Sharing is about consumers providing each other temporary access to a good and not about the transfer of ownership of the good. Thus, the sharing economy does not include the second-hand economy, in which goods are sold or given away between consumers (as occurs on online platforms such as eBay or Facebook).
- Sharing is about the more efficient use of physical assets and not about private individuals developing each other a service. After all, physical goods can go unused, but people cannot. Internet platforms that bring consumers together to provide each other with service represent the on-demand economy. An example of such platforms is Task Rabbit, through which you can hire people to carry out work around the house.

The combination of low sunk costs of entry, a larger number of potential entrants, downward-moving incumbent's prices, and the entrant's lower production costs, would suggest that entry or the threat of entry would be eliminate the economic rent earned by the incumbents. This should mean that economic rents would disappear, not so much from the reduction in the retail price of a yellow cab, as from the reduction in the amount of business that drives up average production costs and reduces total revenue" Gabel (2008: 530). In South Africa, there is a high failure of businesses and the low levels of entrepreneurship further undermine efforts to drive economic growth, innovation and job creation. Furthermore, Business Partners (2018) noted that 31% of respondents indicated fear for failure as one of the reasons that prevent them from being entrepreneurs, according to the 2017/2018 Globa Entrepreneurship Monitor Report. The built-in business sytems within companies in the sharing economy ensure success for partner drivers in Uber and hosts in Airbnb, ensuring sustainable businesses.

Chua *et al.* (2019) noted that there are several drivers of the sharing economy:

- **Economic Drivers:** Amid economic crisis, consumers' financial constraints and urban problems, sharing economy compliments in the current situation as it gives solutions to the problem more than a substitute. During high unemployment rate, sharing economy can act as a supplementary source of income in times and provide flexibility for individuals to gain financial independence.
- **Social Drivers:** Being connected and able to interact with one another makes sharing economy more effective. Sharing economy compliments the need of the new consumers as it opens a new channel to interact with strangers that leads to authenticity of its experience.
- **Environmental Drivers:** When people start to share, people learn to optimize idle and underutilize assets that reduce production costs.
- **Legal Drivers:** Sharing economy has disrupted traditional companies and has established tough competition since then. Authorities are becoming more sympathetic with conventional companies and try to mitigate the disruption by regulating sharing companies. Sharing companies should work with local governments to educate them about their real business model and comply with what is required to make their operations formal and abide by the regulations and taxation.

Airbnb becomes a platform that meets the needs of millenials who get to book authentic experiences and choose hosts according to their interests. "The millennials have changed how travel should be. Traveling has become more affordable and engaging as millennials want to part take and experience more authentic ways of traveling. Studies have shown that millennials can easily trust new innovations, even trusting their belongings and safety to a stranger. This is why Airbnb's growth has been increasing exponentially over the years. Furthermore, social media, ratings and electronic Word of Mouth (eWOM) has become the medium of trust as this become the quality basis of sharing economy users" Chua *et al.* (2019: 32).

Airbnb in South Africa

"Since 2010, Airbnb has been helping South African households and communities create economic opportunities for themselves through innovative and sustainable travel. Six hundred thousand guests arrived at Airbnb listings in South Africa in the past year, representing year-on-year growth of 144%" Airbnb (2017: 14). Ever since Airbnb launched, hosts have been able to make on average R28 000 per year by hosting tourists as noted by Alfreds (2016). Destiny (2018) noted that hosts around South Africa have earned over $260 million (about R3.8 billion) and supported over 22 000 jobs. South Africa is the leading market for Airbnb on the African continent, with Cape Town being the leading city in terms of listing. "The company has reported 190% growth in local hosts to 7 500 with 134 000 inbound guests, representing 250% growth and guest staying an average 4.9 nights" Alfreds (2016: 1). The concept is relatively new in South Africa. For instance, according to the Holmes (2015), Airbnb was officially launched in July 2015. Around 2016, international visitors outsold locals in terms of Airbnb bookings,, representing 72% of guests (Alfreds, 2016).

Shevel noted that (2016) Airbnb Guests in South Africa, represent the following characteristics:

- **134 000:** Number of inbound guests in the past year, up 250% over 2014;
- **4.9:** Average length of stay (measured in nights) per guest;
- **37:** Average age of the booking guests;
- **91%:** Proportion of trips involved four or fewer guests;
- **2.5:** Number of people in the average travel group; and
- **4.7:** Average host rating, out of five.

However, it is argued that early adopters started posting their properties on the site a few years ago. Perhaps that's the reason the San Francisco-based company identified South Africa as its hub for expansion into Africa. Interestingly, South Africa represents about a third of all the Airbnb listings in Africa, with Cape Town leading the Airbnb listings in South Africa. "In Cape Town, there were 17 600 active listings on Airbnb, earning a combined income of R762 million over the past 12 months. In the past year, Airbnb hosts in Cape Town welcomed almost 290 000 visitors, which resulted in an estimated economic boost of R2,4 billion for the city" Tourism Update (2018: 3). Nkanyeni (2017) noted that Cape Town became the first African city to sign an agreement with the City of Cape Town to promote the benefits of people-to-people tourism for the residents and their communities. The agreement was signed on 18 October 2017 by the former Cape Town Executive Mayor, Patricia de Lille and Airbnb Global Head of Public Policy and Public Affairs, Chris Lehane. The agreement between Airbnb and the City of Cape Town, makes sense as Cape Town had the most leading listings on the African continent, would allow for Cape Town to be promoted as a unique travel destination (Tourism Update, 2018).

Figure 1. Airbnb consumption across the African continent in 2017
Source: Wee Trackêr (2019)

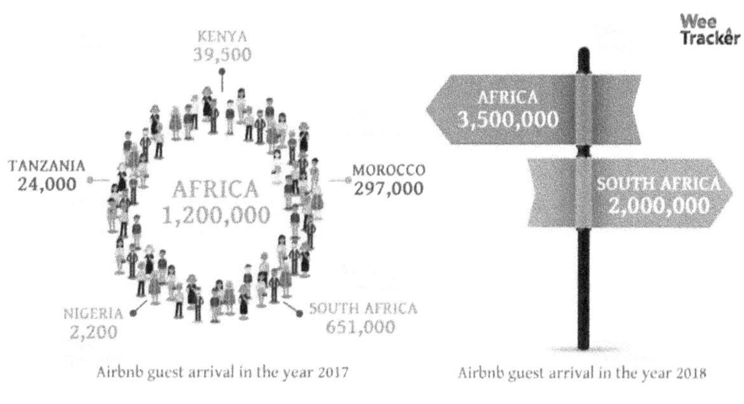

With the figure (depicted above) in mind, it could be surmised that Airbnb does not have as much a stronghold or as loud clout in Africa as it does in some other parts of the world. However, as per the diagram above, SA enjoys a competitive advantage compared to its African counterparts. SA, Airbnb's largest African market, is still waxing strong and going steady at a growth rate of 65%; a feat that is made all the more glaring by the fact around 2 million of the total of 3.5 million guest arrivals in Africa have been in SA alone. With multiple initiatives in Africa and policies that might be suggestive of an onslaught on the market, Airbnb's posture depicts an interest making the most of the tremendous opportunities that are available in African hospitality industry. A good illustration of this improved posture can be cited in its collaboration with the Mother City, Cape Town. This example is unique in that it represents Airbnb's first partnership with an African city, even though the company is known to have already collaborated with over 300 governments worldwide.

"In 2016, 63% of Airbnb hosts in South Africa were women. A recent study of Airbnb's global women host community found that the typical women hosts in South Africa earned nearly US$2,000 last year, more income than earned by the typical women hosts in Brazil, China or India. More than 60% of women hosts in South Africa are Super hosts, hosts who are specifically designated by Airbnb as hosting guests frequently, receiving a high number of five-star reviews and being exceptionally responsive to guests and committed to reservations" Airbnb (2017: 15).

Senior citizens in South Africa have jumped on the Airbnb bandwagon, and have listed their properties and are curating Airbnb Experiences, as a means to earn an additional income. Senior citizens are usually "time-rich" because they are retirees, and can therefore dedicate themselves fully. "Senior South African hosts are also "much loves", receiving better reviews than any other age group, with 85% boasting a five-star rating" De Villiers (2018: 1).

Airbnb Disaster Response Program

Airbnb has a Disaster Response Program that offers free accommodation by Airbnb hosts to those in need during times of global crisis. "Airbnb hosts can respond to crises in their communities by offering housing to displaced neighbors and relief workers deployed to help. During an emergency, Airbnb may email local hosts with information about how to help and how to offer their extra space to affected community members. These hosts are still covered by the Host Guarantee, and Airbnb's fees are waived" Airbnb (2019: 3). Airbnb hosts that respond to participate in the Disaster Response Program after activation, would list their spaces for free of charge and Airbnb will waive all booking fees. The Disaster Response Program would be actived by Airbnb, and the tool automatically contacts hosts in the affected areas for assistance in hosting. Airbnb would achieve the Disaster Response Program by partnering with governments, disaster relief organizations and Non-Government Organizations (NGOs). During Hurricane Sandy that devastated New York in 2012, Airbnb hosts opened their homes and warm hospitality. "The concept began organically back in 2012 in the wake of Hurricane Sandy, and has since grown into a global disaster response initiative that makes it easy for Airbnb hosts to provide space for people in need when disasters like this strike" Jamal (2017: 2).

According to Airbnb (2019), when the Disaster Response Program is activated, the partnership includes the following:

- Providing disaster and emergency preparedness education material to Airbnb hosts so they can become the most prepared residents on the block.

- Help arrange community emergency response training with local preparedness expert to train Airbnb hosts as response leaders within their neighborhoods.
- Use Airbnb communication channels to notify hosts and visitors about significant emergencies.
- Depending on the emergency and where Airbnb can have a positive impact, activate the disaster tool to help hosts open their homes as temporary housing for relief workers deployed for the response or displaced neighbors.

Cape Town is not just the leading tourism destination in South Africa, it is also leading in terms of Airbnb listings on the African continent. Cape Town is the seat of Parliament and the Western Cape Province is popular for the diversity of tourism experiences it can offer. Towns and cities such as Knysna, Stellenbosch, Hermanus, Ceres, George, Plettenberg Bay, and Still Bay are major tourism destinations, that have experienced an inward migration of tourists, and those that purchase second homes. Furthermore, emigration has changed the character of the towns and cities in the Western Cape as residents of the Gauteng Province, purchase homes instead of migrating out of the country, and migrate to become permanent residence in the Western Cape, and commute to Gauteng and other provinces for business and employment. Knysna is one the most popular listed destinations in the Western Cape, and in 2017 a major fire engulfed the town, destroying properties. Knysna is surrounded by the Knysna Forest, which according to Mgwadlamba (2017), is South Africa's largest Afrotemperate Forest spanning across 65 000 hectars of land from George to Tsitsikamma. "Knysna located along South Africa's Garden Route in the Western Cape, is famous for the Oyster Festival and magnificent views from Leisure Isle to the Knysna Heads, among other things, which has made it a popular destination for local and international travellers" Bizcommunity (2017: 1).

The Pick 'n Pay Knysna Festival happens for ten days usually from 7-16 July yearly, is a food and wine-related festival. The festival was threatened when the Knysna area was ravaged by fires in late June 2017, leading to an evacuation of nearly 100 000 people. Firefighters from other regions of South Africa had their flight bookings sponsored by British Airways and low cost carrier (LCC) kulula.com according Henama (2017a) to help in efforts to stop the fire. The Knysna Forest was particulally affected by the blaze that destroyed houses and some commercial tourism accommodation providers such as Phantom Forest Lodge, Bradach Manor, Kanonkop House, The Big Tree House Lodge, Bridgewater Manor, and Four on Faure (Henama, 2017a). The Knysna Forest remained intact. For the first time in the South African history, the Disaster Response Tool was activated. Jamal (2017) noted that Airbnb offered temporary housing for displaced people in Knysna at no charge until 30 June 2017. The messaging after the Knysna fires was that the public must support the Pick 'n Pay 2017 Knysna Oyster Festival as an economic revitalisation strategy for the devastation after the Knysna fires. "Some of the properties destroyed by the Knysna fire included several tourism accommodations, which meant that there was a decrease in the number of accommodation establishments that would offer accommodation for the Knysna Oyster Festival. Airbnb once again provided a platform for an increase in the number of beds in Knysna when more residents, became Airbnb hosts" Henama (2018b: 11).Tourism has once again picked up in Knysna as local and international tourists returned, and the 2017 Knysna Oyster Festival was a resounding success.

Airbnb Africa Academy

Airbnb in the South African context had developed the Airbnb Africa Academy, as an additional measure for potential hosts to learn how to become hosts (for accommodation) and experiences (Airbnb Experiences). Airbnb usually operates remotely from hosts, allowing the technology platform as the only interface between Airbnb and hosts. The Airbnb Africa Academy can be regarded as a form of pro-poor Community-Based Tourism initiative as the vast majority of participants are from previously disadvantaged groups (PDIs) who were dispossessed by apartheid and colonialism. The Airbnb African Academy therefore creates an opportunity to achieve the following:

- Small business creation with Airbnb's technology platform; and
- Upskilling the participants to become hosts and after-service and back of the house support for a period of up to six months to ensure that hosts meet Airbnb standards, so that they become sustainable businesses.

The Airbnb Africa Academy is therefore a small business development success story that was developed to disproportionally benefit the poor. The Airbnb Africa Academy according to was succesfully piloted in collaboration with Open Africa, the South African College for Tourism and the Cape Innovation and Technology Initiative (CiTi), with the aim of empowering underpriveleged communities in South Africa (Tourism Update, 2018). This would be an inclusion strategy for underpiveleged communities into the tourism value chain provided by Airbnb and the sharing economy. The free of charge training provided as part of the Airbnb Africa Academy intentionally aims to level the playing field by training potential hosts, to ensure that they become succesful when they get listed, thereby achieving "Access with Success". The Airbnb Africa Academy is closely associated with the ideals of Airbnb, which seeks to make travelling more democratic not just for visitors, but by also attracting hosts who would have never thought in their wildest idea that they would operate and offer services within the sharing economy. Airbnb developed an Airbnb Africa Academy Toolkit that is downloadable, which provides a step-by-step guide to assist hosts and co-hosts in rural and township communities to understand the tourism opportunity presented by Airbnb. The project is the first of its kind in the history of Airbnb and was piloted in Cape Town, to ensure that would be hosts-get additional training and support to ensure that they meet the Airbnb standards, to ensure their "Access with Success".

The Airbnb Africa Academy is premised on having would be hosts apply for the two-day training, with the help of a locally based technology partner, that would interface in addition to providing additional support to Airbnb in getting the would be hosts ready for "Access with Success". The two days training would be divided into a practical component and possibly their firsthand interface with the Airbnb platform. Some of the would be hosts may have never experienced, a stayover at a commercial tourism accommodation establishment and/or been on an excursion tour before, therefore Airbnb has a responsibility of mitigating that experience and knowledge gap. This is a teaching and learning opportunity as they become hosted overnight and/or participate in a tour (Airbnb Experiences) with Super hosts, who are perennial hosts with exceptional reviews. According to Airbnb (2019), the Airbnb Africa Academy took place in Johannesburg and Cape Town in 2018, working with12 communities in Gauteng and the Western Cape.

"Our work now underway with women hosts in Cape Town's townships offers an especially compelling example of the economic empowerment homes sharing can bring. In Langa Township, Airbnb is partnering with iKhaya le Langa, a nonprofit dedicated to revitalizing the area through social enterprise and tourism" (Airbnb, 2017: 15). Airbnb has emerged as an opportunity to include PDIs into the tourism value chain as product owners. The Airbnb (2017) noted that the work of Airbnb in South Africa has demonstrated that Airbnb can be a powerful way for women in emergeing economies to independently achieve greater economic, professional and social empowerment.

CONCLUSION

Airbnb has provided many South Africans with "access" into the tourism value chain in SA. Ensuring that Black South Africans benefit from the tourism industry is important to correct one of the legacies of apartheid, a tourism industry that was "anti-developmental" that excluded Black South Africans. Interesingly, the Tourism White Paper (1996) elucidates that tourism in SA has been deemed as a "White man's thing". The existing tourism value chains, which are vertically and horizontally integrated, were very concentrated, with high barriers of entry that prevented disruption by competition. Airbnb hosts demographically are older, over the age of 60 years, and this provides an avenue for entrepreneruship to earn money and supplement their retirement savings. The tough economic conditions in South Africa, which include a declining economy, can be mitigated if more time rich retirees use Airbnb to earn an income, whilst maintaining their lifestyle.

The price premium associated with accessing the Internet and have mobile wifi network in South Africa remains high, and this increases the cost of business. Poor connectivity, especially in townships and rural areas has a negative impact on accessing the benefits of the Internet and the sharing eocnomy. The success of the sharing economy, therefore depends on lowering the costs of data and access to fibre, which depends on government regulations. The declining economy in South Africa has seen a decline in travel budgets by citizens and corporates, and the sharing economy services of Airbnb can reduce the cost of travelling. The post-Corona virus pandemic may lead to more tourism consumption going to Airbnb, as tourists will come out to visit, with less money due to the economies that would have suffered due to stagnation caused by the pandemic. This means that it can start to increase its appeal to businesspeople and hopefully government, and quasi-government agencies, considering that government is the largest procurer of goods and services within the South African economy.

It is important for the tourism industry to embrace the sharing economy as it has a potential to bring much needed economic spin-offs. Moreover, Apleni (2017) underscores that tourism is valued as one of the leading contributors to a sustainable South African economy. It needed to be borne in mind that the tourism industry is one of the fastest growing sectors as well as one with the best chance to create significant job opportunities that can further nurture patriotic pride and reduce poverty. The industry employs many thousands directly whilst creating even more indirect employment opportunities and playing an important part in the further stimulation of the stagnating economy (Apleni, 2017).

It is not only the local small businesses that benefit from the Airbnb sharing economy. The grim reality is that in many countries the economic outlooks are bleak. There are many people who are struggling to earn enough money to live without acquiring debt. Though, thanks to the growth of the Airbnb sharing economy, hundreds of thousands of Airbnb hosts are able to earn a much-needed extra income. As a matter of fact, according to data gathered by Airbnb, more than half of all Airbnb hosts have admitted that if it were not for the financial benefits of the Airbnb sharing economy, they would not have been able to afford to remain in their homes.

REFERENCES

Airbnb. (2016). *Airbnb and the Rise of the Millennial Travel.* Retrieved from: https://www.airbnb.com/

Airbnb. (2017). *Across the BRICS: How Airbnb connects the emerging economies.* Retrieved from: https://www.airbnb.com/BRICS

Airbnb. (2019). *Disaster Response Program.* Retrieved from: https://www.airbnb.com/disastor-response

Alfreds, D. (2016). *Here's how much money SA Airbnb hosts earn, Fin24.* Retrieved from: https://www.fin24.com/Tech/News

Almeida-Alonso, M. M., & Millán-Borrajo, F. (2019). Are Social Media Data Pushing Overtourism? The Case of Barcelona and Chinese Tourists. *Sustainability, 11*(12), 3356. doi:10.3390u11123356

Apleni, L. (2017). 'Residents' perceptions on urban tourism as a catalyst for economic development: A case study of Buffalo City, South Africa. *African Journal of Hospitality, Tourism and Leisure, 6*(3), 1–12.

Apleni, L., Vallabh, D., & Henama, U. S. (2017). Motivation for tourist's participation in religious tourism in Eastern Cape: A case study of Buffalo City, South Africa. *African Journal of Hospitality, Tourism and Leisure, 6*(2), 1–14.

Arival Travel. (2018). *Airbnb.* www.arival.travel

Badenhorst-Weiss, J., & Nel, J.D. (2011). A conceptual framework to analyse supply-chain designs. *Acta Commercii*, 1-16.

Benner, M. (2019). From overtourism to sustainability: A research agenda for qualitative tourism development in the Adriatic. *Munich Personal RePEc Archive*, 1–18. Retrieved from https://mpra.ub.uni-muenchen.de/92213/1/MPRA_paper_92213.pdf

Bizcommunity. (2017). Airbnb Experiences one year later: Foodies and millennials lead the way. *Tours News*. https://www.bizcommunity.com/Article/1/

Boston Consulting Group. (2013). *Traveling with Millennials.* Retrieved from: https://www.bcg.com

Business Partners. (2018). *Fear of failure hinders South Arica's aspiring enterpreneurs.* Retrieved from: https://www.businesspartners./en-za/media-centre/media-release/

Capocchi, A., Vallone, C., Pierotti, M., & Amaduzzi, A. (2019). Overtourism: A Literature Review to Assess Implications and Future Perspectives. *Sustainability, 11*(12), 3303. doi:10.3390u11123303

Cheer, J. M., Milano, C., & Novelli, M. (2019). Tourism and community resilience in the Anthropocene: Accentuating temporal overtourism. *Journal of Sustainable Tourism, 27*(4), 554–572. doi:10.1080/09669582.2019.1578363

Chua, E. L., Chiu, J. L., & Bool, N. C. (2019). Sharing Economy: An Analysis of Airbnb Business Model and the Factors that Influence Consumer Adoption. *Review of Integrative Business and Economic Research, 8*(2), 19–37.

Claudio, M., Joseph, M. C., & Marina, N. (2018). Overtourism: a growing global problem. *The Conservation*. Retrieved from https://theconversation.com/overtourism-a-growing-global-problem-100029

Cruz, M. S., & Zaragoza, M.P.P. (2019). Analysis of the accommodation density in coastal tourism areas of insular destinations from the perspective of overtourism. *Sustainability, 11*(11). doi:10.3390u11113031

De Villiers, J. (2018). There are now 40% more Airbnb hosts over the age of 60 in SA than a year ago. *Business Insider*. Retrieved from: https:www.businessinsider.co.za/-south-african

Destiny. (2018). Airbnb hosts across SA made R3.8bn in the past year. *Destiny Connect*. Retrieved from: https://www.destinyconect.com/

DigitalTrends. (2018). *Airbnb targets crious travelers as 'experiences' expands to 200 U.S. cities*. Retrieved from: https://www.digitaltrends.co.za/web

Dordevic, M. (2010). Supply Chain Management in tourism. In *4th International Quality Conference*. University of Kragujevac.

Duan, W., Gu, B., & Whinston, A. B. (2008). Do online reviews matter? An empirical investigation of panel data. *Decision Support Systems, 45*(4), 1007–1016. doi:10.1016/j.dss.2008.04.001

Expedia. (2018). *Millennial Traveller Report: Why millennials will shape the next 20 years of travel*. Future Foundation.

Font, X., Tapper, R., Schwartz, K., & Kornilaki, M. (2008). Sustainable Supply Chain Management in Tourism. *Business Strategy and the Environment, 17*(4), 260–271. doi:10.1002/bse.527

Forno, F., & Garibaldi, R. (2015). Sharing Economy in Travel and Tourism: The Case of Home-Swapping in Italy. *Journal of Quality Assurance in Hospitality & Tourism, 16*(2), 202–220. doi:10.1080/1528008X.2015.1013409

Gabel, D. (2016). Uber and the persistence of market power. *Journal of Economic Issues, 50*(2), 527–534. doi:10.1080/00213624.2016.1179060

Ghaderi, Z., Hatamifar, P., & Khalilzadeh, J. (2018). *Analysis of tourist satisfaction in tourism supply chain management*. Anatolia. doi:10.1080/13032917.2018.1439074

Goodwin, H. (2017). *The Challenge of Overtourism*. Responsible Tourism Partnership. doi:10.1016/S0140-6736(84)91114-0

Guttentag, D. (2006). *Airbnb: Why tourists choose it and how they get it*. University of Waterloo.

Haywood, J., Mayock, P., Freitag, J., Owoo, K. A., & Fiorilla, B. (2017). *Airbnb & Hotel Performance: An analysis of properietary data in 13 global markets*. Retrieved from: https://wwww.str.com

Henama, U. S. (2013). Attracting Indian Outbound Tourists to South Africa: A BRICS Perspective. *India Quarterly, 69*(3), 229–247. doi:10.1177/0974928413489466

Henama, U. S. (2014a). Health tourism in South Africa: Opportunities and possibilities. *African Journal of Hospitality, Tourism and Leisure, 3*(1), 1–14.

Henama, U. S. (2014b). The demise of 1Time Airline and the reaction of various interest groups. *African Journal of Hospitality, Tourism and Leisure*, *3*(20), 1–11.

Henama, U. S. (2017a). Airbnb a solution for lack of tourism stock in Knysna. *Cape Argus*. Retrieved from://www.pressreader.com/south-africa/cape-argus

Henama, U. S. (2017b). International Migration and National Tourism Industry Development: The Case of South Africa. The EUrASEANSs. *Journal of Global Socio-Economic Dynamics*, *2*(3), 72–82.

Henama, U. S. (2018a). A literature review into South Africa's attempt to create a nudist beach at Ray Nkonyeni Municipality, KwaZulu-Natal. *African Journal of Hospitality, Tourism and Leisure*, *7*(3), 1–14.

Henama, U. S. (2018b). Disruptive entrepreneurship using Airbnb: The South African experience. *African Journal of Hospitality, Tourism and Leisure*, *7*(1), 1–13.

Henama, U. S. (2019a). The Sharing Economy in South Africa's Tourism Industry: The Case of Uber E-Hailing Taxi Services. In O. Oncioiu (Ed.), *Improving Business Performance Through Innovation in the Digital Economy*. IGI Publishing.

Henama, U. S. (2019b). *Tourism as a Neoliberal Economic Messiah: The Case of South Africa. Neoliberalism in the Tourism and Hospitality Sector*. IGI. doi:10.4018/978-1-5225-6983-1.ch003

Henama, U. S., & Sifolo, P. P. S. (2017). Tourism Migration in South Africa: Current Dynamics, Immediate Challenges and Future Prospects. *International Journal of Innovation in the Digital Economy*, *8*(1), 47–60. doi:10.4018/IJIDE.2017010103

Henderson, L. (2016). Innovation or Rule Breakers? Regulating Uber, Airbnb & The Sharing Economy (Bachelor's Dissertation). Otago: University of Otago.

Holmes, T. (2015). Airbnb in SA selling trust to the wary. *Mail and Guardian*. Retrieved from: https://www.mg.co.za

Hong, J. (2018). Rise of the sharing economy and the future of travel and tourism industry. *Journal of Hotel & Business Management*, *7*(2), 5–11. doi:10.4172/2169-0286.1000180

IOL. (2018). *Airbnb opens experiences platform for submission from hosts across SA*. Retrieved from: https://www.iol.com.za/news/south-africa/

Issacs, L. (2017). Fedhasa demands Airbnb be regulated. *IOL*. Retrieved from: https://www.iol.co.za/capetimes/news/

Jamal, S. (2017). *Airbnb offers free accommodation for people displaced by Western Cape disasters*. Retrieved from: https://www.timeslives.co.za/news/south-africa/

Jaremen, D. E., & Nawrocka, E. (2015). Relations in the tourism supply chain. *Seria. Administracja i Zarzadzanie*, *104*, 98–110.

Jungleworks. (2019). *How Airbnb works – insights into business and revenue model*. USA Clicklabs.

Koens, K., Postma, A., & Papp, B. (2018). Is overtourism overused? Understanding the impact of tourism in a city context. *Sustainability*, *10*(12), 1–15. doi:10.3390u10124384

Kuščer, K., & Mihalič, T. (2019). Residents' Attitudes towards Overtourism from the Perspective of Tourism Impacts and Cooperation–The Case of Ljubljana. *Sustainability, 11*(6), 1823. doi:10.3390u11061823

Lalicic, L., & Weismayer, C. (2017). The role of aunthenticity in Airbnb Experiences. In R. Schegg & B. Stangl (Eds.), *Edited. Information and Communication Technologies in Tourism*. Springer Publishing.

Lane, L. (2016). Are millennial travel trends shifting in 2016? *Forbes*. Retrieved from: https://www.forbes.com/sites/lealane

Marais, M., Du Plessis, E., & Saayman, M. (2017). Critical success factors of a business tourism destination: Supply side analysis. *Acta Commercii, 17*(1), 1–12. doi:10.4102/ac.v17i1.423

Martín, J. M. M., Martínez, J. M. G., & Fernández, J. A. S. (2018). An analysis of the factors behind the citizen's attitude of rejection towards tourism in a context of overtourism and economic dependence on this activity. *Sustainability, 10*(8), 2851. Advance online publication. doi:10.3390u10082851

Mgwadlamba, N. (2017). Letter: Iconic Kynsna forest still standing. *Cape Time*. Retrieved from: https://www.iol.co.za

Molefe, L. P., Tauoatsoala, P., Sifolo, P. P. S., Manavhela, P., & Henama, U. S. (2018). 'The effects of tourism supply chain management practices on tourism operations in Pretoria, South Africa'. *African Journal of Hospitality, Tourism and Leisure, 7*(2), 1–15.

Nielsen. (2017). *Young and Ready to Travel (And Shop): A look at Millennial Travellers*. Retrieved from: https://www.nielsen.com

Nkanyeni, U. (2017). Airbnb signs Africa's first city agreement with Cape Town. *Traveller24*. Retrieved from: https://www.traveller24.com/FindYourEscape/

Oskam, J., & Boswijk, A. (2016). Airbnb: The future of networked hospitality business. *Journal of Tourism Futures, 2*(1), 22–42. doi:10.1108/JTF-11-2015-0048

Pantazatou, K. (2018). Taxation of the Sharing Economy in the European Union. In Cambridge Handbook of Law and Regulation of the Sharing Economy. CUP. doi:10.1017/9781108255882.028

Parliamentary Monitoring Group. (2019). *Report of the Portfolio Committee on Tourism for Budget Vote No. 33*. Retrieved from: https://pmg.org.za/tabled-committee-report/3843/

Perkumienė, D., & Pranskūnienė, R. (2019). Overtourism: Between the Right to Travel and Residents' Rights. *Sustainability, 11*(7), 2138. doi:10.3390u11072138

Pimonratanakan, S., & Ayasanond, C. (2018). The Supply Chain Management of Pomleo Production, Process and Tourism Sustainability in Nakhon Chaisri, Nakhon Pathom Province, Thailand. *African Journal of Hospitality, Tourism and Leisure, 7*(5), 1–14.

Polzin, S. E., Chu, X., & Godfrey, J. (2014). The impact of millennials' travel behaviour on future personal vehicle travel. *Energy Strategy Reviews, 5*, 59065. doi:10.1016/j.esr.2014.10.003

PWC. (2015). *The Sharing Economy. Consumer Intelligence Series*. Retrieved from: https://pwc.com/CISsharing

Queensland Tourism Industry Council. (2014). *The Sharing Economy: How it will impact the tourism landscape and what businesses can do*. Queensland Tourism Industry Council.

Rawlings, L. K. (2016). Airbnb takes on more travel pie. *IT Web*. Retrieved from: https://www.itweb.co.za/content

Roblek, V., Stok, Z. M., & Mesko, M. (2016). Complexity of a Sharing Economy for Tourism and Hospitality. *Tourism & Hospitality Industry 2016, Congress Proceedings*, 378-387.

Schofield, J. (2017). How the sharing economy has changed the supply chain. *System ID Barcode Solution*. Retrieved from: https://www.systemid.com/learn/hhh/

Séraphin, H., Zaman, M., Olver, S., Bourliataux-Lajoinie, S., & Dosquet, F. (2019). Destination branding and overtourism. *Journal of Hospitality and Tourism Management, 38*, 1–4. doi:10.1016/j.jhtm.2018.11.003

Shevel, A. (2016). The fast facts for those considering Airbnb. *Business Day Live*. Retrieved from: http://www.bdlive.co.za

Shrikant, A. (2018). *Can guided tours save Airbnb? How Airbnb is ducking governent regulations with pottery classes and nature walks*. Retrieved from: https://www.vox.com

Sifolo, P. P. S. (2015). A tourism theoretical gap: The case of the Northern Cape Province in South Africa. *African Journal of Hospitality, Tourism and Leisure*, *4*(1), 1–12.

Sifolo, P. P. S., & Henama, U. S. (2019). Implications of Climate Change for Tourism in Africa. *Geo Journal of Tourism and Geosites*, *2*(20), 191–198.

Sifolo, P. P. S., Maimela, M. J., & Tladi, M. P. (2014). Exploitation of Technology by Tshwane Residents for Tourism Development Purposes. World Academy of Sciences, Engineering and Technology. *International Scholarly and Scientific Research & Innovation*, *8*(2), 386–389.

Strydom, A. J., Mangope, D., & Henama, U. S. (2019). Making Community-Based Tourism Sustainable: Evidence from the Free State Province, South Africa. *Geo Journal of Tourism and Geosites*, *24*(1), 7–18.

Tauoatsoala, P., Henama, U. S., & Monyane, M. (2015). Service quality as a fundamental cornerstone for visitor satisfaction: Lessons learned from Pretoria, South Africa. *African Journal of Hospitality, Tourism and Leisure*, *4*(1), 1–8.

Tourism Update. (2018). *First Airbnb city collaborative agreement in Africa signed*. Retrieved from: https://www.tourismupdate.co.za/article

United Nations World Tourism Organisation. (2018). *'Overtourism'? Understanding and Managing Urban Tourism Growth beyond Perceptions*. Author.

Van Hull, P. (2016). Supply Chain-The Next Big Winner from the Sharing Economy. *SDC Executive*. Retrieved from: https://www.sdcexecutive.com/sourcing-procurement/

Venktess, K. (2017). South Africans rake in R24.37m from Airbnb. *Fin24*. Retrieved from: https://www.fin24.com/Tech/News/

Visit Utah. (2019). *Airbnb Signup*. Retrieved from: https://www.visitutah.com/airbnb-signup/

Walker, A. (2019). *Airbnb believes it has a R10bn impact on South Africa's economy*. Retrieved from: https://memeburn.com/2019/07/

Walmsley, A. (2017). Overtourism and underemployment: a modern labour market dilemma. In *Responsible Tourism in Destinations 13*. Tackling Overtourism.

Wee Tracker. (2018). *Can Airbnb master the ABCs of Africa?* Wee Media.

White Paper. (1996). *The Development and Promotion of tourism in South Africa*. Department of Environmental Affairs & Tourism.

Zhang, Z., Chen, R. J. C., Han, L. D., & Yang, L. (2017). Key factors affecting the price of Airbnb listings: A geographically weighted approach. *Sustainability*, *9*(9), 3–13. doi:10.3390u9091635

ADDITIONAL READING

Blal, I., Singal, M., & Templin, J. (2018). Airbnb's effect on hotel sales growth. *International Journal of Hospitality Management*, *73*, 85–92. doi:10.1016/j.ijhm.2018.02.006

Böcker, L., & Meelen, T. (2017). Sharing for people, planet or profit? Analysing motivations for intended sharing economy participation. *Environmental Innovation and Societal Transitions*, *23*, 28–39. doi:10.1016/j.eist.2016.09.004

Bridges, J., & Vásquez, C. (2018). If nearly all Airbnb reviews are positive, does that make them meaningless? *Current Issues in Tourism*, *21*(18), 2057–2075. doi:10.1080/13683500.2016.1267113

Cheng, M. (2016). Sharing economy: A review and agenda for future research. *International Journal of Hospitality Management*, *57*, 60–70. doi:10.1016/j.ijhm.2016.06.003

Gurran, N., & Phibbs, P. (2017). When tourists move in: How should urban planners respond to Airbnb? *Journal of the American Planning Association*, *83*(1), 80–92. doi:10.1080/01944363.2016.1249011

Gutiérrez, J., García-Palomares, J. C., Romanillos, G., & Salas-Olmedo, M. H. (2017). The eruption of Airbnb in tourist cities: Comparing spatial patterns of hotels and peer-to-peer accommodation in Barcelona. *Tourism Management*, *62*, 278–291. doi:10.1016/j.tourman.2017.05.003

Guttentag, D. (2015). Airbnb: Disruptive innovation and the rise of an informal tourism accommodation sector. *Current Issues in Tourism*, *18*(12), 1192–1217. doi:10.1080/13683500.2013.827159

Hall, C. M., & Williams, A. M. (2019). *Tourism and innovation*. Routledge. doi:10.4324/9781315162836

Hassanien, A., Dale, C., Clarke, A., & Herriott, M. W. (2010). *Hospitality business development*. Routledge. doi:10.4324/9780080884981

Hobson, K., & Lynch, N. (2016). Diversifying and de-growing the circular economy: Radical social transformation in a resource-scarce world. *Futures*, *82*, 15–25. doi:10.1016/j.futures.2016.05.012

Liang, L. J., Choi, H. C., & Joppe, M. (2018). Understanding repurchase intention of Airbnb consumers: Perceived authenticity, electronic word-of-mouth, and price sensitivity. *Journal of Travel & Tourism Marketing*, *35*(1), 73–89. doi:10.1080/10548408.2016.1224750

Lutz, C., & Newlands, G. (2018). Consumer segmentation within the sharing economy: The case of Airbnb. *Journal of Business Research*, *88*, 187–196. doi:10.1016/j.jbusres.2018.03.019

Oskam, J., & Boswijk, A. (2016). Airbnb: the future of networked hospitality businesses. *Journal of Tourism Futures*.

Teubner, T., Hawlitschek, F., & Dann, D. (2017). Price determinants on AirBnB: How reputation pays off in the sharing economy. *Journal of Self-Governance & Management Economics*, *5*(4).

Tussyadiah, I. P., & Zach, F. (2017). Identifying salient attributes of peer-to-peer accommodation experience. *Journal of Travel & Tourism Marketing*, *34*(5), 636–652. doi:10.1080/10548408.2016.1209153

Wang, D., & Nicolau, J. L. (2017). Price determinants of sharing economy based accommodation rental: A study of listings from 33 cities on Airbnb. com. *International Journal of Hospitality Management*, *62*, 120–131. doi:10.1016/j.ijhm.2016.12.007

KEY TERMS AND DEFINITION

Aribnb: It is an online marketplace which deals with offering and demand of lodging, homestays at a convenient price.

Hospitality: It is a rite of passage oriented to welcome and receive strangers as long as their sojourn.

Price Determinants: Elements of the markets which boost or constrain tourism demand and offering.

Reputation: It means a social entity resulted from the evaluation of others or other groups.

Tourism: It is defined as the act of traveling for pleasure or business out for home more than one day and less than six months.

Chapter 3
Heritage Consumption in the Neo-Liberal Agenda:
El Abasto, Almagro, Buenos Aires, a Study-Case

Maximiliano Emanuel Korstanje
https://orcid.org/0000-0002-5149-1669
University of Palermo, Argentina

ABSTRACT

Buenos Aires city in the days post convertibility crisis appealed to the adoption of tourism as a fresh economic alternative to struggle against poverty and the financial crises. The currency exchange, as well as the abandonment of the convertibility system, favoured Argentina in many ways. For example, Buenos Aires was selected by neighbouring countries as a main tourist destination. This chapter reviews the effects of tourism in the urban landscape, such as real estate speculation, gentrification process, only to name few. The authors stress the role played by urban heritage in the process of touristification—as some voices adhere—which is conducive to the transformation of old industrial neighbourhoods.

INTRODUCTION

Without any doubt, the tourism industry has considerable growth over recent years. This untrammeled growth was accompanied by a strong ideological culture that punctuates tourism is vital for in the revitalization of under-developed economies. In part it was true but the fact was that some nations failed in their efforts to reach a sanitized economy. The theory of development showed some limitations to explain why some nations are rich while others remain poor. What seems to be clear is that under some conditions the tourism industry generates positive changes which are conducive to an economic recovery (De Kadt 1979; Richards & Hall 2003; Pigram & Wahab, 2005). The stimulation of consumption associated to foreign investment has durable positive consequences for the local economies. Having said this, some experts agree that the economic resources of a society and tourism consumption seem to be inextricably

DOI: 10.4018/978-1-7998-4330-6.ch003

intertwined (Korstanje & Seraphin 2017). Based on this perspective, the current chapter studies the role of culture in post-convertibility Argentina just after 2001. In so doing, this research focuses strictly on the theory of urban planning and international tourism trends within the constellations of cultural studies. As George Yudice (2002) puts it, there exists a valorization of cultural consumption as the best (developing) factor that leads societies towards growth and development. Per his viewpoint: "culture can generate income through tourism, crafts and other cultural activities." Such urban transformations in Buenos Aires city have increased since the end of convertibility and become the essential *resource* for development. We have worked on secondary statistical sources administered by some local administration institutions such as the Ministry of Culture (Buenos Aires City Government) and CEDEM (Studies Center for Metropolitan Economic Development, Secretary of Economic Development, Buenos Aires City Government). The selected methodology helps us analyzing the evolution and maturation of artistic activities as well as the cultural consumption associated to the tourism industry. In the global market, cities are commoditized, embellished and exchanged as cultural products while natives are considered cultural commodities to be unilaterally gazed by foreign tourists (Korstanje 2012). In their seminal book, *Economies of Signs and spaces,* Urry & Lash (1994) call the attention to the rise of new global economies where cultures are packaged and sold under the auspices of the sign. Cities are nowadays offered as cultural products which are consumed in the regional and international markets. Recently, managerial disciplines such as Management and Marketing have devoted considerable attention in the production of destination image (Beerti & Martin, 2004; Echtner & Ritchie 1991). Paradoxically, after the turn of the century, political riots, economic crises and violence flourished affecting not only the international destination but also placing the industry in jeopardy (Avraham 2015). The post conflict destination theory emphasizes on the potential of tourism to revitalize and recycles to areas whipped or destroyed by disasters, a state of durable conflict or war (Vitic & Ringer 2008; Morgan & Pritchard 2011). Sites of the caliber of Belfast, New Orleans, Sri Lanka, Bali and even New York (after 9/11) have been fully recovered from their days of violence and darkness (Winter 2007). This means that tourism situates as something more complex than a resilient industry but as an instrument that helps societies to recover from the adversity. Above all tourism seems to be a mechanism of resiliency inherently ingrained in the societal order (Korstanje & Ivanov, 2012).

As the previous backdrop, Buenos Aires has changed its image in the days after 2001 deploying countless resources to position internationally as a preferred destination for the demand. The policies adopted in the 1990s ushered the nation to poverty and higher rates of unemployment. In this case, the urban industrial sectors populated of factories set the pace to a new urban landscape ghettoized by the lack of stable employment, drug addiction and alcoholism. These spaces of segregation and precaritization are dubbed by Urry & Lash (1992) as *wild spaces.* In this context, the neoliberal discourse emphasizes the economic revitalization of these types of spaces is transformed into remarkable leisure-spots. This was particularly probed in different works which take part of what scholars name as "the theory of gentrification" (Gotham 2005; Maitland 2010; Skoll & Korstanje 2014) but what is more important there is little evidence how the theory of gentrification applies in Latin American cities. In such a direction, the present chapter intends to be a significant contribution. At a closer look, some areas and neighbourhoods of the city have been re-functionalized to attract foreign investors creating the conditions for new business deals. That is to say, this is an "urban renewal" which tries to turn obsolete, abandoned areas in strategic places in the city into attractive spots. As a toolbox usefull in the task of promoting destinations, the used strategies are certainly drawn to recover harmed economies –like in Argentina- to be widely fostered through the adoption of tourism as a fertile ground of activity. This was particularly true in the case of

Argentina in 2001. It is important to remark that "convertibility," the international relation the country adopted in which the Argentinian peso was equalled to the US dollar, led to fluent, periodic foreign trade and international relations mostly favourable to the other countries: Argentina imported products at such low cost that local manufacturers could not compete against that. This situation drastically changed with a financial market crisis that led gradually the country in bankruptcy. As a result of this, the successive period of devaluation in Argentina affected the indicators of economy but at the same time it fostered the competitiveness of some urban cities such as Mar del Plata, Buenos Aires, and Bariloche. After years of convertibility Argentina becomes a tourist destination for many segments coming from Brazil and Chile. Buenos Aires city, as a cheap destination, has been chosen as an international tourist destination. Thus, the city's situation changed from being in a national crisis to becoming one of the cheapest tourist destinations in the world. New strategies were therefore designed to attract international tourism, trying to position Buenos Aires as the "Latin American Capital of Culture." Inevitably after the crisis, Argentina "benefited" from the tourism boom which installed the formula of "culture as an economic resource"; this was studied by Harvey (1990) and Yudice (2004). This formula was set in a bigger context: urban culture as merchandise was proposed for the administration of local governments with a strong influence of culture management experience and Spanish theoretical production: Ballart Hernández and Tresserras (2001), Cortés Puya (2005), Santana (1997), Prats (1997), Puig (2004), etcetera).

SOCIAL-TERRITORIAL TRANSFORMATIONS IN BUENOS AIRES CITY: THE STATUS OF THE QUESTION

The decade of the 90s was marked by a clear territorial transformation of global cities as never before. The neoliberal agenda promoted tourism as a valid and rapid way of economic revitalization in the under-developing world (Ayikoru, Tribe & Airey 2009; Mosedale 2016). As Mosedale (2016) observes, the tourism industry has been re-framed under the lens of neoliberals while neoliberal policies were placed at the hands of governments. Per this project, the cultivation of liberal ideals, as well as democracy as the main form of government, would inevitably lead to prosperity and well-being. For those cultures which were historically incompatible with democracy and individual rights, tourism would help to open the doors to more tolerant and more credible institutions. Finally, the failures of Keynesianism to give efficient solutions to the problems of capitalism, adjoined to the end of the welfare state, paving the ways for the rise and installation of liberalism as the dominant ideology. In this conjuncture, the supporters of neoliberal thinking embraced enthusiastically unregulated policies which decentralized the monopoly and intervention of the state in the domestic economy. Mosedale interrogates furtherly on the role of tourism in this liberal process, as a vehicle towards de-regulation. As he notes, the tourist enterprise is embedded on the urgency of assisting –at least in the economic field- locals whereas the involving actors are trained according to neoliberal rationality. In other terms, some disadvantaged groups which were historically relegated form the economic productive system can be re-integrated to the society. However, in so doing, the responsibility of the state for citizen protection is endorsed to the local actor. Undoubtedly, this happens because nation-states are systematically integrated into a liberal and global marketplace taken together with other private actors. Argentina, of course, was not an exception (Skoll & Korstanje 2014). For the turn of the century, the commercial exploitation of Argentinean heritage, as well as historical patrimony, played a crucial role in the urban renewal process. Buenos Aires was symbolically re-imagined as the cradle of Tango music, positioning as a highly-demanded tourist city. But these changes altered radically the territoriality of the city and its respective neighborhoods. For the sake of clarity, the city was diversified into fragmented territorial pieces expressing different material asymmetries, inter-class struggles and socio-economic landscapes. The disappearances of factories –through 90s decade- and the decline of industrial neighborhoods during neoliberal days, were fertile grounds for the arrival of different tourist enterprises. This is exactly the case of La Boca, San Telmo, and El Abasto-Almagro dwelled Carlos Gardel (the well famous Tango singer). Here we raise a question, is tourism and international demand a solution to a local crisis, some policymakers started to design to face the economic crisis in 2001? One of them was doubtless the transformation of tertiary sector. Among these proposals, urban renewal while boosting degraded areas into richer spaces of consumption. In this sense, we can see that the general movement of people and their activities was not only fostered and modeled by the market but also promoted and monitored by local governments, in their role of ensuring and managing the city's fragmentation according to its diverse territorial uses (Marcuse, 1986). The tourist city showed the interplay of recent urban shifts aimed at the revitalization of the neighborhood and the ways of living and interpreting the city (Judd & Fainstein 1999; Carman 2006). But far from being naive, these new sensibilities revolving around the city re-inscribe into the neoliberal dominant discourse which is linked to the economic ways of production. Gorelik (2006) makes the following comments about Buenos Aires:

The urban boom now simply implies the idea of an explosive and successful reactivation to be observed in the tourist circuit and the real estate market [...]. For the first time ever, tourism has become an evident phenomenon in Buenos Aires: tourism is normal now [...], whose only perception of the crisis is low international prices; such economic difference has managed to make the miracle of awakening two corpses from the tomb: tango and Florida Street. On the other hand, there is also local tourism, looking for diversity and complexity in theatre plays, film festivals or the art market move with its typically urban, global expressions, such as gallery nights. With a mix of paternalism and interest in the city's innate picturesque aspects to foreign eyes, local tourism values a double difference in Buenos Aires: higher social energy compared to developed cities and higher cultural and urban density in comparison to other Latin American cities.

This above-noted "urban boom" (touristic and cultural), as Gorelik adheres, ascribes to a stereotyped and singled image of the city, with strong emphasis in expressions as football, the "conventillo," tango, the "gaucho", etcetera. The problem, precisely, lies in the fact that the fabricated image is ultimately internalized by *porteños* (Buenos Aires city-dwellers). As debated the logic of neo-liberalism operated into two different but complementary sides. On one hand, the classic-established industrial areas declined to consolidate pauperized neighborhoods and slums in the affected urban areas. These (wild) spaces, citing Urry & Lash, corresponded with spaces of high criminality, unemployment and unfair wealth distribution –only to name few of social maladies city-dwellers faced). As a result of this, the land prices slumped down in comparison with other parts of the city. Just in the post-2001 context, these lands were occupied and bought by foreign investors interested in developing tourism as the main economic alternative. It is safe to say, the nation-state, as well as the imaginary revolving around the national heritage, played a marginal role not only allowing the investors to pay for the lands but also increased the taxes to force the neighbor to migrate to other places. Paradoxically, though tourism was seen as a part of the solution to alleviate communities from the extreme poverty, and social inequality, no less true was that it implemented a policy of relocation that excluded many actors from the so-called benefits of tourism.

The city finally recovered while a picturesque of poverty was created through the articulation of new consuming landscapes. Degraded areas which were subject to crime and poverty were symbolically and materially transformed in rentable tourist attractions.

[...] Indeed, Buenos Aires, the "most European" South American city once, became more "Latin American" in the past years at the same time as the social-economic collapse got deeper and deeper and its streets filled with wanderers, indigents, "cartoneros" [...].However, both internal and external imaginary worlds share a much more contradictory representation. There are still traces of the city of lights [...] lights which sometimes change into blackouts, moments in which the city looks sombre and frightens its inhabitants [...].

CRISIS OF THE CONVERTIBILITY MODEL: ARGENTINA 2001

To some extent, the crisis of convertibility model jettisoned many of the liberal promises and programs, as well as the hopes of prosperity. As stated, the global trend to commoditize the cultural lifestyle in cities was proportional to the upsurge of economic crises in Latin America and other regions. In his book, *Development and Social Change,* Phillip McMichael (2011) has exerted a radical criticism to the theory of development as well as globalization as a leading cultural project, or at the best an ideal to reach in the developing economies. Basically, we cannot understand the term development without re-approaching the influence of globalization in local economies. He traces back the origins of development theory and its evolution up to date. It is important not to lose the sight of the fact that development theory reinforced a long-dormant colonial discourse which situated the European ideals of free-trade and democracy over other cultural values. The colonial powers, in the eighteenth and nineteenth centuries, launched to other overseas territories to conquest and index emerging economies to reinforce their war-machine. Colonialism was based on much controversy. On one hand, democracy was cultivated in the core of European citizens and respected only for European citizens, but on another, the colonial armies moved with cruelty and cynicism in the peripheral colonies. What is important to discuss, here, seems to be that paradoxically the ideals of democracy which were systematically neglected for Non-western natives inspired a strong counter-insurgency which led to the political independence from the central authority. Sooner or later, natives stripped down the veil of European colonialism as well as its double moral standard. The former imperial powers, just in the end of WWII, signaled to the development theory to maintain the dependency of their colonies. The existent social inequality can be explained by the combination of three key factors: a) the acceptance of financial loans to higher interests to accrue, b) the capital intense methods in agriculture which enlarged the dependency of developing economies and c) a new division of labor based on the access to technology. As the previous argument is given, McMichael argues convincingly that globalization succeeded in expanding thanks to the lack of protective barriers of the Third world where the capital investors were welcomed. This, in consequence, provoked two alarming situations. An increase in unemployment and the decline of unionization in the North was accompanied by the arrival of international business corporations seduced by the low-cost of workers in the South. The doctrine of "free enterprise" was presented as a superior ladder in the evolutionary process. Each state should adopt a specialized role in a much wider "world factory" where some provide with the raw-materials and others with elaborated products. This trend which characterize the 90s decade created a new asymmetry between skilled (located in the first world) and under-skilled human resources (situated in the periphery). The recession produced by oil-embargo pressed First World to borrow a massive influx of money to the Third world, but now it will be carefully selected by two organisms, GATT and WTO. Both curtailed the protective measure of local economies by consolidating of a new model which combined the reduced public capacity with the needs of governance. If Nationalism showed the importance of nation-state to protect the citizen from Market's arbitraries, now neo-liberalism focused on the inefficiency public administration to regulate the economy.

In short, the making of a free trade regime reconstructed food security as a market relation, privileging and protecting corporate agriculture and placing small farmers at a comparative disadvantage. Food security would now be governed through the market by corporate, rather than social criteria. (p. 136).

In this token, the World Trade Organization was relentless by charging or applying sanctions over those countries who affected the new system of import-exports. Less interesting in freeing trade than in consolidating their hegemony, main powers prompted the discourse that Third World had not the right towards "self-sufficiency" any more. Once again, globalization accelerated the accumulation of profits (beyond the boundaries and control of nation-states) but enlarging the levels of poverty as never before. Per some calculations, the UN declared that only 20% of the world population is situated within 20% of the richest people, whereas the rest is facing serious economic problems. It is important not to lose the sight that almost 1 billions of people are living in slums. In parallel with this, the crisis of the economic convertibility model in Argentina accelerated the end of a neoliberal agenda, which far from disappears, mutated to new more subtle forms. To a closer look, the convertibility system was over-valorized –over other models- because it would boost the economy situating Argentina as a competitive actor regarding its neighbors as Brazil, Chile or Uruguay. Originally the peso-Dollar parity not only reduced the inflation, which wreaked havoc in the local economy but also stimulated the purchasing power of middle classes. This aspect facilitated the re-election of former president Carlos Saul Menem. But thing came worse to worst, the parity gradually altered the productive system affecting the competition powers of the local firms respecting to the coming exports. Since convertibility was held based on direct foreign investment, privatization occupied a central position in the policies of Carlos Menem´s administration. The opening up for exports, associated with an urgency to privatize the state firms, created a financial stock crisis that ends with the collapse of the banking system in 2001. To here, we have summarized the phenomenon of convertibility to skip to the role of cultural tourism as an instrument to refurbish former industrial neighbor in tourist destinations.

CULTURAL TOURISM IN CITIES: A PHENOMENON

In the last years, there has been growing interest in visiting urban spaces typical of other times and cultures, leading to the denominated "the city tourist or urban tourism" (Gotham, 2005, Judd and Fainstein 1999, Judd 2003). All these voices hold the thesis that the process of deindustrialization, which whipped many economies during 70s and 80s decades, adopted a rapid program of segmentation to recycle the deteriorated urban space. This benefited the life of many citizens but affected the poorer (outcast) classes. These examples apply for Las Vegas, New Orleans, Boston and Miami to cite a few only, where leisure and entertainment are deemed as essential values not only for visitors but for locals (Judd 2003). Lacarrieu (2002) argued convincingly that cultural tourism is creating new places and policies oriented to cultivate the respect for the "Other". However, the guest-host encounter has its own contradictions and limitations. The quest for novelty, which marks the tourists´ expectancies, ushers local in the urgency to see their culture as a commodity, as a product to be gazed and consumed. Such a process, unless regulated, recreates the conditions to situate tourists as spectators and local as players. Under some conditions, these types of fabricated spectacles are conducive to real speculation and business where local stakeholders are certainly excluded. Starting from the premise culture is defined and re-negotiated as a resource originally disposed to draw the city´s image, Lacarrieu turns the attention of how new tourist circuits and capital take the same ways.

TOURISM AS AN ECONOMIC STRATEGY FOR RENOVATION AND ITS GROWTH IN NUMBERS

In 1996, Secretary of Tourism of Buenos Aires City conducted an exhaustive research to decipher the evolution of tourist demands. The results dropped out interesting numbers. Almost a 46.6% of consulted persons were foreign tourists while only 50% was domestic. In Buenos Aires city, the gastronomic offer is very wide locating mainly in the neighborhoods of Buenos Aires downtown, (311 restaurants); Recoleta, (60 restaurants); San Telmo, (22 restaurants); Palermo, (49 restaurants); Puerto Madero (15 restaurants) and Costanera, (20 restaurants). At that instance, la Boca scarcely had 3 restaurants only. The main destination elected by foreign tourists was: Recoleta 42.6%, Palermo 44.4%, La Boca 36.7% and San Telmo 33%. As stated, La Boca and San Telmo attracted tourists focusing on the Tango Music and their heritage linked to Tango and the figure of Gaucho (arrabalero). Some years later, in a survey-poll administered by the Department of Tourism of Buenos Aires City (DTBAC), the Statistical Report of 1997-1998 reads with no substantial differences. La Boca and San Telmo remained as marginal destinations in the fields of gastronomy and tourism in comparison with other neighborhoods. Palermo and Recoleta still had the most influential infrastructure according to bars, restaurants and pubs in the city (14. 6% y 14.4%, respectively), La Boca had 1.4% and San Telmo, 2.14% gastronomic establishments. This was complemented by other investigation opened in 1999 by DTBAC gives a fresh snapshot of the situation. Now things were radically different. Now, La Boca and San Telmo located as one of the most visited neighborhoods while Palermo and Puerto Madero kept their hegemony. Such trends increased notably in the years to come because of the state investment in infrastructure and the constructions of new ways, ship terminals and transport-hubs. Statistically speaking, tourism has grown according to a clear-cut tendency; a crucial point which was widely favored by the peso devaluation in the post-2001 days. There was no before the crisis in 2001 that Argentina overweighed the tourism balance receiving further inbound tourism than other decades. For example, In the 2002-2003 tourism report carried out by the PEIET (Program of Statistics and Economic Impact of Tourism) and the CEDEM (Studies Centre for Metropolitan Economic Development), surveys were made in different terminals of access to Buenos Aires (Ezeiza International Airport, Retiro Bus Terminal, J. Newbery Metropolitan Airport and the City's fluvial harbor). As long as 2002, the visitors came from Europe (22%), Latin America 11%), US (10%) and others (5%) (CEDEM, 2002). The 90s decade witnessed the arrival of international hotel chains such a Four Seasons, Arcor, Marriott and Howard Johnson. The international cruise companies were an important source for profits for the city. In 2002 and 2004, the cruise arrivals not only increased but the estimate length of state was simply triplicated. During the 1996-2004 period, we can see that from 1996 to 2000, 149 cruises arrived in Buenos Aires city, while from 2001 to 2004, 216 cruises were reported. These numbers evince how the increase of cruise arrivals was a palpable reality respecting the 1996-2000 period.

Table 1. Cruises arrivals in Buenos Aires city

Season	Cruises	Season	Cruises
1996-1997	25	2000-2001	62
1997-1998	39	2001-2002	57
1998-1999	38	2002-2003	47
1999-2000	47	2003-2004	50

Source: Own elaboration based on data from CEDEM (Studies Center for Metropolitan Economic Development. Department of Production, Tourism and Sustainable Development. 2004).

In the table shown above, it is clear how the cruise arrivals stem from the US and Europe. At a closer look, within the European visitors, Germans and Britons prevail. Needless to say that these segments are highly interested in consuming Tango Music and heritage associated to the history of Buenos Aires city. Among the activities, these visitors do are the following: 52.9% people go on city tours, 45% go to restaurants and 37% go to tango shows identifying Carlos Gardel (tango singer) and Caminito Street (in La Boca) as the icons of the city. According to this, the latest CEDEM report shows two important things. On one hand, the number of hotels and restaurants in San Telmo and La Boca has risen. On another, Gay-friendly circuits developed together with the tourism boom of San Telmo, Palermo and la Boca. There are presently over 200 restaurants, bars and clothing shops oriented to this segment, regarded as DINK ('double income no kids') by tour operators. Indeed, San Telmo is considered "gay friendly." Besides, an important five-star hotel is being built in San Telmo for gay visitors only with a USD3.5 million investment. Although the statistical presentation does not say much on the qualitative impact of tourism in Buenos Aires city, we present a rapid snapshot for improving the current understanding of the tourism effects in urban cities. In the next, we offer a qualitative discussion about Tango and its interpretations for the European demands. This begs the question of what happens when heritage is commoditized to be gazed by an international segment?

CULTURE AS A RESOURCE: FOR WHOM AND FOR WHAT?

Urban planning and cultural studies have much to say or explain in the transformation of urban landscapes, above all in periods of crises and political violence. The revalorization of neighborhoods through the adoption of tourism seems not to be new, but what is more than important to debate is its intersection with global modernity. In the earlier section, we cited George Yudice´s contributions who argued convincingly that heritage consumption plays a crucial role in the recovery of urban areas harmed by financial crisis or the decline of industrial neighborhoods. Tourism, after all, opens the doors to foreign tourists who cultivate different lifestyle and cultural values. This process makes not only more tolerant and open-minded but resilient cities. Nonetheless, he cautions on the problems of globalization imposing externally-designed ideals or stereotypes which are automatically internalized by locals. This is a growing use made for "the social-political and economic improvement in this age," among other issues, "due to the appearance of what Jeremy Rifkin (2000) called ´cultural capitalism´ (2002: 23). This author claims that cultural development is a "holistic perspective of development" in which culture acts as a vehicle towards human development: "Culture can generate income through tourism, crafts and other cultural activities." For Buenos Aires city, culture has become the axis of the local government's official campaign: *Buenos Aires, Latin American Cultural Capital*, which marks its "comparative advantage" at the regional level. In the field of urban studies, culture includes new multicultural-related phenomena which, in the face of the tourism phenomenon, highlight local culture's identity, diversity and development values. As Gorelik (2004) claims, these cultural studies have taken over the political discourse of local governments of Latin America, emphasizing culture and sifting politics; cultural studies look as if they have replaced opinion polls through urban imaginary worlds, making identity, memory and cultural diversity part of a political horizon, displacing political debates about urban imagination, that is, about the city project. It is noteworthy that culture should not be contemplated as singled or a closed concept, rather, it signals to lifestyles and minds in constant movements (Williams, 2000). To wit, Raymond Williams (1981) defines culture as a created significant system, which in practice can be identified as language, system of thoughts (or consciousness) and ideology (in a broad sense as modern anthropology takes it). This system of meanings, beliefs and values is inherent in the economic or political sphere, spheres which belong to a social system. In this sense, we think it is of utmost importance to regard the use of culture as a *resource* from a critical point of view and its symbolic and material component as a *cultural mercantilization* phenomenon that therefore makes it a cult. It is thus necessary to understand urban transformations also as an ideological phenomenon of late capitalism (Jameson 1991), in which case culture and economy are closely related in the production of "places" as merchandises with highly cultural content.

CONCLUSION

In a nutshell, the successive social-territorial transformation processes suffered Buenos Aires city -since the 1990s- seems not to be pretty different than those situations faced by European, Latin American or American cities. Tourism ignites the necessary force to open the doors to a gentrification process which helps but at the same time can affect local life. In the 90s decade, urban heritage was imagined as a *resource* for economic development, the conservation of heritage which enhances very well the solidarity, associated to the formula of development gave to policymakers sufficient reasons to believe in tourism as the best cures for the problems of unemployment and local crime. As the specialized literature holds, tourism allows a rapid change for the urban landscape that in theory would assist the poorer classes. As this chapter holds, this was partially true. While Buenos Aires city becomes an international tourist destination, the city went through one of its worse economic crises in history. Some southern urban area (i.e. La Boca and San Telmo) were refurbished and subject to much deep transformations and interventions in which case, there is no further to object to the specialized literature. At a closer look, anyway, these changes brought some undesirable or negative effects associated to the re-localization of established families or the rise of real estate speculation. Of course, the peso devaluation was a key factor that increased the number of arrivals to the city. Likewise, tourism passed to be a utopia to a major economic strategy applied for the local administration. To put the same in other terms, tourism not only assisted but triggered the recovery of some neighborhoods during the 2001 economic crisis. As claimed at the introduction, the urban transformation ignited by tourism was conducive to the selection of some narratives and stereotypes carefully gathered to give to visitors a one-sided discourse. Buenos Aires city was incorporated to the global marketplace through a mere symbolism linked to Tango Music and a commoditized history. We, in this chapter, stress on the concept of heritage as an economic resource those re-dynamize spaces whipped by the neoliberal crisis. In this way, Carlos Gardel Museum or el Abasto was designed as an iconic symbol of culture while La Boca –and Caminito were packaged as landscapes of international cinema. The city continues to offer a "for export" seduction which, since the end of convertibility, has been an opportunity for public policies to generate money income because foreign visitors can access high-level cultural goods and services for a low economic cost. The way such foreign capital enters the country and the protector and regulatory role public policies have are yet to be debated and reconsidered. The same neoliberal agenda that shipwrecked Argentina to a chaotic situation was reformulated to offer a new aesthetic dynamic more oriented to what Veblen called "conspicuous consumption". The crisis was an opportunity for some private groups, but a real turning point to introduce tourism as a form of modern reformation. The dominant (neoliberal discourse) keeps both, the diagnosis of the illness and the cure, where the agonizing economies are next to die.

REFERENCES

Avraham, E. (2015). Destination image repair during crisis: Attracting tourism during the Arab Spring uprisings. *Tourism Management*, *47*, 224–232. doi:10.1016/j.tourman.2014.10.003

Ayikoru, M., Tribe, J., & Airey, D. (2009). Reading tourism education: Neoliberalism unveiled. *Annals of Tourism Research*, *36*(2), 191–221. doi:10.1016/j.annals.2008.11.001

Ballart Hernandez, J., & Tresseras, J. (2001). *Gestión del patrimonio cultural*. Ariel.

Beerli, A., & Martin, J. D. (2004). Factors influencing destination image. *Annals of Tourism Research*, *31*(3), 657–681. doi:10.1016/j.annals.2004.01.010

Benko, G. (2000). Estrategias de comunicación y marketing urbano. EURE, 26(79), 67-76. doi:10.4067/S0250-71612000007900004

Capanegra, A. (2007). *La Política Turística Argentina del Siglo XX; emergencia e institucionalización del turismo como Política y Planificación pública 1930-2001*. Jornadas Nacionales de Sociología 2007, Mesa 37: Sociología del Turismo, Facultad de Ciencias Sociales, Universidad de Buenos Aires.

Carman, M. (2006). *Las trampas de la cultura: los intrusos y los nuevos usos de la cultura en el barrio del Abasto*. Editorial Paidós.

Carrion, F. (2000). *Lugares o flujos centrales: los centros históricos urbanos, Santiago de Chile, CEPAL, Serie Medio Ambiente y Desarrollo. CEDEM*. Anuarios Estadísticos. Centro de Estudios para el Desarrollo Económico Metropolitano. Secretaría de Desarrollo Económico. Gobierno de la Ciudad de Buenos Aires.

Ciccollela, P. (1999). Globalización y dualización en la Región Metropolitana de Buenos Aires. Grandes inversiones y reestructuración socio-territorial en los años noventa. *Revista Eure*, *25*(26), 5–27.

Cortez Puya, T. (2005). *Recuperación de patrimonio cultural urbano como recurso turístico*. Universidad Complutense de Madrid.

De Kadt, E. (1979). *Tourism: Passport to Development. In Perspectives on thesocial andcultural effects of tourism in developing countries*. World Bank Press.

Echtner, C. M., & Ritchie, J. B. (1991). The meaning and measurement of destination image. *Journal of Tourism Studies*, *2*(2), 2–12.

Fainstein, S., & Judd, D. (1999). *The Tourist city*. Yale University Press.

Garreton, M. (2001). *El espacio cultural latinoamericano. Bases para una política cultural de integración*. Santiago de Chile, Fondo de Cultura Económica.

Gorelik, A. (2006). Buenos Aires, de la crisis al boom. *Revista Punto de Vista*, *84*, 1–8.

Gotham, K. F. (2005). Tourism gentrification: The case of new Orleans' vieux carre (French Quarter). *Urban Studies (Edinburgh, Scotland)*, *42*(7), 1099–1121. doi:10.1080/00420980500120881

Guattari, F., & Rolnik, S. (2005). *Micropolíticas*. Editorial Tinta Limón.

Harvey, D. (1990). *The Condition of Posmodernity*. Blackwell.

Jameson, F. (1991). *Postmodernism, or, the cultural logic of late capitalism*. Duke University Press.

Judd, D. (1999). *Constructing the Tourist Bubble. In The Tourist city*. Yale University Press.

Judd, D. (2003). Urban tourism and the geography of the city. *Eure magazine. Santiago de Chile*, *29*(87), 51–62.

Korstanje, M. (2012). Reconsidering cultural tourism: An anthropologist's perspective. *Journal of Heritage Tourism*, *7*(2), 179–184. doi:10.1080/1743873X.2011.639883

Korstanje, M., & Seraphin, H. (2017). Revisiting the sociology of consumption in tourism. In R. K. Dixit (Ed.), *The Routledge Handbook of Consumer Behaviour in Hospitality and Tourism* (pp. 16–25). Routledge.

Korstanje, M. E., & Ivanov, S. H. (2012). Tourism as a form of new psychological resilience: The inception of dark tourism. *CULTUr-revista de Cultura e Turismo*, *6*(4), 56–71.

Maitland, R. (2010). Everyday life as a creative experience in cities. *International Journal of Culture, Tourism and Hospitality Research*, *4*(3), 176–185. doi:10.1108/17506181011067574

Marcuse, P. (1986). Abandoment, gentrification and Displacement: Linkage in New York City. In N. Smith & P. Wiliams (Eds.), *Gentrification of the city*. Allen & Unwin.

McMichael, P. (2011). *Development and Social Change*. Sage.

Morgan, N., Pritchard, A., & Pride, R. (2011). *Destination brands: Managing place reputation*. Routledge.

Mosedale, J. (2016). Neoliberalism and the political economy of tourism. In J. Modesale (Ed.), *Neoliberalsm and the political economy of tourism* (pp. 1–19). Routledge. doi:10.4324/9781315597782

Pigram, J. J., & Wahab, S. (Eds.). (2005). *Tourism, development and growth: the challenge of sustainability*. Routledge. doi:10.4324/9780203975138

Prats, L. (1997). *Antropología y Patrimonio*. Ariel Antropología.

Richards, G., & Hall, D. (Eds.). (2003). *Tourism and sustainable community development* (Vol. 7). Psychology Press.

Rifkin, J. (2000). *La era del acceso*. Editorial Paidós.

Skoll, G. R., & Korstanje, M. (2014). Urban heritage, gentrification, and tourism in Riverwest and El Abasto. *Journal of Heritage Tourism, 9*(4), 349–359. doi:10.1080/1743873X.2014.890624

Urry, J. (1990). *The Tourist Gaze: Leisure and Travel in Contemporary Societies*. Sage Publications.

Urry, J., & Lash, S. (1994). *Economies of signs and space*. Sage.

Vitic, A., & Ringer, G. (2008). Branding post-conflict destinations: Recreating Montenegro after the disintegration of Yugoslavia. *Journal of Travel & Tourism Marketing, 23*(2-4), 127–137. doi:10.1300/J073v23n02_10

Winter, T. (2007). *Post-conflict heritage, postcolonial tourism: tourism, politics and development at Angkor*. Routledge. doi:10.4324/9780203946381

Wortman, A. (1996). Repensando las políticas culturales de la transición. In *Revista Sociedad*. Buenos Aires: Facultad de Ciencias Sociales, UBA.

Yudice, G. (2002). *El recurso de la cultura*. Gedisa.

Yudice, G., & Miller, T. (2004). *Política cultural*. Editorial Gedisa.

ADDITIONAL READING

Causevic, S., & Lynch, P. (2011). Phoenix tourism: Post-conflict tourism role. *Annals of Tourism Research, 38*(3), 780–800. doi:10.1016/j.annals.2010.12.004

Gant, A. C. (2015, August). Tourism and commercial gentrification. In *Proceedings of the RC21 International Conference on "The Ideal City: Between Myth and Reality, Representations, Policies, Contradictions and Challenges for Tomorrow's Urban Life", Urbino, Italy* (pp. 27-29).

Garrod, B., & Fyall, A. (2000). Managing heritage tourism. *Annals of Tourism Research, 27*(3), 682–708. doi:10.1016/S0160-7383(99)00094-8

Gravari-Barbas, M., & Guinand, S. (Eds.). (2017). *Tourism and gentrification in contemporary metropolises: International perspectives*. Taylor & Francis. doi:10.4324/9781315629759

Hines, J. D. (2010). Rural gentrification as permanent tourism: The creation of the 'New' West Archipelago as postindustrial cultural space. *Environment and Planning. D, Society & Space, 28*(3), 509–525. doi:10.1068/d3309

Korstanje, M. E. (2011). Detaching the elementary forms of dark-tourism. *Anatolia*, *22*(3), 424–427. doi:10.1080/13032917.2011.620800

Korstanje, M. E. (2016). *The rise of thana-capitalism and tourism*. Routledge. doi:10.4324/9781315457482

Poria, Y., Butler, R., & Airey, D. (2003). The core of heritage tourism. *Annals of Tourism Research*, *30*(1), 238–254. doi:10.1016/S0160-7383(02)00064-6

Prentice, R. (1993). *Tourism and heritage attractions*. Routledge.

Seraphin, H., Gowreensunkar, V., & Ambaye, M. (2016). The Blakeley Model applied to improving a tourist destination: An exploratory study. The case of Haiti. *Journal of Destination Marketing & Management*, *5*(4), 325–332. doi:10.1016/j.jdmm.2016.07.004

Seraphin, H., Sheeran, P., & Pilato, M. (2018). Over-tourism and the fall of Venice as a destination. *Journal of Destination Marketing & Management*, *9*, 374–376. doi:10.1016/j.jdmm.2018.01.011

Seraphin, H., Yallop, A. C., Capatina, A., & Gowreesunkar, V. G. (2018). Heritage in tourism organisations' branding strategy: The case of a post-colonial, post-conflict and post-disaster destination. *International Journal of Culture, Tourism and Hospitality Research*, *12*(1), 89–105. doi:10.1108/IJCTHR-05-2017-0057

Viladrich, A., & Baron-Faust, R. (2014). Medical tourism in tango paradise: The internet branding of cosmetic surgery in Argentina. *Annals of Tourism Research*, *45*, 116–131. doi:10.1016/j.annals.2013.12.007

Weed, M. (2003). Why the two won't tango! Explaining the lack of integrated policies for sport and tourism in the UK. *Journal of Sport Management*, *17*(3), 258–283. doi:10.1123/jsm.17.3.258

Welch, M. (2013). Penal tourism and a tale of four cities: Reflecting on the museum effect in London, Sydney, Melbourne, and Buenos Aires. *Criminology & Criminal Justice*, *13*(5), 479–505. doi:10.1177/1748895812474660

KEY TERMS AND DEFINITIONS

Gentrification Theory: It is understood as an urban process where the landscapes of neighborhoods are radically changed. While gentrification allows the increase of the economic value of a neighborhood, low-income residents are forced to abandon the urban area.

Heritage: It signals to the physical and cultural legacy which shapes and identify a society or a group as such.

Tango: It is a popular music genre born in Buenos Aires, Argentina and Montevideo (Uruguay) during 1880s.

Tourism: It is a growing industry where travels are marked by pleasure or businesses. Tourism should be defined as the act of traveling to know further of new peoples, cultures, and landscapes.

Chapter 4
Destination Attributes and Tourist Satisfaction in Jaipur as a Culture Tourism Destination

Lakhvinder Singh
Government College, Kurukshetra University, Kaithal, India

ABSTRACT

The chapter is an attempt towards measurement of tourist satisfaction towards different selected destination attributes from responses of four hundred randomly collected sample data through a structured questionnaire. The collected data were analyzed with the help of suitable statistical tools. The result of the chapter identified different attributes of the destination and classified them into satisfying, neutral, and dissatisfying attributes. The chapter also provides an overview of overall satisfaction of the tourists as well as discovers the important factors of destination attributes. The findings conclude some recommendations and suggestions to develop strategies to increase tourist arrival to ensure repeat visit.

INTRODUCTION

Cultural tourism is the subset of tourism concerned with lifestyle of the people, history, art, architecture, religion(s), fairs, festivals and other elements that helped shape their way of life in a particular geographical area. It also includes the travelling to heritage places like famous monuments, museums, archeological sites etc. India is known for its rich cultural heritage and an element of mysticism, which is why tourists come to India to experience it for themselves.

India is one of the oldest civilizations with variety of rich cultural and natural heritage. It has the world's youngest mountain chains from east to west, to be called The Great Himalayas. Its rich architectural legacy dating to an ancient past with a great philosophical thought, timeless monuments, magnificent temples, breathtaking beaches, nature-majestic mountains, wonderful wildlife make India never ceases to lure the travelers. India has fascinated people from all over the world with her geographical and cultural locale. In the north India, there are different culture tourism destinations which attract large number of foreign and domestic tourists (Singh, 2002).

DOI: 10.4018/978-1-7998-4330-6.ch004

The Rajasthan state of India is having number of culture tourism destinations including famous pink city 'Jaipur' which enriched with its handicraft, fairs, folk dances, music forts, palaces, museums and people of different religions. It is an enthralling and memorable experience to one and all. India is also a tourist paradise and has all unique features of modernity and traditional hospitality (Dasgupta, 2011). Due to rich cultural resources, Jaipur has registered phenomenal growth and projected as ultimate culture tourism spot.

In this chapter, an endeavor has been made to provide an overview of tourism in Jaipur with particular reference to experience of tourist towards different destination attributes (facilities and services). The chapter starts with brief introduction of culture tourism, overview of India as tourist destination followed by description of Jaipur tourism potential number of tourist arrivals and focus of the chapter.

The city of Jaipur looks vibrant and colorful with its unique architecture and natural surroundings that makes it important destination of India. Jaipur is well-connected by road, rail and air. The city railway station and airport are the major points of entry and exit to the state. The Jaipur Airport is located at a distance of 13 km from the main city. Most of the domestic and international airlines operate to and from Jaipur. Flights to Jaipur from Delhi and Mumbai are frequent. The regular international flights operate mostly from Jaipur to Sharjah, Muscat, Dubai, Bangkok and other destinations. The Jaipur Railway Station is one of the main stations of Rajasthan. Most of the cities and towns in India are connected to Jaipur by railway. There are a number of trains that run almost on a daily basis to and from Jaipur. Some of the important trains to Jaipur are; Shatabdi, Mandore Express, Jammu Tavi Express, Ashram Express, Ahmadabad Mail, Ibadat Express etc.

Jaipur is a major tourist destination with many places of historical and religious importance. Some of the important tourist attractions include Wind palace as one of the important hallmarks of Jaipur that attracts tourists from all over the world. The City palace in Jaipur is one of the architectural wonders that reflect a perfect blend of the Rajasthan and Mughal architectural style. The famous observatory knows as Jantar Mantar also attracts several tourists throughout the year particularly those interested in study the movements of stars and planets. The Albert Hall museum in city is a magnificent building that served as the store house of artifacts and paintings of the past. The fort of Jaigarh provides a bird's eye view of Jaipur and frequently touched by tourists from country and overseas territories. Another famous and crucial attraction of the city is known as Rambagh palace which is the former residence of Maharaja of Jaipur, and now one of the luxurious hotels in Jaipur. During the vacations, large influx of visitors can be seen in the palace. Amber Fort built by Meena dynasty and later ruled by Raja Man Singh, is fort witnesses numerous tourists all through the year. From religious point of view, the Birla temple also situated in the city as a wing of several Birla temples located across the country. The temple looks stunning during night.

Jaipur is a place you can plan for a weekend leisure trip from any point of the country. Jaipur featured with monuments, palaces, variety of foods, shopping avenues and so on. The smart city of Jaipur has huge potential and prospects for tourism and continuously moving towards receiving increasing tourists' footfalls both domestic tourists visits (DTV) as foreign tourist arrivals (FTA) as presented in the table 1 indicated that tourists in Jaipur are growing year by year which indicted positive growth of tourism in the area. The domestic tourism registered an enhanced increase over the previous years in contrast to foreign tourist arrival in Jaipur.

Table 1. Tourist traffic in Jaipur

Year	Domestic Tourist Visits	Foreign Tourist Visits
2014	1170152	568234
2015	1201152	596756
2016	1544730	565978
2017	1702665	633990

Source: Tourism Department Website of Rajasthan Government
(http://www.tourism.rajasthan.gov.in/annual-progress-report.html)

Taking a note from this growing phenomenon of tourism in Jaipur, it is essential to observe tourist behavior towards different products and services in the city. The present study tries to determine this aspect that the whether the tourists really focus on destination attributes for ensuring their satisfaction in study area. The presented study has following objectives:

1. To measure the satisfaction of the tourists from the destination attributes in Jaipur.
2. To understand the overall experience of tourists in Jaipur.
3. To identify the important factors of tourists satisfaction of destination attributes in Jaipur.

THEORETICAL BACKGROUND

A destination is popular place with attraction and number of facilities and services. So, the measurement of satisfaction of tourist is depends on experience perceived by tourists from different service elements at the destination or attributes (Turner et al., 2001; Prayag, Hosany & Ode, 2013; Arowosafe & Emmanuel, 2014; Lai & Li, 2016). These attributes are used as parameters to understanding the overall experience of tourists from their trip and thus providing a ground to the tour planner and destination managers to design better package including different attributes of the destination (Pizam et al., 1978; Rust et al., 1993; Battour et al., 2014; Lin & Cai, 2016). So, it is crucial to consider the quality of different service elements or attributes to understand the satisfaction or dissatisfaction of tourists for planning or success of the destination (Akama & Kieti, 2003; Hassan & Shahnewaz, 2014;Prayag, Hosany, Muskat, & Del, 2017).

There are many studies that applied the above approach, such as, Sofield & Li (1998) discovered attraction such as culture, history, religious and monumental buildings as significant attributes of destination. Yuksel, (2001) consider service delivery, cleanliness, attitude of locals, helpfulness of employees, communication facilities, as crucial attributes of the destination. Larry & Munhtuya (2005) noticed cost effectiveness of different facilities and services as important parameters of destination. Woodside & Mazanec (2004) identified lodging, foodstuff, retail or souvenir, infrastructure, basic amenities as important attributes for the destination. Some more studies were conducted to find out important attributes like accommodation, transportation, tourist information centers, cultural and musical events, recreation opportunities, safety and security as essential at destination (Reisinger,2009;Assefa, 2011).

For measurement of tourist satisfaction with different attributes of the destination, several studies have been conducted so far. These studies were reviewed by author in order to development of conceptual part of the research study. Huh (2002) assessed the relationship of destination attributes and satisfaction of the tourists visiting the Virginia Historic Triangle (Williamsburg, Jamestown, and Yorktown) from twenty five attributes of the destination. By using factor analysis attributes were extracted into four factors namely general tour attraction, heritage attraction, cultural attraction and maintenance factor. The identified factors significantly co-related with the overall satisfaction of the tourists. Tosun, Temizkan, Dallen, & Fyall (2007) measured the tourist's level of satisfaction at Cappadocia destination in Turkey on the basis of retail, shopping, service quality, staff credibility, ambience, payment of different services attributes. The measurement of attributes resulted in to variant satisfaction level of the tourists and concluded to enhanced tourist satisfaction in the area. In a similar study, Aggarwal, Handa, & Singh (2010) identified satisfaction of tourists visiting India on the basis of selected attributes namely facilities, visual appealing, information, lodging facility (hotel/resort, etc.), eating and drinking facilities and safety and security arrangements. In this vein, it came to noticed that the satisfaction of tourists fails to match with their expectation in responses to all the selected attributes. Hence, suggested to step up concrete strategies for improving facilities and services to bring more satisfaction index.

The present study focuses on culture tourism and hence inclusion of some recent studies particularly conducted in pursuit of understands tourist satisfactions towards different attributes of culture tourism. On global side the research undertaken by Putri (2017) in Saung Angklung in West Java (Indonesia) and Damayanti, Ekasari, & Syaodih (2019) in Sundapolis area of Bandung city and from Indian perspective, the studies of Kumaran, Kannan, & Milton (2013) in Madurai of Taminladu and Dar (2018) in Gulmarg area of Jammu and Kashmir indicated the measurement of tourist satisfaction were considered and reviewed to formulate a strong theoretical background to the chapter. The analysis of review of earlier studies identifies limited attention of researchers towards topic particularly in study area. Hence, it is justified to conduct a meaningful study to cover this gap.

RESEARCH APPROACH

This present chapter aims at investigating the satisfaction of tourists regarding various set of destination attributes in Jaipur. Different attributes of destination were identified by author from the review of previous studies considered in chapter. The study adopted expectation-disconfirmation paradigms to accomplish the objective of the chapter. The selection of model asserted from the study of Chon & Olsen (1991) that support the relationship between tourist expectation and satisfaction are related to each other. The study of Clemons & Woodruff (1992) also in favor to use the model as identified the customer (tourist in the present study) first form expectations regarding product/service before experiencing product/service customer to make comparison between expectation and actual performance of to find the result of the comparison. Some authors argue that (Churchill & Surprenant, 1982; Oliver & Beardon, 1985; Patterson, 1993) in terms of expectation-disconfirmation theory further elaborated that customer expectations are confirmed when performance matches its expectation, positively disconfirmed (satisfaction arises) when performance exceeds expectation and negatively disconfirmed (dissatisfaction arises) when performance fails to match expectations.

The author used conduct field survey to get the responses of tourist and used self-administered questionnaire to note the rating of expectation and satisfaction regarding selected destination attributes on the basis of five point Likert scale highly satisfied-5 to highly dissatisfied-1). The questionnaire was divided in four different parts. The first part of the instrument covered questions related to socio-demographic profile of the tourists. The second part contained questions related to measurement of expectation and satisfaction regarding attributes of the destination.

The present study is cross-sectional in nature thus limited to primary data collected from tourist during June 2019 to December 2019 in Jaipur as a subgroup of study population. The study used simple random sampling method to collect the responses of tourists and the survey form in the form of questionnaire distributed to 700 tourists, but only 400 tourists completely filled up the questionnaire and returned to author with a response rate of 57.14 percent. A variety of statistical techniques have been used through application of Statistical Software Package (SPSS) to analyze the collected data in order to make the study more meaningful. A brief description of tools and the formulas used for data analysis includes descriptive statatistics such as frequency, percentage, means and standard deviation and also inferential statistics like paired t test and factor analysis.

MAIN FOCUS OF THE CHAPTER

The previous section of the chapter provided details on the research techniques adopted for analysis of the data. The present section will discuss the result of the data analysis and various statistical tests used. The result of the study presented in tables for providing clear understanding of the data.

Demographic and Travel Profile of Tourists

The demographic profile of the respondents is presented with the help of frequency and percentage analysis in the below tables.

Above table highlighted that the out of 400 respondents, 77 percent were male where as 23 percent were female. It indicates that male respondents are more active in responding during field survey. The most of the tourists participated in the survey were from India only and poor responses from female tourists could be due to several other reasons such as male dominated region and hesitation of female tourists. In terms of age wise classification, 28.25 percent were in the age group of 21-40, followed by 27.5 percent in 41-60, 23.5 above 60 years and 20.75percent up to the age of 20 years. It means that senior tourist groups above 40 years are more frequently visited study area as compared to their younger counterparts. According to marital status classification of the respondents which indicated that 71.5 percent were married and 28.5 percent unmarried respondents. According to place of residence, 92.25 percent were Indian or domestic visitors as compared to 7.75 percent international or foreign visitors. Out of total 31 foreign respondents participated in the survey indicated a wide coverage of tourist generating regions for Jaipur. The foreign respondents from United Kingdom indicated highest number (8) of participant in the study followed by France (5 respondents). Almost equal number of tourist respondents was noted as 3 each from United States of America, Germany, Canada and Japan. 2 tourists each from Australia and Bangladesh, followed by 1 each from Italy and Switzerland. This shows that there is lack of marketing and promotion strategies to attract foreign tourists in the area. Poor response of foreigner tourists in survey is also due to language and communication barrier. Education wise distribution highlighted that 32.5 percent respondents were graduate followed by 26 percent with senior secondary, 25 percent with post graduate or above, and 16.5 percent respondents were higher secondary education only. This depicted that the more than half of the respondents are in the highly educated group i.e. above graduation. On analyzing the occupation of the respondents, it came to notice that 31 percent serviceperson, 26 percent businessperson, 23.75 percent involved in other occupations (such as retirees, farmers, freelancers etc.) and 19.25 percent were students. It shows that the also every segment of the society visited the selected culture area. Income wise classification highlighted that 33 percent were in the group of 2-4 lakhs followed by 28.5 percent up to 2 lakhs, 22.5 percent in 4-6 lakhs and remaining 16 percent in above 6 lakhs group.

Tourist travel profiles are presented in the below table with the help of frequency and percentage analysis.

Table 2. Demographic profile of the respondents

Features	Variables			
Gender	Male		Female	
Frequency	308		92	
Percentage	77		23	
Age (in years)	Up to 20	21-40	41-60	Above 60
Frequency	83	113	110	94
Percentage	20.75	28.25	27.5	23.5
Marital Status	Married		Unmarried	
Frequency	286		114	
Percentage	71.5		28.5	
Residence	Indian		Foreigner	
Frequency	369		31	
Percentage	92.25		7.75	
Education	H. Secondary	Sr. Secondary	Graduation	P. Graduation or above
Frequency	66	104	130	100
Percentage	16.5	26	32.5	25
Occupation	Student	Serviceperson	Businessperson	Other
Frequency	77	124	104	95
Percentage	19.25	31	26	23.75
Income	Up to 2 Lakhs	2-4 Lakhs	4-6 Lakhs	Above 6 Lakhs
Frequency	114	132	90	64
Percentage	28.5	33	22.5	16

In terms of the travel behavior features of the respondents mentioned in the above table, 34.25 percent visit for second time followed by 27.25 first time visitors, 22.75 percent third visit and remaining 15.75 visits fourth or more times to study area. It depicted that more than half of the respondents are repeat visitors to the selected culture sites. The largest portion of respondents i.e. 42.75 used rail transportation to reach the destination as compared to 30.75 percent through road transport services. The remaining 26.5 percent were travelled through their own vehicle. It shows that the study area has the easy accessibility. The foreign respondents participated indicated travelling through air transport to reach India, but on arrival the services of rail and road transportation were used to reach Jaipur. In this way, none of the respondents used air transportation to reach the Jaipur.

Table 3. Travel profile of the respondents

Features	Variables			
No. of Visit	1st	2nd	3rd	4th or more
Frequency	109	137	91	63
Percentage	27.25	34.25	22.75	15.75
Transportation to reach Jaipur	Road Transport	Rail transport	Own vehicles	Air transport
Frequency	123	171	106	-
Percentage	30.75	42.75	26.5	-
Accommodation	Hotel	Guest House	Dharmshalas/ Gurudwara	Friends/ relatives houses
Frequency	147	109	83	61
Percentage	36.75	27.25	20.75	15.25
Length of stay	1 day	2 days	3 days	4 or more days
Frequency	111	155	89	45
Percentage	27.75	38.75	22.25	11.25
Sources of Information	Print Media	Electronic Media	Travel Agent/ Tour Operator	Word of Mouth
Frequency	90	106	84	120
Percentage	22.5	26.5	21	30
Companion of Tour	Alone	Family	Friends	Group tour
Frequency	85	142	105	68
Percentage	21.25	35.5	26.25	17
Purpose	Culture	Leisure	Business	Education and research
Frequency	211	91	52	46
Percentage	52.75	22.75	13	11.5

According to the classification of accommodation used, 36.75 percent used hotels, 27.25 percent used guest house, 20.75 percent used dharmshalas/gurudwars and 15.25 percent used to stay at the destination. It highlighted that in study area different types of the accommodation to offer to tourists according to their taste and budget. In terms of length of the stay at the destination, 38.75 percent stay for two days followed by 27.75 percent for at least one day, 22.25 percent for three days and 11.25 percent with four or more days stay at the destination. For the sources of information used to search about the destination, 30 percent were know about the destination from word of mouth publicity, 26.5 percent used electronic media to search the destination, 22.5 used printed media and 21 percent used to search through travel agents or tour operators. With regards to travel companion 35.5 percent visited with family members followed by 26.25 percent visited in friends group, 21.25 percent visited alone and remaining 17 percent traveled in group tour of school/colleges or office tours. On asking about the purpose of visit, 52.75 percent visited for culture tourism, 22.75 percent for seeking leisure (holiday), 13 percent visited for business purpose and 11.5 percent visited for the purpose of some education or research work. It support that more than half of the respondents visited the selected culture sites with religious motivation.

Destination Attributes and Tourist Satisfaction in Jaipur

Table number 4 listed 40 attributes of the Jaipur and result indicated those tourists were showing different satisfaction and expectation from different attributes of the destination. The result of paired t test is presented on next page.

Table 4. Result of paired t-test for identification of satisfaction with attributes

Attributes	ME	S.D.	MP	S.D.	Paired Diff.	t	Sig. (2tailed)
Attractiveness of tourist places	3.98	1.144	4.10	1.018	-.118	-3.447	.001
Attractiveness of fairs and festivals	3.60	1.148	3.78	1.133	-.177	-4.190	.000
Uniqueness of culture and heritage	3.30	1.054	2.78	1.197	.522	11.821	.000
Weather conditions at the destination	3.30	1.194	3.85	1.130	-.545	-11.822	.000
Availability of souvenirs/crafts	3.35	1.260	3.86	1.192	-.507	-13.369	.000
Cost effectiveness of souvenirs and crafts	3.52	1.193	3.21	1.259	.315	5.878	.000
Availability of bus services	2.85	1.182	3.68	1.203	-.830	-13.206	.000
Availability of private taxi/cab services	3.87	1.093	4.08	1.035	-.212	-5.216	.000
Availability of parking area	2.83	1.177	2.66	1.165	.167	4.234	.000
Condition of roads and signage	3.31	1.246	3.16	1.293	.145	3.172	.002
Price of transportation services charged	3.49	1.155	2.88	1.167	.605	10.879	.000
Connectivity of trains from destination	3.16	1.153	3.26	1.244	-.108	-2.682	.008
Availability of accommodation facilities	3.25	1.118	3.96	.980	-.710	-16.477	.000
Quality of services provided at accommodation units	3.78	1.044	2.76	1.182	1.022	26.773	.000
Cleanliness of accommodation unit area	4.08	1.016	2.71	1.153	1.365	26.337	.000
Behavior of the hotel/guest house staff	3.62	1.198	3.40	1.314	.210	5.302	.000
Price charged for accommodation services	3.02	1.317	3.17	1.303	-.152	-3.221	.001
Availability of food and beverage outlets	4.01	.943	4.02	1.036	-.002	-.066	.948
Variety of food and beverages	3.54	1.134	3.79	1.124	-.248	-6.376	.000
Healthy and hygienic condition of the food and drink served	3.80	1.080	3.27	1.352	.533	9.058	.000
Cleanliness at food serving area	4.00	1.000	2.62	1.283	1.382	32.746	.000
Neat and clean appearance of food serving staff	3.61	1.223	2.91	1.337	.700	13.711	.000
Prices of the food and drink served	3.24	1.322	3.37	1.248	-.127	-3.271	.001
Attributes	ME	S.D.	MP	S.D.	Paired Diff.	t	Sig. (2tailed)
Availability of tourist information centers	3.58	1.248	3.04	1.392	.547	11.949	.000
Knowledge of tourist information office staff	3.21	1.321	3.20	1.368	.002	.045	.964
Availability of supporting staff (Guide/escort/pithoo/ porter)	3.16	1.221	3.09	1.314	.075	1.650	.100
Behavior of supporting staff	3.40	1.332	3.58	1.194	-.188	-4.972	.000
Responsiveness of local people	3.63	1.231	3.69	1.285	-.060	-1.098	.273
Behavior of temple/shrine staff	3.56	1.355	3.75	1.158	-.188	-2.950	.003
Communication and Language of the locals	3.47	1.264	3.62	1.243	-.152	-4.377	.000
Availability of banking and ATM facilities	3.52	1.270	3.84	1.223	-.320	-5.569	.000

continues on following page

Table 4. Continued

Attributes	ME	S.D.	MP	S.D.	Paired Diff.	t	Sig. (2tailed)
Availability of medical facilities	3.57	1.216	3.53	1.264	.045	.814	.416
Availability of clean drinking water	3.04	1.397	3.94	1.128	-.900	-15.416	.000
Availability of toiletry facilities	3.22	1.169	3.34	1.183	-.127	-3.446	.001
Availability of cloak room and locker facilities	3.35	1.326	2.80	1.332	.553	10.024	.000
Mobile network availability	3.21	1.347	2.82	1.283	.388	6.632	.000
Overall cleanliness and sanitation at the venues	3.54	1.246	2.90	1.318	.640	9.891	.000
Presence of police posts in the area	3.42	1.295	3.72	1.189	-.308	-6.690	.000
Security of tourists' luggage from theft and robbery	3.88	1.066	3.26	1.328	.620	13.841	.000
Personal safety of tourists in the area	3.88	1.200	3.98	1.122	-.107	-2.314	.021

N=400, df=399(for all cases),
ME= Mean of expectation, MP= Mean of the perception, S.D.= Standard deviation
Expectation level ranges from 1(Very Low Expectation) to 5(Very High Expectation)
Satisfaction level ranges from 1(Highly Dissatisfied) to 5(Highly Satisfied)

The satisfaction of the tourists analyzed from above was divided into three categories i.e. satisfactory attributes, dissatisfactory attributes and neutral attributes. The satisfied attributes are explained as those attributes with satisfaction score above expectation scores and with t-value significant at the 0.05 level. Analysis of t-test revealed that tourists were satisfied with attractiveness of religious places, fairs and festival attractiveness, weather condition, availability of souvenirs and crafts, availability of bus service, presence of taxi/cab service, train connectivity to and from destination, availability of accommodation facilities, price charged for accommodation, variety of food and beverages, price of the food and drinks served charged, behavior of supporting staff, behavior of temple/shrine staff, communication and language ability of locals, banking and ATM availability, availability of clean drinking water, availability of toiletry facilities, presence of police posts in the area and personal safety of tourists in the area. The respondents' satisfaction with these attributes was positively disconfirmed with their expectation level, which led to satisfaction in relation to those attributes.

The most satisfying attribute is availability of clean drinking water (paired difference value= 0.900), availability of bus service (paired difference=0.830) and availability of accommodation facilities (paired difference=0.710) whereas, the least satisfying attribute is personal safety of tourists in the area (paired difference=0.107).

Neutral attributes were defined as those attributes which are having non-significant t-value ($p^{3}0.05$), regardless of a positive or negative mean difference. The neutral attributes identified from t-test analysis are availability of food and beverages (paired difference=0.002), knowledge of tourist information centre staff(0.002), availability of supporting staff(0.075), responsiveness of local people(0.060) and availability of medical facilities(0.045).

Dissatisfied attributes are indicated as the attributes with expectation score outweighing satisfaction scores, regardless of a significant or non-significant t-value at the 0.05 level or below. Analysis in the table 3 noted that tourists were dissatisfied with uniqueness of culture-heritage, price charged for souvenirs and crafts, availability sufficient parking space, condition of roads and signage, price charged for transportation, quality of accommodation facilities, cleanliness of accommodation area, behavior of hotel/guest house staff, hygienic condition of food and drink served, cleanliness at food serving area, neat and clean appearance of food serving staff, availability of tourist information centers, availability of cloak room and lockers, mobile network availability, overall cleanliness in the area, and security of tourist luggage. The most dissatisfying attribute is cleanliness at food serving area (paired difference=1.382), cleanliness of accommodation area (paired difference=1.365) and quality of services provided at accommodation unit (paired difference=1.022).

Overall Satisfaction of the Tourists

Respondents' were also questioned about their overall level of satisfaction with the study area. The result indicated that 59.8% of the respondents were highly satisfied and satisfied with their visit to the study area. 19.2% of the respondents neutral in their opinion and 21% of the respondents were dissatisfied and highly dissatisfied. The mean value of overall satisfaction level of the respondents derived from their visit was 3.66, which tended the moderate end of the satisfaction scale. This suggests that the study area provides tourists with a satisfactory experience.

Table 5. Overall satisfaction

Satisfaction Level	Frequency	Percentage	Valid Percent	Cumulative Percent
Highly Dissatisfied	28	7.0	7.0	7.0
Dissatisfied	56	14.0	14.0	21.0
Neutral	77	19.25	19.25	40.25
Satisfied	103	25.75	25.75	66.0
Highly Satisfied	136	34.0	34.0	100.0
Total	400	100.0	100.0	

Note: Overall satisfaction mean ranges from 1(Highly dissatisfied) to 5(Highly Satisfied)

Factors of Destination Attributes

For identifying the underlying factors of tourists' perception of attributes principal component analysis was used to generate the initial solution. The eignvalue suggested that a five factor solution explained 76.16% of the overall variance before the rotation. The factors with eignvalues greater than or equal to 1.0 and attributes with factor loading 0.30 were reported. Hair et al. (1995) also suggested the factor loading should not be less than 0.3 for consideration in factor analysis. The table 5.13 illustrates the results of the factor analysis. The five factors were: Amenities and basic services, services and responsiveness, hospitality services, tour attraction and accessibility.

The Kaiser-Meyer Olikin value was 0.886, the variables were interrelated and share common factor. In addition to Bartlett test of Sphericity was significant as p value is less then0.05 (p£0.05) which indicated that the correlation between items were enough to accommodate principal component analysis and the use of factor analysis was appropriate. The Kaiser-Meyer-Olkin which is the overall measure of sampling adequacy was 0.886, which was commendable. The acceptable value of KMO must to be 0.60 or more than it (Hinton et al., 2004). The value of Kaiser-Meyer-Olkin in the present study is more than its acceptable value; hence KMO value indicated that sampling adequacy was suitable to conduct factor analysis.

From the varimax-rotated factor matrix, five factors with thirty three (33) variables were defined by the original 40 variables that loaded most heavily on them (loading^{3}0.30). Seven attributes were dropped due the failure of loading any factor at the level of 0.30 (or higher). These were price of the souvenirs/crafts purchased, price charged for transportation services, availability of parking area, Quality of services provided at accommodation unit, Behavior of hotel/guest house staff, healthy and hygienic condition of the food and drink served and presence of police posts in the area.

To test the reliability and internal consistency of each factor, the Cronbach's alpha of each was determined. The result showed that the alpha coefficients ranged from 0.710 to 0.839 for the five factors. The result was considered more than reliable, since 0.50 is the minimum value for accepting the reliability test. The values of Cronbach's alpha for each underlying factor is also greater than minimum acceptable value. Some authors (Sekaran, 2000) suggested that the value of Cronbach's alpha less than 0.60 considered as poor and more than 0.60 considered as good for considering the internal consistency of the scale as adequate.

Table 6. Factor analysis result of the satisfaction of attributes in Jaipur

Rotated Component Matrix						
	Component					Communalities
	1	2	3	4	5	
Factor 1: Amenities and Basic Facilities						
Personal safety and security of tourist in the area	.528					0.294
Mobile network availability	.505					0.288
Availability of cloak room and locker facilities	.502					0.395
Overall cleanliness and sanitation in the area	.487					0.325
Availability of medical facilities	.483					0.238
Availability of clean drinking water	.390					0.387
Security of tourists' luggage from theft and robbery	.332					0.342
Availability of banking and ATM facilities	.317					0.156
Availability of toiletries facilities in the area	.301					0.103
Factor 2: Services and Responsiveness						
Availability of tourist information centers		.563				0.528
Knowledge of the tourist information centre staff		.557				0.351
Availability supporting staff (guide/escort/pithoo/purport etc.)		.512				0441
Behavior of supporting staff		.465				0.490
Responsiveness of local people		.462				0.389
Communication and language of the locals		.452				0.233
Behavior of temple/shrine board staff		.340				0.290
Factor 3: Hospitality Services						
Availability of accommodation in the area			.525			0.308
Price charged for accommodation			.519			0.312
Cleanliness of accommodation area			.482			0.401
Availability of food and beverages centers			.478			0242
Variety of food and beverage available			.400			0.202
Price of the food and drinks served			.399			0.257

continues on following page

Table 6. Continued

Rotated Component Matrix						
	Component					Communalities
	1	2	3	4	5	
Neat and clean appearance of food serving staff			.399			0.162
Cleanliness of food serving area			.372			0.286
Factor 4: Tour Attraction						
Attractive tourists places				.822		0.706
Attractive fairs and festival				.742		0.596
Uniqueness of culture and heritage				.394		0.306
Good weather at the destination				.332		0.237
Availability of souvenirs and crafts				.323		0.163
Factor 5: Accessibility						
Availability of buses					.809	0.677
Connectivity of trains					.725	0.568
Availability of taxi/cab service					.468	0.270
Condition of roads and signage					.404	0.281
Eignvalue	5.17	3.64	4.32	2.25	1.68	
Percent Variance	27.78	12.53	15.76	10.44	9.65	
Cumulative Variance	27.78	40.31	56.07	66.51	76.16	
Cronbach's Alpha	0.839	0.827	0.774	0.730	0.710	
No. of items per factor	9	7	8	5	4	

Extraction Method: Principal Component Analysis.
Rotation Method: Varimax with Kaiser Normalization.

The five underlying factors of tourists' perception of attributes in Jaipur were described in the chapter. Amenities and basic facilities (Factor 1) contained nine attributes and explained 27.78% of the variance in the data, with an eignvalue of 5.17 and a reliability of 0.839. The attributes associated with this factor dealt with personal safety of the tourists in the area, mobile network availability, availability of cloak room and locker, overall cleanliness in the area, availability of medical facilities, availability of clean drinking water, security of tourist luggage from theft and robbery, availability of banks and ATMs in the area and availability of toiletries facilities in the area. Services and responsiveness (Factor2) accounted for 12.53 percent of the variance, with an eignvalue of 3.64, and a reliability of 0.827. This factor was loaded with seven attributes that were Availability of tourist information centre, knowledge of the tourist information centre staff, availability of supporting staff (guides/escort/pithoo/porter etc., behavior of supporting staff, responsiveness of local people, communication and language of the locals and behavior of temple/shrine board staff in the area. Factor 3 referred as Hospitality services and loaded with eight attributes. This factor represents the 15.76 percent of the variance, with an eignvalue of 4.32, and a reliability of 0.774. The attributes falls under this attributes were availability of accommodation facilities in the area, price charged for accommodation, cleanliness of the accommodation area, availability of food and beverage counters, variety of food and eatables available, price of the food and drink served, neat and clean appearance of food serving staff and cleanliness of food serving area. Factor 4 labeled as tour attraction and contained five attributes namely attractive tourist places in the area, attractiveness of fairs and festival, uniqueness of culture and heritage, Good weather at the destination and availability of souvenirs and crafts. This factor accounted for 10.44 percent of the variance, with an eignvalue of 2.25, and a reliability of 0.730. Factor 5 designated as accessibility, accounted for 9.65 percent of the variance, with an eignvalue of 1.68, and a reliability of 0.710.This factor was loaded with four attributes such as availability of buses in the area, connectivity of trains, availability of taxi/cab services and condition of road and signage in the area.

SOLLUTIONS AND RECOMMENDATION

The study also draw some suggestions that can help destination managers and service providers to provide quality services according to needs and requirements of the tourists and increase the number of satisfied tourists. Suggestions based on the result of the study and analyses are measures to develop variety of attractions, adopt tourist circuit phenomena, capacity building, and infrastructure development, improve cleanliness, and formulate suitable marketing strategy to different tourist generating regions. Organization of further research o uncovers the different perspective of destination survival and competitiveness to ensure sustainable development of tourism.

DIRECTIONS FOR FUTURE RESEARCH

The present study was conducted to investigate the tourist satisfaction with different destination attributes. Different stakeholders of tourism industry can also undertake further research as a extension of the findings of the chapter. The academicians and researcher can conduct further studies to identify the relationship between motivation to visit and satisfaction at tourism destination. Tourism planners and marketers can also used the findings of the study and can try to learn about the effect of satisfaction on behavioral intentions in order to extend the loyalty and repeat visits of tourists for survival of their business. The policy makers can also planned further investigation of application of findings of study to formulate suitable tourism policy and strategies in order to boost the competitiveness of destination on global level.

CONCLUSION

The culture tourism is growing rapidly in India. Most of the domestic tourists in India specifically travel for culture tourism purposes. The study area occupies a good range of culture tourism attractions. Therefore, it become essential to ensure the comfortable stay and better services to tourists and to measure the performances of services through identifying the level of satisfaction is also crucial for the success of the destination. Jaipur city of Rajasthan is among the important culture tourist destinations of the country. This frequently visited by number of tourist from different nations as well as domestic tourists. Furthermore, it is the need of the hour today is to understand experience gained by tourist from different attributes of destination in the area. The present chapter observed that Jaipur has been designated as culture tourism destination. The availability of sufficient attractions and facilities in the area provided a boon to tourism industry to excel its growth and improve the image of Jaipur as global tourism destination. Due to this,

Jaipur proved an added advantage in beautification of their attractions, improvement in mobility, tourist and general infrastructure enhancement, better communication and technology and improved utilities.

The demographic and travel behavior profile of the tourists reveals that there is a dominance of male tourists over female tourists. The most of the respondents were among the senior age groups as compared low age groups. While considering the nationality of the tourists, the majority were the Indians in comparison to foreign tourists. The chapter noticed overseas tourists in study area visited mostly from developed regions of the world; hence the stakeholders need to consider this point while marketing strategies for the destination in pursuit of increasing number of international tourists to the area. In terms of education, all types of people from uneducated to highly educated professionals also (graduate and above) visited the study area. In occupational status, people from almost all categories visited the area. While considering economic status of the respondents, the tourists with low to high disposable income visited to the destinations of study area. The travel profile of the respondents reveals that the majority of the respondents were repeat visitors to the area. The respondents used rail transportation more frequently to reach the study area as compared to other means of the transportation. The most of the respondents used to stay in hotels and guest houses to stay at the destination. More than half of the respondents used to stay at least two or more days at the destination visited. The respondents used all sources of information to search about the destination with word of mouth more frequently as compared to other means. Mostly respondents visited with family, friends, and official or student groups in comparison to visiting alone. A wide majority of the respondents visited for culture tourism purpose followed by for holiday or leisure seeking purpose.

The present study draws a linear relationship between pre-purchase and post-purchase behavior of the tourists by comparing their expectation with their satisfaction. Assessing the difference indicated that positive gap between expectation and satisfaction with attributes weather condition, availability of souvenirs and crafts, availability of bus service, availability of accommodation facilities, availability of clean drinking water. On the other hand, some of the major causes of dissatisfaction that include lack of cleanliness at food serving area, lack of cleanliness of accommodation area and poor quality of services provided at accommodation, lack of neat and clean appearance of food serving staff, overall cleanliness in the study area, poor security of tourists' luggage, price charged for transportation service with low satisfaction and negative gap between expectation and satisfaction of attributes. The remaining attributes availability of food and beverages, knowledge of tourist information centre staff, availability of supporting staff, responsiveness of local people and availability of medical facilities are rated as neutral in terms of satisfaction by tourists. In terms of overall experience of the tourists, the high level of overall satisfaction was found. The study also used to discover important factors of destination attributes from application of factor analysis technique. The result identified five factors from selected forty attributes of the destination. The five factors were designated as amenities and basic services, services and responsiveness, hospitality services, tour attraction and accessibility. The Kaiser-Meyer-Olkin (0.886) for measure of sampling adequacy and Cronbach's alpha of each factor (0.710 to 0.839) was also considerable and reliable. These factors were observed as important parameters to ensure the extension of satisfactory experience to tourists.

The present chapter also provides some implications for concerned stakeholders including government at central and state level and tourism marketers. Most of the policy and schemes pertaining to tourism in India are formulated by central government. Hence, the present study through investigation of the satisfactory and dissatisfactory attributes of culture tourism is providing support to formulate suitable tourism policies to attract more tourists to the destination. Moreover, the local government by understanding about tourists' satisfaction during their visit to Jaipur may be helpful to proficiently comprehend present and potential tourists to the area for survival of the destination. In terms of tourism planner, the identification of satisfied tourist segment group is helpful in reducing the marketing cost that aid to increase revenue from tourism and finally achievement of sustainable development of area.

REFERENCES

Aggarwal, V. S., Handa, M., & Singh, K. A. (2010). Expectation and Satisfaction of Adventure Tourists for Land, Air and Water based Sports in India. *DIAS Technology Review*, *7*(2), 9–19.

Akama, J. S., & Kieti, D. M. (2003). Measuring tourist satisfaction with Kenya's Wildlife safari: A case study of Tsavo West National Park. *International Journal of Tourism Management*, *24*(1), 73–81. doi:10.1016/S0261-5177(02)00044-4

Arowosafe, F. C., & Emmanuel, A. A. (2014). Investigating indicators for tourist satisfaction at Mole National Park, Ghana. *American Journal of Tourism Management*, *4*(1), 1–6.

Assefa, B. (2011). *Assessing Tourist Satisfaction in Ethiopia* (Master Thesis). Addis Ababa University.

Battour, M., Ismail, N. M., Battour, M., & Awais, M. (2014). Islamic tourism: An empirical examination of travel motivation and satisfaction in Malaysia. *Current Issues in Tourism*. Advance online publication. doi:10.1080/13683500.2014.965665

Chon, K. S., & Olsen, M. D. (1991). Functional and Symbolic Approaches to Consumer Satisfaction/Dissatisfaction. *Journal of the International Academy of Hospitality Research*, *28*, 1–20.

Churchil, G., & Surprenant, C. (1982). An investigation into the determinants of customer satisfaction. *JMR, Journal of Marketing Research*, *19*(4), 491–504. doi:10.1177/002224378201900410

Clemons, S.D. & Woodruff, R.B. (1992). Broadening the view of Consumer (Dis) satisfaction: A proposed Means – end Disconfirmation model of CS/D. *American Marketing Association*, *13*(5), 413-421.

Damayanti, V., Ekasari, A. M., & Syaodih, E. (2019). The Determinants of Cultural Tourism Attractions Based on Tourist Satisfaction in the Sundapolis Area, Bandung City. *Advances in Social Science. Education and Humanities Research*, *409*(1), 538–541.

Dar, H. (2018). Satisfaction of Domestic Tourists Visiting Gulmarg in Jammu and Kashmir, India. *International Journal on Recent Trends in Business and Tourism*, *2*(1), 16–22.

Dasgupta, D. (2011). *Tourism Marketing*. Pearson Education.

Department of Tourism. Rajasthan. (2019). *Tourism Department Annual Progress Report 2018-19*. Published By Govt. of Rajasthan. Retrieved from http://www.tourism.rajasthan.gov.in/annual-progress-report.html

Hair, J., Anderson, R. E., Tatham, R. L., & Black, W. C. (1995). *Multivariate data analysis* (4th ed.). Prentice-Hall Inc.

Hassan, M. M., & Shahnewaz, M. (2014). Measuring tourist service satisfaction at destination: A case study of Cox's Bazaar sea beach, Bangladesh. *American Journal of Tourism Management*, *3*(1), 32–43.

Hinton, P. R., Brownlow, C., McMurray, I., & Cozens, B. (2004). *SPSS Explained.* Routledge. doi:10.4324/9780203642597

Huh, J. (2002). *Tourist Satisfaction with Cultural/ Heritage Sites: The Virginia Historic Triangle* (Thesis). Virginia Polytechnic Institute and State University, Blacksburg, VA.

Kumaran, D. S., Kannan, D. R., & Milton, T. (2013). Tourist Satisfaction with Cultural/Heritage Sites at Madurai, *Asia Pacific. Journal of Marketing Management*, *2*(3), 62–68.

Lai, K., & Li, X. (2016). Tourism destination image: Conceptual problems and definitional solutions. *Journal of Travel Research*, *55*(8), 1065–1080. doi:10.1177/0047287515619693

Larry, Y., & Munhtuya, G. (2005). A Comparative Analysis of International Tourist Satisfaction in Mongolia. *Tourism Management Journal*, *27*(6), 1331–1342.

Lin, L. F., & Cai, X. W. (2016). Research on influencing factors of tourist satisfaction in Hengshan Mountain. *Journal of Hunan Finance and Economics University*, *1*(1), 130–134.

Oliver, R. L., & Bearden, W. O. (1985). Disconfirmation Processes and Consume Evaluations in Product Usage. *Journal of Business Research*, *13*(3), 235–246. doi:10.1016/0148-2963(85)90029-3

Patterson, P. G. (1993). Expectations and product performance as determinants of satisfaction for a high-involvement purchase. *Psychology and Marketing*, *10*(5), 449–465. doi:10.1002/mar.4220100507

Pizam, A., Neumann, Y., & Reichel, A. (1978). Dimensions of tourist satisfaction with a destination. *Annals of Tourism Research*, *5*(3), 314–322. doi:10.1016/0160-7383(78)90115-9

Prayag, G., Hosany, S., Muskat, B., & Del, C. G. (2017). Understanding the relationships between tourists' emotional experiences, perceived overall image, satisfaction, and intention to recommend. *Journal of Travel Research*, *56*(1), 41–54. doi:10.1177/0047287515620567

Prayag, G., Hosany, S., & Odeh, K. (2013). The role of tourists' emotional experiences and satisfaction in understanding behavioral intentions. *Journal of Destination Marketing & Management*, *2*(2), 118–127. doi:10.1016/j.jdmm.2013.05.001

Putri, M. (2017). Tourist Satisfaction at Cultural Destination-A Case Study of Saung Angklung Udjo Bandung, West Java (Indonesia). *International Journal of Tourism & Hospitality Reviews*, *4*(1), 35–43. doi:10.18510/ijthr.2017.415

Reisinger, Y. (2009). *International Tourism: Cultures and Behavior.* Butterworth - Heinemann Publications.

Rust, R. T., Zahorik, A. J., & Keininghan, T. L. (1993). *Return on Quality*. Probus Publishing.

Sekaran, U. (2000). *Research method for business: A skill building approach.* John Wiley and Sons, Inc.

Singh, S. (2002). Tourism in India: Policy Pitfalls. *Asia Pacific Journal of Tourism Research, 7*(1), 45–59. doi:10.1080/10941660208722109

Sofield, T. H. B., & Li, F. M. S. (1998). Tourism development and cultural policies in China. *Annals of Tourism Research, 25*(2), 362–392. doi:10.1016/S0160-7383(97)00092-3

Tosun, C., Temizkan, S. P., Dallen, J. T., & Fyall, A. (2007). Tourist Shopping Experiences and Satisfaction. *International Journal of Tourism Research, 9*(2), 87–102. doi:10.1002/jtr.595

Turner, L. W., Reisinger, Y., & McQuilken, L. (2001). How cultural differences cause dimensions of tourism satisfaction. *Journal of Travel & Tourism Marketing, 11*(1), 79–101. doi:10.1300/J073v11n01_05

Woodside, A., & Mazanec, J. (2004). *Consumer Psychology of Tourism. Hospitality and Leisure*. CAB International.

Yuksel, A. (2001). Managing Customer Satisfaction and Retention: A Case of Tourist Destinations. Turkey. *Journal of Vacation Marketing, 7*(2), 153–168. doi:10.1177/135676670100700205

ADDITIONAL READING

Aksu, A., Icigen, E. T., & Ehtiyar, R. (2010). A Comparison of Tourist Expectations and Satisfaction: A Case Study from Antalya Region of Turkey. *Turizam, 14*(2), 66–77. doi:10.5937/Turizam1002066A

Canny, I. U. (2013). An Empirical Investigation of Service Quality, Tourist Satisfaction and Future Behavioral Intentions among Domestic Local Tourist at Borobudur Temple. *International Journal of Trade. Economics and Finance, 4*(2), 86–91.

Goswami, B. K., & Raveendran, G. (2003). *Text Book of Indian Tourism*. Har Anand Publications.

Jauhari, V., & Sanjeev, G. M. (2010). Managing Customer Experience for Spiritual and Cultural Tourism: An Overview. *Worldwide Hospitality and Tourism Themes, 2*(5), 467–476. doi:10.1108/17554211011090094

Kent, R. (2007). Marketing Research: Approaches, Methods and Applications in Europe (1st ed.). London: Thomson Learning.

Kozak, M., & Rimmington, M. (2000). Tourist's satisfaction with Mall orca, Spain as an off-season holiday destination. *Journal of Travel Research, 38*(3), 260–269. doi:10.1177/004728750003800308

Morales, M., & Ladhari, R. (2011). Comparative cross-cultural service quality: An assessment of research methodology. *Journal of Service Management, 22*(2), 241–265. doi:10.1108/09564231111124244

Morris, W. (1995). The American heritage dictionary of the English language (4th ed.). Boston: Houghton Mifflin.

KEY TERMS AND DEFINITIONS

Attributes: The service elements that pull and satisfied the tourists and force to revisit.

Destination Image: The formation of perception about a tourist destination in the psychology of tourists.

Expectation: Expectation is the rational affirmation of destination, products/service to be used during stay at the destination.

Revisit: To visit the destination repeatedly after performing at least one visit.

Satisfaction: Satisfaction is the evaluation of the experience or conclusion drawn from the actual consumption of different attributes of tourist destination.

Sustainable: Current use of economic, social, cultural, and physical resources in such a manner to provide maximum utilization without compromising needs of future generations.

Tourist Traffic: The total number of tourist's foreign and domestic tourist's visits in a particular.

Chapter 5
Problems of Monetary Rewards in Tourist Organizations:
Rent-A-Car Sectors in Buenos Aires, Argentina

Maximiliano Emanuel Korstanje
https://orcid.org/0000-0002-5149-1669
University of Palermo, Argentina

ABSTRACT

Over years, management and marketing have been influential and useful to organize the working condition in modern corporations in tourism fields. The classical literature suggests that motivated workers enhance their commitment to the goal of the corporation. In this process, the assistance of managers is vital to achieve the success. In doing so, the program of incentive (monetary awards) gives to workers a reason to internalize management decisions. This not only is not truth, but also, we have found in this case-study opposite evidence. The capital somehow disorganizes the human relations and harmony in service organizations. Of course outcomes of this review should not be extrapolated to other unit of analysis; this is valid only for this organization.

INTRODUCTION

In recent years, specialized literature stresses the needs of implementing efficient methods to motivate workers and desk-staff. Studies published in these fields indicate that psychological motivation occupies a central position in the enhancement of production. Of course, the industry of tourism seems not to be an exemption. The customer´s satisfaction plays a leading role in the configuration of a commercial campaign oriented to gain and retain loyalties. For this reason, tourist companies devote considerable resources and time to keep their desk-staff motivated. In many cases, desk-staff is pressed or subject to countless deprivations such as low-salaries, excessive working hours and distress (Ross, 2006; Deery

DOI: 10.4018/978-1-7998-4330-6.ch005

& Jago 2009). Not surprisingly, some voices have alerted on the negative consequences of monetary rewards when they are not combined with non-monetary rewards (Lafont & Tirole 1991).

The present chapter explores the contradictions and limitations monetary rewards in tourist organizations with strong foci in rent-a-car services, a particular sub-sector of the tourism industry which remains unchecked. The chapter focuses on two auto-ethnographies conducted in a well-known rent-a-car company geographically located in Buenos Aires city, Argentina. Because of confidentiality, real names of interviewees, as well as the company, will not be revealed. Ethnography was the correct method selected in this research because of two main motives. On one hand, motivation is a complex phenomenon which is associated with fear and retaliation, above all when managers realize the desk staff's discontent. On another, desk-staff often lies to protect their interests or are unfamiliar with their inner-life.

We start from the original premise that corporations often adopt monetary rewards to incentives their personnel but in so doing, they trigger a climate of extreme competition, greed and fear which directly or not affects the organization. At a closer look, management history shows workers and capital-owners´ relationships have varied on time and culture, but the inter-class conflict remains stable in the time. Originally, those policies oriented to enhance economic production were centered on a paternalist viewpoint and culture, which paved the ways for the rise of quantitative methods and instruments to control and measure –rather than understanding the worker. In a post-Fordist world, the philosophy of measuring, which is encrypted in tourism management, speaks us of the needs of optimizing labor relations to improve efficiency. In this process, sociology and psychology were of paramount importance for capital-owners to understand the workforce (Davis 2006).

Having said this, I want to remark the auto-ethnography was conducted during 2004 and 2008. In so doing, the role of ethnographer was covert. We gathered substantial evidence which described the complexity of human behavior, above all in a climate of competence and conflict. The fear of retaliation or labor sanctions by the side of managers was the common-thread argument in interviewees. We headed the research with basis on a snow-ball method and interviews were finely tape-recorded and verbatim transcribed.

MOTIVATION AND ORGANIZATIONS

Today´s experts valorize psychological motivation as a key factor to engage with pleasurable experiences in consumers. But to some extent, these experiences are mutually constructed and negotiated. In tourism fields, many scholars have certainly explored the factors and conditions that determine the loyalty of the internal client, which means the worker. Desk-staff workers are more active, and of course creative when they feel part of something bigger. Hence, these studies punctuate on inter-department cooperation as a fertile ground towards more stable working environments (Zamora Gonzalez et al, 2004; Arkursus and Tarkan, 2002; Lindroth, 2008; Zehrer et al, 2008). What is more important, other complementary approaches suggest that monetary rewards notably contribute to better the connection of workers and consumers (Charles and Marshall, 1992; Hall, 1995; Rodriguez and San Martin, 2008; Mckercher and Lau, 2007; Um, Chon and Ro, 2006; Fuller, Matzler and Faullant, 2006; Ball, 1988; Brymer, 1991; Bigne and Andreu, 2004; Szivas, Riley and Airey, 2003; Muller and Wyss, 2007; Moller et al, 2007; Alonso Ferreras, 2002). Although the number of publications on workers´ motivation abounds, less attention was paid to the negative effects of rewards in tourism organizations. This is the main reason for this research. Centered on an empirical-based investigation, Bochner & Coulong show how complainers not only had a bad experience from the received service but they adopt previously-based cultural narratives which encourage or discourage them to present the complaint. Authors use a Culture Assimilator Technique administered to understand the cultural interaction between Australian airlines´ employees and Japanese tourists. Their conclusion holds that tourists often behave according to their cultural background (software). For example, Japanese tourists are not accustomed to complaint personally at the desk rather they prefer to write the claim once returned home. Our cultural values indicate what is a lack of respect or an accepted course of action, these researchers remark. In this respect, those tourists who repeat the visit have fewer probabilities to present a claim than the tourists who visit the destination by first time (Bochner & Coulong 1997). These results are coincident with the recent advances in social psychology. As Geert Hofstede (1991) puts it, in multicultural organizations the interpersonal conflict is given by the tolerance uncertainness workers bear as well as the cultural background of each organization. While some organizations face serious frustrations with certain stability others have problems to survive in similar conditions. In this vein, Earley (1989) offers a more than interesting comparison which deserves surely our attention. Chinese and American managers develop different (if not contrasting) patterns to the adaptation to social conflict. While Chinese managers worked harder when they are in loneness, American managers incur often in social loafing. This coincides with the outcomes found in Yamagishi (1988) who argues that American managers are culturally educated to promote individual performance, avoiding penalties and maximizing the proper gains where Japanese managers are inclined to enhance the tacit social contract. The quest for excellence and competition are inextricably intertwined. In a landmark paper, Lillo Bañuls (2007) calls the attention on the volatile nature of tourism as well as the inter-firm competition in a global marketplace, where training and motivations are vital to surviving. The decline of a company is given for numerous reasons, but motivation seems to be an invisible force sometimes experts overlook. This point leads Alexandru Nedelea to hold that organizational consultants should work with managers delineating efficient actions to improve the tourist destination attractiveness. In so doing, they ought to adopt flexible programs since tourism not only rests on a multicultural logic but it evolves in a climate of high competition given by the numerous stakeholders and their interests. To avoid the Hobbesian War of all against all, firms should adopt innovative campaigns supported by marketing and management research (Nedelea 2010). Last but not least, British Philosopher, Geoffrey Klempner acknowledges that ethics help to resolve not only the social conflict but cements the social trust which is the touchstone of modern organizations.

Capitalism, Conflict and Competition

The collapse of communism not only evinces the end of Cold war but introduced radical shifts in the labor organization worldwide. As a result of this, lower-cost destinations (peripheral economies) received the investment of global companies which looked to maximize their profits at minor costs. The process of globalization created a new flat world where the industrial worker was subject to a climate of precaritization a never before. In the liberal market, each worker should compete for the basic resources in question, a job. The welfare state, as well as the Fordist doctrine, sets the pace to a new emerging Darwinist climate of extreme competition (Gottdiener 1994). As Michael Burawoy (1979) brilliantly observes, the capitalist system expanded not only through the technological breakthrough that revolutionized the economic production but by the cultural legitimacy of Science. Sociology, as a critical discipline, failed to explain the nature and logic of *the organization-factory* which is erroneously portrayed as a harmonic home for workers. This conception led founding parents of sociology to understand the factory in analogy to a family. The figure of the father was equated to the capital-owner while workers were comparable to children. From its inception, as Burawoy adheres, the idea of social conflict –always present in these spaces- was juxtaposed to harmony. In this way, conflict was seen as a social malady, management should locate and eradicate. The introduction of more efficient programs would lead to a climate of cooperation and reduced conflict. In some respect, since conflict is an obstacle towards prosperity it should be avoided. Contrariwise, Burawoy argues convincingly that social conflict –far from being negative or disruptive, is part of system reproduction. For some reason, this is the point that sociology and the managerial perspective misjudged. He goes on to write,

Conflict and consent are neither latent nor underlying but refer to directly observable activities that must be grasped in terms of the organization of the labor process under capitalism. Conflict and consent are not primordial conditions but products of the organization of the work. (1979: p. 12).

This above-noted citation reveals two significant things. On one hand, the specialized literature demonized conflict as an irrational force which leads an organization toward social disorganization. For that reason, conflict should be eradicated from the core of the factory. On another, the idea of a conflict-zero organization serves ideologically to calibrate the workers´ exploitation. Workers not only are motivated to compete with others but also they internalize the impossibility to confront with managers or capital-owners. Being a wayward worker is a sign of immaturity and lack of emotional intelligence to deal with emotions. Workers are ideologically educated to serve passively managers no matter than the cost. Instead, conflict is a basic force in the configuration of sociality and any human organization. The power of ideology is not marked by what is being said, but in what remains unsaid. In a seminal book, which entitles *Dialectics of Social Thought,* Professor Geoffrey Skoll reminds that the figure of capital (money) mediates between citizens and their institutions. Capitalism revolves around a manifest problem of accumulation which engenders an inevitable inter-class struggle (conflict). The question of whether Marx interrogated furtherly the reasons why and how society keeps united, he obtained his answer in the role of ideology and dialectics. While the former refers to a rapid solution for the material asymmetries created by capital reproduction, the latter gives citizens a good opportunity to play the game. To put differently, dialectics oppose two asymmetrical objects, introducing the third object in order for the causality between them are ignored. In the Marxian texts, the figure of the city occupied a prominent position as the main recipient of human relations. Metropolis condenses capital accumulation and the necessary manpower to keep the machine functioning. The capital exhibits a dialectical relation between the worker and the machines disposed to amass the surplus. In this respect, Skoll mentions the *"dialectics of triadic thought"* to denote the juxtaposition of two complementing modes of thinking. In a nutshell, the (real) conflictive relation between workers and capital-owners camouflages through the introduction of capital, a concept which never is critically interrogated.

As the previous argument is given, Richard Hofstadter (1992) accepted that the quest for amassing profits and competition are inevitably entwined in the capitalist ethos. The doctrine of social Darwinism, which stems from the Puritan culture, has considerable influence on the grassroots level in the US. Social Darwinism historically defended off two important axioms. First and most important, the US is touched by a sentiment of exceptionalism which marks the pace of other cultures nation across the globe. The US should be distinguished as something else than an example to follow but as a source of international legislation. While the law emanates from the US, the law does not apply for the US. Secondly, Darwinism certainly centers on the idea only the fittest –or strongest men- should survive. The material asymmetries or inequalities are not proper of capitalism, but by a struggle among different actors. The richest, the strongest citizen is an example of hard-work and self-enhancement. Thus, "primitive man, who long ago withdrew from the competitive struggle and ceased to accumulate capital goods, must pay with a backward and unenlightened way of life" (p 58). Millionaires have not resulted from the greed or avarice but the evolution of natural selection. They have been selected by their strengths, tested in their success in business, and abilities to achieve adaptation to the environment. Rather, others have been relegated to occupy pour conditions of existence or to disappear. Per Zygmunt Bauman (2007), the liberal man has been disposed from all his rights, subject to a climate of competence, where he is the manager of his own fate. In view of this, in the liberal marketplace, people are commoditized and constantly exchanged to be consumed by others. The industrial logic that constructed the scaffolding for the well-fare state has been gone forever, leaving the doors open to new liquid modernity and consumerism prevails.

The Psychology of Managers

Unlike the classical sociologists, Miklas Luhmann and Anthony Giddens focused their attention on the figure of money in our modern style of life. Luhmann clarifies that money serves as a symbolic mediator (among many other else as power and beautiness) for the self to reduce the uncertainty in society (Luhmann, 1979). Complementarily, Anthony Giddens (1991) explains that capitalism expanded through what he dubbed as "disembedding process", a new stage where presences connect with absences. Corporate capitalism rests on two pillars: the risk and the net of expertise. While the former creates the contingency as a new exploration towards the future, the later signals to the necessary protocols and guidelines on how these risks should be addressed (Giddens 1991)).

A more than a pungent book, in this discussion, sheds light on the passage from family to corporate capitalism. Michael Useem (1984), in *the inner Circle,* points to the problems to understand if nations are governed by a powerful business-elite class or by an atomized cells of corporations which only exerts pressure over governments. This debate is far from being closed. The psychological portraits of American and British managers vary on by many factors. While Americans are oriented to profit-generations and private welfare, the British are less prone to productivity issues. However, over the years, sociologists portrayed a benefactor character of managers. Unlike the "predatory" capital-owners, managers should be esteemed as workers who want the best for the company. They not only care for the good function of the company but also are motivated by non-profit concerns. One of the founding fathers of organizational psychology, Harry Levinson (1976) writes that the business-related leadership is important to achieving a good performance in the lines of production. The decision making the process as it is made by top-management molds the destiny of workforce. Leadership marks the start for the total authority of executives over workers. But this does not resolve an old problem: greed. Levinson says overtly that the problem of greed associates to the education levels. Less educated people not only are unable to cultivate other arts but are unfortunately limited to the obsession of amassing wealth. No less true seems to be that the efficacy of a good chief consists in the articulation of motivational programs which alternate monetary with non-monetary rewards. Like parents motivate and care, children, in the same way, chiefs should protect employees.

The superego incorporates the moral and spiritual value of culture in which a person lives, the rules and regulation within the family, and the attitudes toward himself which the person has acquired from those around him. Thus the superego is at a policeman, a judge, and a preceptor. The superego represents the law, telling us what we should or should not do. It represents the judiciary, in judging how well we conform to the rules which it has set up, and it represents the preceptor, incorporating values and aspirations and goals. (1976: p. 11).

Following Levinson´s models, employees and managers agree on a tacit contract which draws and contains the expectances, fears and goals of each organization. This contract, needless to say, varies from organization, culture and time. Other interesting studies in the field stress on the recruitment process to incorporate highly-skilled staff. Whether the recruitment stage is a success, a peaceful internal climate is ensured. For these scholars, competition and conflict are not resulting from the capitalist system but it derives from the insecurities of subjects (Rogelberg & Books-Laber 2002; Miner 1992). What is equally important, managers exert a direct influence in the formation of labor climate in the organization as well as the communicational processes which often are the precondition to conflict and mistrust (Daft 2011). At a closer look, the conflict has advances and disadvantages which directly affect the economic production cycle and job-satisfaction levels. Doubtless, there is a gap between conflict management and sociology of organization which was not dully filled.

One of the first methodological problems to study conflict is how it can be measured. What one person may perceive as conflictive atmosphere, others experience as harmonic one and vice-versa. In this discussion, the role played by networks in the reduction or inflation of conflicts is of significant importance for social analysts. Under some conditions, the conflict triggers team performance or the commitment to a certain task. Ching Tsung Jen (2013) conducted an empirical-research to validate the above-noted axioms,

- In groups where the authority is centralized, the conflict influences negatively in job satisfaction.
- Team-structured organizations managers and people who are in better position are prone to conflict than others.
- Psychological problems as anxieties or any type of disorder lead people not to understand social change.
- Centrality in the task, not authority, is benefited by the exacerbation of conflict.
- The distress caused by discrepancies, in the task-oriented group, fosters the productivity and the competition among workers.

In the fields of hospitality, Sanchez Cañizares (2007) et al. discuss to what extent workers are more productive and efficient when they feel motivated. Per their viewpoint, interaction and fluid genuine communication is essentially an aspect that correlates directly with job-satisfaction (Sanchez Cañizares et al, 2007:244-246). Last but not least, there is a consensus to suggest that workers move usually maximizing gains while avoiding risks. This means that the quest for power, competence and benefit-cost balance plays a crucial role in the motivation of workers in tourism and service sectors (Gursoy & Rutherford, 2004; Franch et al 2008; Dwyer 2008; McNaughton, 2006; Pearce, 2008). In this respect, Tran and Phillip emphasized that the nature of human beings is based on the need for achievement (prizes), power (control) and affiliation (sentiment of belonging). The pattern of behavior is often determined by the typology of subjective needs (Tran and Phillip, 2010).

Study-Case: Buenos Aires

The present section is a result of our previous fieldworks conducted in a rent-a-car company, located in Buenos Aires Argentina. The literature revolving around monetary rewards in tourism sectors abounds, but in rent-a-car, there is little research done. At a preliminary glimpse, the findings can be summarized as follows:

The organization in question recruited and trained its employees to ensure high-quality levels of service. For that, the company invests considerable resources to motivate employees. The pyramid of this organization was based on three levels. At the top, Chief Executive Officer (CEO), who makes the day-to-day strategic decisions that involve the company. Top-managers, almost ten, who is situated at a second middle level in the hierarchy. They are often in charge of controlling the quality of the service. At the bottom, the front-desk staff who is in direct contact with consumers, renters and tourists. While the managers have no financial or monetary incentive program, front line workers have paid depending on their sales. Originally, the monetary rewarding system was applied to the employees´ discontent but the problem –far from being solved- aggravated in the threshold of time. The recruitment process is carefully articulated to select skilled-staff who have fluency in foreign languages. We have interviewed almost 35 front-desk staff and 10 top managers who range from 20 to 55 years old. A whole portion of employees was discontent with the monetary rewards, and only 4 were comfortable with the money.

Because of time and space, we are not able to give more specification about all interviews, but only the most relevant are transcribed: Romina (female 32 years old) was upset stating that:

The current form of incentivisation does not work in our company; I believe this is because managers apply penalizations without any previous notice. Quite aside from this, we realize at the end of the month how much money will earn. Sometimes the penalizations are based on fact invented to debit from arbitrarily that is not a serious system; however, the system has been designed properly the problem lies in the application if you ask me.

Similarly, another agent Agustin (Male, 27 years old) claimed:

I am unhappy with our incentive programs for two reasons, penalties are not applied in a fair manner and it is an invention of Managers to boost the performance but save money. If the monetary penalties over us are not clear, it is because we bill more and more each month. As a result of this, they (Managers) like no to pay us and invent errors in procedures for the application of debits and pay less than due.

The above-noted interviewees reveal not only how the conflict between front-desk staff and managers remain covert, but also the limitations of open or closed-ended questionnaires often applied by Human resources Departments. One of the problems monetary rewards engenders, associate to the clerks´ resistance to accepting (discounts) penalties when some protocols were omitted or error detected. Let´s clarify readers that front-desk staff was paid of money –over the wage- according to their levels of sales, but involuntarily errors were discounted from that gross monetary amount. Still further, sometimes managers manipulate the monetary rewards as a form of control over their employees.

Managers were reluctant to support their subordinate in view of the increasing of monetary awards given. Eager and envy were two emotions that flourished in this organization while the incentive program applied. The gap between Managers and workforce's wages reduced considerably to the extent that some managers earned a less salary than their subordinates. In view of this, workers were unduly penalized by errors that were fabricated by their managers. This generated a climate of conflicts or tension as never before. Managers not only were not paid of rewards, but they were zealous to witnesses how front-desk staff received further money by their sales. To set an example of this which helps to illustrate the point, Esteban (male 35 years old)

Employees are very ambitious when more money they earn more like. I understand that they are receiving at hand more than $ 500/600 for productions but claim that that is not enough. Truthfully, Managers do not inform the penalties in due course, but this is a consequence of all problems employees have brought. In past, when we reported to agents that a penalty of $ 150 was debited from there accounts in accordance to some errors, they responded aggressively and threatened us to mistreat to clients because of their unsatisfaction. Please figure out that we do not manipulate the incentive program if this measure was taken it is due to diverse problems program of this nature has been caused; up-to-date we have modified this program more than twice but the unhappiness still remains.

The conflictive atmosphere in this rent-a-car does not authorize to affirm that monetary rewards do not boost sales. In fact, during 2005 the volume of sales reached $ 2.000.000 ARS, followed by 2006 of 2.500.000 ARS, and finally $ 3.500.000 for the period of 2006-2007. These promising numbers were not accompanied by the number of clients´ complaints. In 2005, the customer department received and processed almost 65 complaints about bad services in the month. This number notably increased to 125 for 2006 and 140 in 2007. Although sales volume was not hurt by monetary rewards, the customer relation declined. According to the gathered interviewees and the reaction of staff to the application of monetary reward, a table, which is shown below, was drawn.

Table 1 shows the types of reactions noted as long as ethnography which varied according to the type of punishment (penalty) or recognition. Broadly speaking, when the employee is penalized with a monetary discount, the answer was hostile and violent, aggravating the working climate. They were allowed to write and present a letter of discharge, explaining the situation that led them to commit the error. These errors were calculated in poor quality services, failures in the car cleanup process, missing cash, and errors in the issuance of a bill or invoices. When employees were charged with a discount, they claimed the responsibility was on the side of managers who were blamed of manipulating the system at their discretion. When the penalty does not account any monetary deduction, the attitude associated with indifference or passive acceptance.

Table 1. Types of reaction based on punishments

Type of Punishment	
Monetary Punishment	Negation and Hostility
Non-Monetary Punishment	Acceptance or Indifference

Table 2 gives an opposite landscape. Front-Desk staff manifested to be motivated only when monetary rewards was applied or ensured. What follows this, their reactions changed radically if the award were non-monetary such as more days of holiday, ticket to fly, or training programs. At a closer look, at least in this organization, once the monetary program was implemented potential alterations or changes which can be seen as threatening to the social contract was rejected by front-desk staff. The conflict was a mix-balanced gridlock between two contrasting forces: the staff working to gain further money and managers forcing penalties because voluntarily or involuntarily errors. Managers were moved by envy while front desk staff for greed. Once the monetary reward program was implemented, the gap between staff and managers enlarged.

Table 2. Types of reactions based on awards.

Type of Award	
Monetary	Motivation and commitment
non-monetary	Acceptance or Indifference.

At first glance, the dichotomy, acceptance or rejection of punishments depends on the type of incentive, award or punishment at stake. Whenever rewards are strictly based on monetary payments, the engagement with the job increases at a short-run, but in the case of penalty-application, this commitment becomes angry. To some extent, front-desk employees are situated in a privileged position because of their nearer and fluid contact with costumers. They give them further liberties to negotiate further wages and monetary rewards directly to CEO while managers are paradoxically relegated to a marginal position, so to speak, their authorities are finally undermined. Front-desk workers are trained and motivated to cultivate a genuine relationship with renters at the time they are pressed to give high-quality service. This leads them in a paradoxical position which facilitates further gains. Symbolically speaking, renters are hosted by front-desk staff speculating with Managers´ fears when the quality service declines. The idea articulates revolving around this axiom: if I am not motivated, the quality service will be poor!

To validate this assumption, we come across with the following interview which is self-explanatory: Marcelo (male, 25 years old) said us the following,

I do not care if I had right or not, if the mistake is mine or not, I will always complain and fight when an economical penalty is being applied on me simply because that is money, and the money worth my time and efforts.

Other consultants agreed to the same question voicing: *"whatever the case may be, I will be in disagreement with this award programs only for one reason, the money".*

This latter excerpt seems to be in sharp discordance with previous managerial literature that outlines stakeholders tend to accept punishments whenever they consider they are fair. To this point of view, one might conclude in two main ideas:

Monetary inceptive programs create a paradoxical situation. On one hand, sales are notably increased but this does not mean that the high-quality service maintains. This particularly prompts a climate of hostility and conflict between managers and front-desk staff. To some extent, the system leads to a climate of individualism and speculation. The privileged situation of front-desk employees who daily deal with customers and renters allow them further negotiations otherwise would be impossible. This includes putting the CEO directly in negotiations with lower-staff, while managers are pressed to occupy a marginal position. It is important not to lose the sight of the fact that monetary rewards mold and determine employee behavior in many dimensions. In the last resource to reduce the conflict, CEO cancelled the program and proposed a new one, but this was in vain. The levels of discontent arose to the extent to place many of the workers in the strike.

The pressure of workforce boycotted the income and profits of organizations from many perspectives. Some agents and front desk workers aborted the activities echoing a strike which was supported by worker unions. In other cases, they refused to offer cheaper rates to their clients, affecting seriously the organizational profits. CEO lost his political arm wrestling against workforce and opted to return the award-system. This measure woke up the fury of managers who quitted in protest. They passed to other rent-a-car companies migrating vital information for local competence in the market.

CONCLUSION

It is noteworthy that the introduction of monetary rewards in a rent-a-car company in the study revealed negative effects on the working climate, above all in the relationships between front-desk staff and top-management. Renters are carriers of capital, and this leads front desk staff to a privileged position. The figure of motivation seems to have a dark face, managers promptly witnessed. One of the premises that guided this ethnography was the needs to confirm if monetary rewards enhance engagement and commitment with the job. At first look, we need to divide the conclusion in two. On one hand, monetary campaigns boost the volume of sales and motivation in a short-run, but on another; they lead to a climate of conflict when they are not combined with non-monetary alternatives. This is a point unexplored by the specialized literature, and where this chapter wants to make a contribution. Paradoxically, capital subverts the lines of authority. While top managers reserved the right to penalize front-desk workers, they were paradoxically relegated since workers negotiated directly with the CEO. The obsession to achieve more motivated workers to enhance the quality service seems to be counter-productive to a coherent political organization of the company.

REFERENCES

Alonso Ferreras, V. H. (2002). Quality, competition and client Satisfaction in Organisation Offering tourism Sciences. *Studies and Perspective in Tourism.*, *11*(3-4), 253–268.

Arksus, A., & Tarkan, E. (2002). Leadership on hospitality industry: An study of case in Turkey. *Studies and Perspective in Tourism.*, *11*(3-4), 253–268.

Ball, R. (1988). *Seasonality in the UK labour market*. Averbury.

Bauman, Z. (2007). *Consuming Life*. Polity Press.

Bigne, E., & Andreu, L. (2004). Emotion in segmentation: An empirical study. *Annals of Tourism Research*, *31*(3), 517–539. doi:10.1016/j.annals.2003.12.018

Bochner, S., & Coulon, L. (1997). A culture assimilator to Train Australian Hospitality Industry Workers serving Japanese Tourists. *Journal of Tourism Studies*, *8*(2).

Brymer, R. A. (1991). Employee empowerment: A guest driven leadership strategy. *The Cornell Hotel and Restaurant Administration Quarterly*, *32*(1), 58–68.

Burawoy, M. (1979). *Manufacturing Consent. Changes in the Labor Process under monopoly Capitalism*. University of Chicago Press.

Charles, K., & Marshall, L. (1992). Motivational preferences of Caribbean Hotel workers: an explanatory study. *International Journal of Contemporary Hospitality Management*, *4*(3), 25-29.

Daft, R. L. (2011). *Leadership* (5th ed.). Cengage.

Davis, P. (2006). Beyond Human Resource management in co-operatives. *Cross Cultural Management*, *13*(1), 69–95. doi:10.1108/13527600610643493

Deery, M., & Jago, L. (2009). A framework for work-life balance practices: Addressing the needs of the tourism industry. *Tourism and Hospitality Research*, *9*(2), 97–108. doi:10.1057/thr.2009.4

Dwyer, L. (2008). Destination compititiveness and policy making for poverty reduction. *Sustaining Quality of Life through Tourism*, 24–28.

Early, P. C. (1989). Social Loafing and collectivism: A comparison of the United States and the People's in the Republic of China. *Administrative Science Quarterly*, *34*(4), 565–581. doi:10.2307/2393567

Franch, M., Martini, U., Buffa, F., & Parisi, G. (2008). 4l tourism (landscape, leisure, learning and limit) responding to new motivations and expectations of tourist to improbé the competitiveness of alpine destination in sustainable way. *Tourism Review*, *63*(1), 4–14. doi:10.1108/16605370810861008

Fuller, J.. Matzler, K., & Faullant, R. (2006). *Asymetric effects in customer satisfaction*. Academic Press.

Giddens, A. (1991). *Modernity and self-identity*. Polity Press.

Gottdiener, M. (1994). *The New Urban Sociology*. McGraw-Hill.

Gursoy, D., & Rutherford, D. (2004). Host attitudes toward tourism: An improved structural model. *Annals of Tourism Research*, *31*(3), 495–516. doi:10.1016/j.annals.2003.08.008

Hall, S. (1995). Hotel's success breeds happy employees. *Hotel and Hotel Management.*, *210*(18), 3–4.

Hofstadter, R. (1992). *Social Darwinism in American Thoughts*. Beacon Press.

Hofstede, G. (1991). *Cultures and Organization: Software of the mind*. Mcgraw-Hill.

Jen, C. T. (2013). The influence of conflict centrality and task interdependency on individual performance and job satisfaction. *International Journal of Conflict Management*, *24*(2), 126–147. doi:10.1108/10444061311316762

Klempner, G. (2004). Ethics and Advertising. *Philosophy for Business*, (9).

Laffont, J. J., & Tirole, J. (1991). Privatization and incentives. *JL Econ. & Org.*, *7*(special), 84–105. doi:10.1093/jleo/7.special_issue.84

Levinson, H. (1976). *Psychological Man*. The Levinson Institute.

Lillo Bañuls, A. (2007). El capital humano como factor estrategico para la competitividad del sector turístico. *Cuadernos de Turismo*, *19*, 47–69.

Lindroth, K., Ritalahti, J., & Soisalon-Soininen, T. (2007). Creative Tourism in Destination Development. *Tourism Review*, *62*(3-4), 53–57. doi:10.1108/16605370780000322

Luhmann, N. (1979). *Trust and power*. John Willey & Sons.

Mckercher, B., & Lau, G. (2007). Understanding the movements of tourists in a destination: Testing the importance of markets in the tourist attraction system. *Asian Journal of Tourism and Hospitality Research.*, *1*(1), 39–53.

Mcnaughton, D. (2006). The host as uninvited guest; hospitality, violence and tourism. *Annals of Tourism Research*, *33*(3), 645–665. doi:10.1016/j.annals.2006.03.015

Miner, J. B. (1992). *Industrial-organizational psychology*. McGraw-Hill.

Mitchell, T. R., & Daniels, D. (2003). Motivation. In Handbook of Psychology (Vol. 12, pp. 225–54). New York: Wiley.

Moller, C., Weiermair, K., & Wintersberger, E. (2007). The changing Travel Behaviour of Austria's Ageing Population and its impact on Tourism. *Tourism Review*, *62*(3-4), 15–19. doi:10.1108/16605370780000316

Muller, H., & Wyss, U. (2007). Aypical Working Hours and their impacts on Leisure – an unofficial Swiss Time Use Study. *Tourism Review*, *62*(1), 14–24. doi:10.1108/16605370780000158

Nedelea, A. (2010). *Comparative Tourism Marketing Case Studies, Dehli*. Abhijeet Publications.

Needle, D. (2004). *Business in Context: An Introduction to Business and Its Environment*. Cengage Learning Business Press.

Pearce, D. (2008). A needs-functions model of tourism distribution. *Annals of Tourism Research*, *35*(1), 148–168. doi:10.1016/j.annals.2007.06.011

Rodriguez, d. B. I., & San Martin, H. (2008). Tourist satisfaction a cognitive-affective model. *Annals of Tourism Research*, *35*(2), 551-573.

Rogelberg, S. G., & Brooks-Laber, M. E. (2002). Securing our collective duture: Challenges facing those designing and doing research in industrial and organization psychology. In S. G. Rogelberg (Ed.), *Handbook of research methods in industrial and organizational psychology* (pp. 479–485). Blackwell.

Ross, G. F. (2006). Tourism industry employee workstress—A present and future crisis. *Journal of Travel & Tourism Marketing*, *19*(2-3), 133–147. doi:10.1300/J073v19n02_11

Sanchez Canizares, S. M. (2007). La satisfaccion laboral en los establecimientos hoteleros. Analisis empirico en la provincia de Cordoba. *Cuadernos de Turismo*, *20*(1), 223–249.

Skoll, G. R. (2014). *Dialectics in Social Thought: The Present Crisis*. Macmillan. doi:10.1057/9781137387066

Szivas, E., Riley, M., & Airey, D. (2003). Labor mobility into tourism: Attraction and satisfaction. *Annals of Tourism Research, 30*(1), 64–76. doi:10.1016/S0160-7383(02)00036-1

Tran, X., & Phillip, S. (2010). Financial Performance and Vision Statements of National Tourism Organizations in Northeast Asia. *E-Ertr Review of Tourism Research.*, *8*(3), 57–70.

Um, S., Chon, K., & Ro, Y. (2006). Antecedent of revisit intention. *Annals of Tourism Research, 33*(4), 1141–1158. doi:10.1016/j.annals.2006.06.003

Useem, M. (1984). *The Inner Circle: large corporations and the rise of business political activity in US and UK*. Oxford University Press.

Yamaguishi, T. (1988). Exit from the Group as an individualistic solution to free rider problema in United States and Japan. *Journal of Experimental Social Psychology, 24*(6), 530–542. doi:10.1016/0022-1031(88)90051-0

Zamora Gonzalez, J. (2004). Proceso de formaci6n de la lealtad del huesped: Teoría y prueba empírica. *Studies and Perspective in Tourism.*, *13*(3-4), 197–221.

Zehrer, A., Pechlaner, H., & Raich, F. (2007). Satisfaction Profiles and Tourism Curricula - Tourism Organization under Study. *Tourism Review, 62*(1), 25–33. doi:10.1108/16605370780000159

ADDITIONAL READING

Blake, A., & Sinclair, M. T. (2003). Tourism crisis management: US response to September 11. *Annals of Tourism Research, 30*(4), 813–832. doi:10.1016/S0160-7383(03)00056-2

Causevic, S., & Lynch, P. (2013). Political (in) stability and its influence on tourism development. *Tourism Management, 34*, 145–157. doi:10.1016/j.tourman.2012.04.006

Chew, E. Y. T., & Jahari, S. A. (2014). Destination image as a mediator between perceived risks and revisit intention: A case of post-disaster Japan. *Tourism Management, 40*, 382–393. doi:10.1016/j.tourman.2013.07.008

Hall, C. M., & Williams, A. (Eds.). (2013). *Tourism and migration: New relationships between production and consumption* (Vol. 65). Springer Science & Business Media.

Heilbroner, R. L. (2011). *The worldly philosophers: The lives, times and ideas of the great economic thinkers*. Simon and Schuster.

Hobbes, T. (1994). Leviathan: with selected variants from the Latin edition of 1668 (Vol. 8348). Cambridge: Hackett Publishing. (1668)

Hurlock, E. B. (1927). The use of group rivalry as an incentive. *Journal of Abnormal and Social Psychology, 22*(3), 278–290. doi:10.1037/h0071601

Korstanje, M. E., & George, B. P. (2012). Falklands/Malvinas: A re-examination of the relationship between sacralisation and tourism development. *Current Issues in Tourism, 15*(3), 153–165. doi:10.1080/13683500.2011.569535

Lambert, E. G., Pasupuleti, S., Cluse-Tolar, T., Jennings, M., & Baker, D. (2006). The impact of work-family conflict on social work and human service worker job satisfaction and organizational commitment: An exploratory study. *Administration in Social Work, 30*(3), 55–74. doi:10.1300/J147v30n03_05

Lash, S. M., & Urry, S. L. J. (1993). *Economies of signs and space* (Vol. 26). Sage.

Locke, J. (1988). *Locke: Two Treatises of Government Student Edition*. Cambridge University Press. doi:10.1017/CBO9780511810268

MacCannell, D. (1973). Staged authenticity: Arrangements of social space in tourist settings. *American Journal of Sociology, 79*(3), 589–603. doi:10.1086/225585

MacCannell, D. (1976). *The tourist: A new theory of the leisure class*. University of California Press.

KEY TERMS AND DEFINITIONS

Conflict: It is tension between two or more parts which needs from a solution. Sometimes conflict is confused or homologized to rivalry.

Labor Relations: This refers to an academic field that often studies different inter-personal relations in working conditions.

Monetary Rewards: A system of motivation oriented to pay workers money in basis of their level of production or performances.

Rent-A-Car: Subindustry within service and tourist sector that offers the renting of cars to clients and drivers.

Tourism Management: IT consists in a net of techniques and disciplines oriented to keep the tourist system functioning.

Chapter 6
Regional Food as the Catalyst for Cultural Tourism in India

Priya Singh
Jamia Millia Islamia, India

Ashaq Hussain Najar
https://orcid.org/0000-0003-3101-4854
Central University of Kashmir, India

ABSTRACT

Culture is playing an increasingly important role in tourism, and food is one of the key elements of culture. Tourists enjoy indigenous food, particularly items of local or ethnic nature. Furthermore, knowledge of the local, regional, and national cuisine has become an interest for tourists. The concept of consuming local food or drink is considered first-hand cultural experience, and it is on top of the tourist attraction list. In India, the promotion of food as a component of its destination attractiveness is in the budding phase. The context of this contribution is to underpin such linkages between tourism and food that can add to the cultural value of the destinations. The present empirical research is aimed to explore the role of food, culture, and tourism in sustaining the tourism of India. The current study will also attempt to address the role of regional food in promoting the cultural tourism of a particular destination. Further, the study will explore how the local food helps the tourist to recall the cultural heritage of a tourist's destination.

INTRODUCTION

The linkage of food and tourism has become powerful in the recent years. The recent trend of experiencing local food is adding value to tourist's experience and it has become one of the favourite activities during travelling. The gastronomic experiences add a new component to the image of the tourist destination as it reñects the culture and tradition of the region. Therefore, regional food can add value to the destination and can enhance the competitiveness of the geographical area. In this view, many researchers highlight the increasing fascinations of the tourists towards the regional cuisines and the local flavours available

DOI: 10.4018/978-1-7998-4330-6.ch006

at their tour spots. In the recent times, gastronomy has become a subject of interest for the tourists as it's a reflection of the people, culture and traditions of the visited place (Ignatov & Smith, 2006).

Food is not just the means of sustenance alone, it also represents the culture of the place. While consuming a particular regional food, tourists can closely associate to the culture of that destination. Food provides an authentic representation of the culture thus enriching the cultural tourism of that particular destination. Through food one can know and relate to the geography, history and people of the visited place. Food restores the regional identity of the place thus contributing towards the regional development (Steinmetz, 2010).

Tourist experience is incomplete without exploring the distinct culture of the destination. Relishing the regional food of the visited destination have always been an entertaining cultural activity. According to Sharpley (1999) "The culture of a society is an amalgamation of its values, morals, behavioural norms, dress, cuisine, artefacts and language". "Food is undoubtedly a significant means of promoting cultural tourism of the destination, where the meal is looked upon as cultural artefact" (Scarpato, 2002). The regional food habits and traditions helps to restore the "culinary heritage" of a destination and is therefore indispensible for identity formation (Bessiere, 1998). Tourist destinations uses regional cuisines as a powerful tool for showcasing the culture of the destination through the culinary activities.

Food plays a vital role in culture and is not just a means of sustenance for tourist, rather it forms the foundation of the culture and plays a major role in overall branding of the destination. "Local cuisines serves as the means of portraying destination's intangible heritage and thus creating truly mesmerizing cultural experience for the tourists" (Okumus, Okumus, & McKercher, 2007; Renko et al., 2010) It is important to note that the culinary experiences sharpen the destination image and guarantees a uniqueness in comparison to other destinations. Where the food can be presented as an icon of cultural identity, it can provide India with a more powerful national dress and can place it among the list of most attractive destinations of the world. Authors, look forward to address the role of regional cuisine in promoting the cultural tourism of a particular destination. The discussions shall take place in the Indian context and shall provide evidences of various regional cuisines served in different destinations across the country. Thus, this chapter shall focus on exploring how the culinary experiences help the tourist to recall the cultural heritage of a tourist's destination.

BACKGROUND

Culinary tourism has emerged in recent decades, although food has been closely associated with travel from ancient times. Food has undergone remarkable changes over the time and is no more confine to a status of basic survival necessity, it is being associated with culture, rituals, wealth and experience (Tannahill,2002). The transfer of food across cultures has contributed towards the formation of cultural identities and social hierarchies (Pilcher, 2006). "Culinary tourism is defined as the desire to explore, experience and relish the food of a particular region (Hall & Sharples, 2003) and attract large number of gastronomic tourists" (Okumus& McKercher,2007).Food has been identified as a means for regional development, strengthening the local produce and it also aids in achieving the sustainable tourism.

Culinary tourism caters to impressive gastronomic experiences of the tourists. It's a new niche that meets both economic and intercultural insights of a destination. Culinary tourism involves experimental eating and encountering new places and culture through food (Long, 2004). Food-based attractions holds special food activities like food festivals or cooking holidays (Hall & Sharples, 2008). According to

Renko (2010) food, like any other phenomena such as religion, events, architecture, craft and festivals is considered indispensable for cultural and heritage tourism. The literature supports the argument that food is no longer a basic necessity for tourist, rather it gives a sense of the regional culture and does a value addition to the destination's image. Regional cuisines give an overview of destination's intangible heritage and offers a rich cultural experience to the tourist (Okumus, & McKercher, 2007).

India's intangible cultural heritage showcases rich ancient cuisines and culinary processes. As a mirror of society, food narrates different stories related to traditions, rituals, religion, festivals, agricultural processes and techniques. India must design a road map on how to promote its food culture to both local as well as international tourists. Food has an infinite capability of touching lives of those who engage in food activities. It breaks down all the religious, cultural and social divide and try to get everybody on the common platform. Thus, food is blessed with an immense power of enriching and showcasing the culture of that region or destination.

Food and Tourism

Food is one of the most important constituent of destination's attribute. The relationship between food and tourism has evolved from "traditional hospitality and cuisine" to the new innovative concept of "food tourism" (Jones & Jenkins, 2002,) also referred to as "culinary", "gastronomic" tourism (Okumus, 2007). Food has always been a motivation for travelling and it adds lot to an overall experience of the tourist. Mitchell and Hall (2003) stated that "food is an integral part of tourist overall travel experience and it becomes highly experiential and "much more than functional". Tourism industry considers food experiences as its prime function and continuously strives to create one for the tourist.

Long (2004) revealed that experiencing food involves all the five senses of the tourist and provides an unforgettable experience. Food has always been a source of satisfaction for tourist as it is being rightly said that "good food is good mood". Tourists love to move around places and for them relishing variety of food is the most preferred and enjoyable activity. "Food consumption involves relaxation, excitement, escapism, education, status and lifestyle" (Corigliano,2002). The cuisine of the destination is one of the key drivers for the selection process of a destination. The food activities creates plethora of everlasting memories which are recalled by the tourist even after long time.

The connection of food and tourism cannot be ignored. Tourist motivation is majorly govern by two factors: "pull factors" and "push factors" (Hudson, 1999). Pull factors are the factors which governs the selection of tourist destination while the push factors motivates the tourist to go for a particular destination. Fields (2002) explained that "gastronomic experiences pull the tourist away from their familiar foods and eating habits and push the tourist to experience new and diversified food options" .McIntosh (1995) further added that food has the capability to satisfy four new travel motivators i.e. physical, interpersonal, cultural, and status and prestige.

The Cultural Dimension of Food

The gastronomic activities of a destination represent the cultural tourism of that particular destination. Cultural tourism is all about embracing the unique cultural attractions as well as the regional food. Food is an important component of intangible heritage which gives the sense of authenticity to the tourist. Local food has always been a motivation for travel for tourists which get them closer to the culture of that destination. The culinary heritage of a destination gives the fair idea about the mentality, behaviour and the character of the society and thus becomes relevant for creating unique destination identity (Bessiere,1998).The different rituals and eating tradition determines the society's culture and this diversity in culture is being well appreciated by the tourists.

Regional food depicts out the tradition, culture and rituals related to food served in that region. Tourists get great opportunities to interact with locals while involved in food experience and this helps them to create a unique experience for them. The food experience at Chowki Dhani Village, Rajasthan is one of such kinds that gives a meaningful cultural exposure to the tourists of witnessing the ambience in which the food is served, the way in which food is served, what all is served in the meal and how is the food supposed to be consumed. Such places exhibits destination's distinct culture and the tourist get to know and appreciate the same. Hence, it is obvious to link food with the identity of the place and food is also used as a place marker while promoting tourism of a destination.

Local cuisine offers an authentic cultural experience through serving the food which is unique in its own way, thus preserving destination's intangible heritage. Regional foods are associated with different localities or regions and are used to market those destinations as food being the major source of tourist attraction. It is important to note that the food tourism activity for any given particular destination will not be viable until it includes the cultural characteristics of the destination. The Gastronomic experience offers tourists to access the cultural heritage of the place through tasting, experiencing and local purchasing.

Figure 1. Food and culture linkage

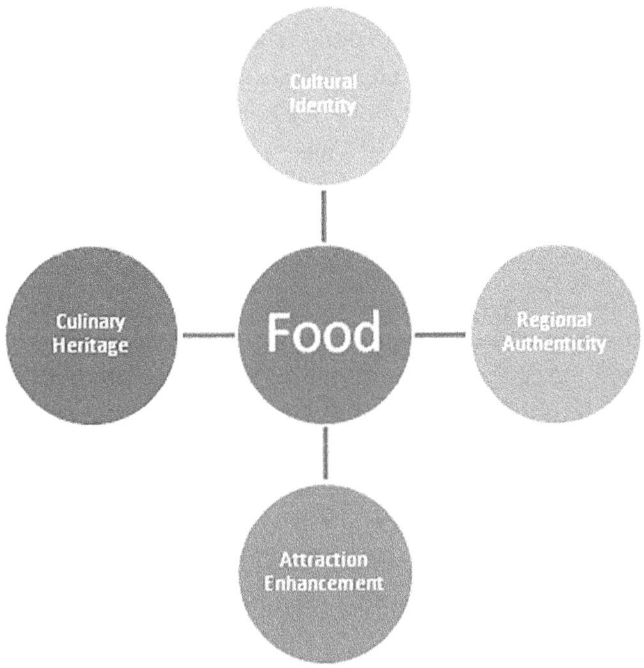

Culinary Tourism

Culinary tourism refers to special interest tourism, where food and beverage is the main motivation for the tourist to visit the destination. It majorly caters to domestic tourist which usually travels to places which are famous for their local delicacies. Food and drink festivals attract the gathering with the sole interest of offering gastronomic experiences to the tourists. In the recent years, there has been a remarkable change in number of travellers looking for culinary tours in India.

Culinary tourism has been in trend from past few years, it offers a wide range of interesting and engaging activities like food tours, food walks, cycling and cultural tours etc. This segment attracts tourists from different segments. Gastronomic tourists are mostly couples between 30 to 50 years with handsome earnings. With an inclination towards the local produce of the area, there has been a recent trend of trying local dishes and drinks and the food festivals, food walks and food tours are catering to such new demands. Growth in this niche segment is expected to grow rapidly and India will soon emerge as one of the popular gastronomic destination in the world.

Gastronomy is becoming an important element of diversification and a tool for defining the brand and the image of a region, a country and a culture (Hjalager, 2010). In recent years, tourism has started to recognize the important power of attraction and communication of food products, to the point of turning them into true tourist attractions (Folgado et al, 2011; Kivela and Crotts, 2006). The peculiarities of the local cuisine are the starting point to create routes, travel packages and thematic proposals that put local food and products at the centre of the development strategies and the promotion of certain territories or destinations.

Food is considered as an element of local culture and as an expression of the history, habits and traditions of a community or a geographical area (Kivela and Crotts, 2006; Lee and Crompton, 1992). Culture determines and shapes the culinary history of a community and, through it, expresses itself and becomes a tangible, sensory and pleasant asset. Thus, traditional dishes and local food turn into new tourist products and motivational elements of a trip

"Gastronomic tourism is rather a new form of tourism (Chaney & Ryan, 2012) and there are many different terms used for food-based tourism such as culinary tourism, gastronomy tourism, and food tourism" (Hall & Sharples, 2003).Culinary tourism gives the glimpse of "food culture" to the tourist and one comes across the tradition and culture of the destination. A country is known for its distinct and unique culinary customs on cooking, dinning and serving styles and this diversification in the food cultures across the world is the prime motivation for the tourist to embrace the culture of the destination. Destinations like Italy, France, Thailand, and India have been emerging as popular gastronomic destinations of the world. Food plays a major role in distinguishing various destinations through the varied cuisines. Different destinations have different regional cuisines which attracts many tourists round the world. The authenticity and uniqueness of regional food has been gaining importance and DMOs uses this as a strength to attract tourists to various budding culinary destinations.

Major Drivers of Culinary Tourism

There have been few key drivers for motivating tourists towards culinary tourism. During any gastronomic experience, the culinary legacy, novelty of cuisine, authenticity and the food culture is being witnessed by the tourist. For a gastro- tourist the experience is only authentic when he/she come across the local produce of the region and participates in the traditional food rituals. Differentiation, authentication and symbolization work as the catalysts for enlightening any tourist's culinary experience. The destination's authenticity is closely linked with social dimensions and creates a unique identity. The food menu and the food culture of a region act as a symbol for a gastro- destination and thus offer distinctiveness to the tourist. The broader purpose of a culinary tourist is to boost the cultural tourism of a destination through the regional food of the locality.

Figure 2. Drivers of culinary tourism

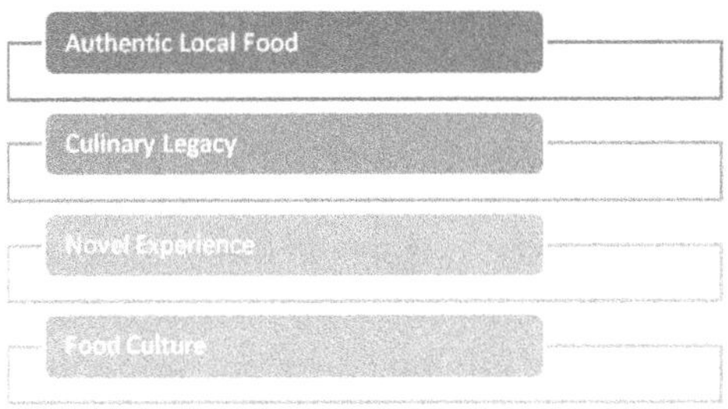

India as a Culinary Destination

Culinary tourism has gained prominence in the recent times. Indian cuisine has been popular globally and known for its unique flavours and spices. Our spices, breads and sweets have attracted not only domestic but international tourists around the world. The visit to India is not complete, if one has not indulged in the culinary activities and has explored the culture through its cuisine. Every state in India offers a distinct cuisine and few of the cuisines are so authentic that a tourist visits a particular destination to satiate their palate.

Indian culinary culture is unique and diverse. It stands out from the rest of the world not only in taste but also the way food is cooked, eaten and served. Indian delicacies are perfect blend of tradition, culture and love. Indian food is region and the gastronomy is as diverse as the Indian population. The regional cuisines of India are more likely influenced by its history, religious, trade and culture of its people.

India: People and Cuisine

India has diverse population and so is the gastronomy. The food of India is heavily influenced by the regional and religious peculiarities. It is interesting to note that, "one-third population of India is vegetarian, dictated by their Hindu, Jain, or Buddhist faiths". Consequently, major portion of the Indian dishes are exclusive of meat. Also, there are few strict dietary restriction followed by the natives based on the religious beliefs which has carved Indian cuisine so unique. Hindu followers abstain from beef, while Muslims refrain themselves from consuming pork. The regional cuisines are basically governed by the religious beliefs of a region and thus offers varied regional foods across the country.

Northern Indian Cuisine

Northern Indian cuisine reflects a strong Mughal culture. The cuisine involves high use of dairy fats like milk, ghee (clarified butter), curd and paneer (cottage cheese).The variety of north Indian snacks display Samosas, aaloo chaat, pakoras etc. Tandoor preparations with charcoal flavour adds uniqueness to the cuisine. Few popular dishes of the cuisine are gravy dishes like paneer makhni, dal makhni, palak paneer, chana masala etc. Korma, is yet another attraction of the menu and portrays the richness of the cuisine .It is served mainly with meats usually lamb or chicken and sometimes with beef too. The sweet dishes offers range of items comprising of kheer, rabri faluda, barfi, rawa halwa etc.

Western Cuisine

Western Indian cuisine primarily features the cuisines of three main regions: Maharashtra, Gujarat, and Goa. The cuisine is mostly dominated by the coastal produce and includes fish in the preparations. Use of coconut-milk is common in the cuisine. Gujarati cuisine mainly comprises of vegetarian preparations and has sweetness in many of its dishes. Due to dry climate in the region is well known for its chutneys and pickles .Goa is a hub of gastronomy and is dominated by the Portuguese culture. Goan cuisine mainly consists of pork and beef preparations and attracts large gastro-tourists from all round the world. Use of vinegar also assures the influence of Portuguese influence on the cuisine. Vindaloo is a traditional Goan dish and baklava is popular among food lovers too.

Eastern Cuisine

Eastern Indian cuisine is popularly known for its desserts. Rasgulla is one among the most common sweet delight of the region. Chenna balls are also in demand and loved by the people relishing the cuisines. Eastern dishes are usually flavoured with mustard seeds and poppy seeds adding light pungency to the dishes. Rice and fish are the main dish of the cuisine. The dishes of this cuisine are less spiced as compared to the dishes of other cuisines.

South Indian Cuisine

South Indian cuisine primarily features the five states ie Karnataka, Andhra Pradesh, Tamilnadu, Telangana, and Kerala. The cuisine is influence by the geography and culture of the region. South Indian cuisine is considered to be the hottest among all other Indian cuisines . Rice is the staple dish of the cuisine and is the centre for all the major preparations of the cuisines. Sambaar, rasam, dry vegetables, meat dishes and coconut chutneys are most popular .South Indian cuisine is a bliss for coffee lovers and produces rich coffee made with chicory .In any South Indian food a special flavour of curry leaves, mustard, heeng, fenugreek seeds, chillies is must.

Indian Food Traditions

Indian culinary heritage is an amalgamation of different cultures, cuisines and customs. Indus valley civilization introduced grains and plants, the Mughals introduced the use of rose water, curd and ghee in Indian cooking. They also changed the way of looking at food and established the idea of eating for pleasure. The Chinese introduced tea and India today treats Tea (Chai) as an emotion. Portuguese gave chillies to India and added spice to the culinary wealth.

India has been a land of rituals and various food traditions can be witnessed across the country. The *prasads*at temples, *langars* at gurudwaras and the *iftaar meals* are reflection of our diverse culture and ethnicity. Consequently, food is believed to be sacred and pure. Also, the eating rituals and traditions create our cultural identity. The following ancient food traditions form the basics of our culinary culture.

Eating With Hands

Eating with hands has evolved from Ayurveda and is believed to offer a sensory experience. According to the Vedas, eating with hands involves emotions and gives a sense of satisfaction. According to the Ayurvedic texts, "every finger is an extension of the five elements. The thumb symbolises space, the forefinger represents air, the middle finger is fire, the ring finer is water and the little finger stands for earth".

Serving Food on Banana Leaf

This tradition is observed in the regions where authentic south Indian meal is served. Banana leaf enriches the food with various nutrients and makes it more healthy for the consumption. The study reveals that "banana leaves contain large amounts of polyphenols; a natural antioxidant found in many plant-based foods."The leaf also adds good natural aroma to the food and also enhances the taste of the food. The leaf is always washed by the users before the use.

One for All

This tradition is practiced by muslim community where the food is served in one big *'Thal'* and the entire family sits together and eats food from the same platter. Every member takes their food portions as per their appetites. Interestingly, the first course served in the thal is usually sweet, followed by the meat appetisers. Starting the meal with the sweet course is considered auspicious and this tradition clearly depicts one of the unique food culture of India.

Great Indian Thali

The *thali* depicts the complete meal of a particular region like in India the most preferred variety of thalis are Rajasthani, Gujarati, and South Indian. The thali offers complete range of light snacks, rich curries, local vegetable preparations, rice and breads. Accompaniments like *chutneys*, pickles and papad are the show of the thali. Thali is most preferred by tourist as it gives the taste of major popular delicacies of a region and is also nutritive when comes to health. Also, the main advantage of trying a thali is that one can get fairly good idea about the main dishes and flavours of the region to which the thali belongs.

Role of Food Tours and Food Walks

Eating habit of an individual reflects the personality of an individual and collectively the community's eating habit showcases the culture of the country. India is rich in culinary and has immense potential to cater the culinary tourists visiting the country. A culinary tour in India will help the tourists to explore the diverse cooking cultures of the nation and appreciate its culinary richness. India is a mix of cultures and has lot to offer when it comes to food, every region has its own unique cuisine and this adds to the culinary wealth of the country.

Indian cuisine is geographically demarked as northern, southern, eastern and western regional cuisines. Each zone is distinct in their cooking styles and culinary traditions. The regional cuisines of different zones are in no way similar when it comes to aroma and taste. Thus, the diversities in the taste and styles motivates the tourist to experience different foods while travelling. The culinary tours across the country can be offered region wise to the tourist and many organizations are working on this niche segment and offering customized culinary tours to the tourists. Companies like India Food Tours offers exciting food tours in numerous cities across India. In every food walk, consumers can relish sample of local delicacies from the region. Presently the foods walks are being offered for destinations like New Delhi, Udaipur, Jaipur, Mumbai, Jodhpur, Kerala and soon will be launched in many other cities of the country.

ISSUES AND CHALLENGES

From the past few decades, tourism industry has been experiencing enormous growth, this calls for an extensive training and research to meet the demands of culinary developments. For the purpose of meeting the demands, tourism industry requires sound fundings from both government as well as renowned private players.

Cuisines are no more pure, multiculturalism has affected the food cultures and traditions. The immigrants have brought a mix of their food traditions to new countries and drastically affected the authentic food culture of the destination. In multiracial destinations, natives are often influenced by different cultures, food habits and lifestyle. The authenticity of the food may get diluted and the traditional preparation techniques may get lost. There has been evidences of fusion recipes, where the traditional recipe of the food has been altered and concept of fusion food has emerged.

Food culture of India should be made popularize across the world. In the recent times, not much emphasis is done in documenting different regional recipes, organizing food festivals and food walks. Serious efforts should be done in showcasing regional cuisines across the country and should formulate various strategies to promote the culture of India through its rich culinary heritage.

In order to boost the culinary tourism in India, it is important to address these issues at the earliest. Government and the private sector should start working for this niche segment and should sooner launch India as a popular food hub.

RECOMMENDATIONS

The initiative should be taken adopts to curate, share and popularize knowledge on Indian Food Culture and engage people to make more informed about the richness of the Indian cuisine. All the stakeholders including government authorities, private players and various food enthusiasts should come on the common platform and promote India as a culinary destination.

Some of the suggestive measures are as follows:

1. **Developing a National Recipe Archive:** The intent of this is to ensure revival of our culinary heritage, promotion and recognition of local and regional recipes. The strategy should include State wise mapping of regional and local cuisines.
2. **Creating and Supporting Local Food Hubs:** Food Hubs are usually located near the tourist attractions or the religious sites. A food hub would not only offer food, it would rather offer a complete cultural experience to the tourists. For example, there could be a food museum, a historical display of utensils, food talks or food trails.
3. **Facilitating and Promoting Food Streets:** Food streets attract major portion of tourist every year and every state has its own famous food street showcasing its popular regional delicacies. From Delhi's Chandni Chowk to Mumbai's Khao Galli, India's 'food streets' are repositories of age-old delicacies and are must-eat locations. However, the hygiene practices at these streets are not up to the mark which needs to be dealt seriously by the concerned authorities.
4. **Supporting and Recognizing Regional Food Festivals:** Food festivals showcase popular dishes, representing preferences of the local community and establishing ecological links. Through the regional festivals one can restore the cultural as well as culinary wealth of the region.
5. **Encouraging Food Trails:** Food Trail can be an effective tool for illustrating the local food of the region .This is an informative and engaging way to share food, its history and the stories behind different dishes and cuisines. To enhance the popularity of food trails, the pre-requisites should be met by the regional food authorities and the certification requirements need to be met by the concerned food authorities.

FUTURE RESEARCH DIRECTIONS

This chapter has emphasised on the role of regional food in promoting the cultural tourism in India. Further chapters can be focus on how the local food aids in generating employment for the locals. The further discussions can include the subjects like tourist perceptions and preferences towards the local food of a tourist destination. In future the perspectives of local communities towards commercialisation of local food can be explored.

CONCLUSION

In today's competitive era, destinations need to be unique and should offer distinguished tourism products. Food promotion can help in attracting huge number of tourists, thus strengthening the region's economy on one hand and retaining the culture of the destination on the other.

This chapter concludes that Indian cuisine has a rich potential and offer range of experiences to the gastro tourists. It is strongly recommended by the author that the regional food should be efficiently linked with the tourist industry so as to enrich with the cultural aspects in tourism. The study ascertains food as a reflection of destination's culture and thus food rebrands the image of the destination. Its not only food but various food traditions and rituals that sets India apart and sooner India will be branded as a popular culinary destination.

In order to motivate tourist to experience the local delicacies proper marketing of local food should be planned and executed. Indian cuisine has immense potential to attract the tourists round the world.

There is huge demand of wider publicity of local food through various food shows, food walks, food tours. This will also encourage the tourists to explore Indian culture through its food.

REFERENCES

Bessiere, J. (1998). Local development and heritage: Traditional food and cuisine as tourist attractions in rural areas. European Society for rural. *Sociology, 38*(1).

Chaney, S., & Ryan, C. (2012). Analyzing the Evolution of Singapore's World Gourmet Summit: An Example of Gastronomic Tourism. *International Journal of Hospitality Management, 31*(2), 309–318. doi:10.1016/j.ijhm.2011.04.002

Corigliano, M. A. (2002). The route to quality: Italian gastronomy networks in operation. In A.-M. Hjalager & G. Richards (Eds.), *Tourism and gastronomy* (pp. 166–185). Routledge.

Fields, K. (2002). Demand for the gastronomy tourism product: motivational factors. In A.-M. G. R. Hjalager (Ed.), *Tourism and gastronomy* (pp. 36–50). Routledge.

Hall, C. M., & Sharples, E. (2003). *The consumption of experiences or the experience of consumption? An introduction to the tourism of taste.* In C. M. Hall, E. Sharples, R. Mitchell, N. Macionis, & B. Cambourne (Eds.), *Food tourism around the world. Development, management and markets* (pp. 1–24). Butterworth Heinemann.

Hall, C. M., & Sharples, E. (Eds.). (2008). *Food and wine festivals and events around the world: Development, management and markets.* Butterworth Heinemann. doi:10.4324/9780080887951

Hudson, S. (1999). Consumer behaviour related to tourism. In A. P. Y. Mansfeld (Ed.), *Consumer behaviour in travel and tourism* (pp. 7–32). Haworth Press.

Ignatov, E., & Smith, S. (2006). Segmenting Canadian Culinary Tourists. *Current Issues in Tourism, 9*(3), 235–255. doi:10.2167/cit/229.0

Jones, A., & Jenkins, I. (2002). 'A taste of Wales - Blas Ar Gymru': Institutional malaise in promoting Welsh food tourism products. In A.-M. Hjalagar & G. Richards (Eds.), *Tourism and Gastronomy* (pp. 114–131). Routledge.

Long, L. M. (2004). *Culinary tourism.* The University Press of Kentucky.

McIntosh, A. (2004). The family meal and its significance in global times. In R. Grew (Ed.), *Food in global history* (pp. 239–263). New Global History Press.

Okumus, B., & McKercher, B. (2007). Incorporating local and international cuisines in the marketing of tourism destinations: The cases of Hong Kong and Turkey. *Tourism Management, 28*(1), 253–261. doi:10.1016/j.tourman.2005.12.020

Pilcher, J. M. (2006). *Food in world history.* Routledge.

Renko, S., Renko, N., & Polonijo, T. (2010). Understanding the role of food in rural tourism development in a recovering economy'. *Journal of Food Products Marketing*, *16*(3), 309–324. doi:10.1080/10454446.2010.485096

Scarpato, R. (2002a). Gastronomy as a tourist product: the perspectives of gastronomy studies. In A.-M. Hjalager & G. Richards (Eds.), *Tourism and gastronomy* (pp. 51–70). Routledge.

Sharpley, R. (1999). Tourism, tourists and society (2nd ed.). Cambridgeshire, UK: ELM Publications.

Steinmetz, R. (2010). *Food, Tourism and Destination Differentiation: The Case of Rotorua, NewZealand.* http://aut.researchgateway.ac.nz/bitstream/handle/10292/1090/SteinmetzR.pdf?sequence=3

Tannahill, R. (2002). *Food in history*. Headline Book Publishing.

ADDITIONAL READING

Du Rand, G. E., & Heath, E. (2006). Towards a framework for food tourism as an element of destination marketing. *Current Issues in Tourism*, *9*(3), 206–234. doi:10.2167/cit/226.0

Getz, D., Robinson, R., Andersson, T., & Vujicic, S. (2014). *Foodies and food tourism*. Goodfellow Publishers.

Hall, C. M., Sharples, L., Mitchell, R., Macionis, N., & Cambourne, B. (Eds.). (2004). *Food tourism around the world*. Routledge. doi:10.4324/9780080477862

Handayani, B., Seraphin, H., & Korstanje, M. (2019). Halal Tourism in Bali, Why Not? In *Special Interest Tourism in Southeast Asia: Emerging Research and Opportunities* (pp. 64–80). IGI Global. doi:10.4018/978-1-5225-7393-7.ch004

Handayani, B., Seraphin, H., Korstanje, M., & Pilato, M. (2019). Street Food as a Special Interest and Sustainable Form of Tourism for Southeast Asia Destinations. In *Special Interest Tourism in Southeast Asia: Emerging Research and Opportunities* (pp. 81–104). IGI Global. doi:10.4018/978-1-5225-7393-7.ch005

Harrington, R. J., & Ottenbacher, M. C. (2010). Culinary tourism—A case study of the gastronomic capital. *Journal of Culinary Science & Technology*, *8*(1), 14–32. doi:10.1080/15428052.2010.490765

Henderson, J. C. (2009). Food tourism reviewed. *British Food Journal*, *111*(4), 317–326. doi:10.1108/00070700910951470

Hughes, G. (1995). Food, tourism and Scottish heritage. *Food, tourism and Scottish heritage.*, 109-120.

Lee, K. H., Scott, N., & Packer, J. (2014). Where does food fit in tourism? *Tourism Recreation Research*, *39*(2), 269–274. doi:10.1080/02508281.2014.11081770

McKercher, B., Okumus, F., & Okumus, B. (2008). Food tourism as a viable market segment: It's all how you cook the numbers! *Journal of Travel & Tourism Marketing*, *25*(2), 137–148. doi:10.1080/10548400802402404

Mitchell, R., & Hall, C. M. (2004). Consuming tourists: Food tourism consumer behaviour. In *Food tourism around the world* (pp. 72–92). Routledge.

Montanari, A., & Staniscia, B. (2009). Culinary tourism as a tool for regional re-equilibrium. *European Planning Studies*, *17*(10), 1463–1483. doi:10.1080/09654310903141656

Privitera, D., Nedelcu, A., & Nicula, V. (2018). Gastronomic and food tourism as an economic local resource: Case studies from Romania and Italy. *Geo Journal of Tourism and Geosites*, *21*(1), 143–157.

Steinmetz, R. (2010). *Food, tourism and destination differentiation: The case of Rotorua, New Zealand* (Doctoral dissertation). Auckland University of Technology.

KEY TERMS AND DEFINITIONS

Cuisine: It is a style of cooking a particular food of a particular region. For ex: Chinese Cuisine, French Cuisine, Indian Cuisine, etc.

Culinary Tourism: It is a type of tourism in which tourist travels with a purpose of exploring food of a destination. It is also known as food tourism or gastro tourism.

Cultural Tourism: It is a type of tourism in which tourist travel to a destination to experience cultural attractions like fairs, festivals, food, arts, etc.

Gastro Tourists: They are the food enthusiasts who travel to places for exploring food of the regions.

Gastronomy: It is the art of preparing the food of a particular region. It also studies the relationship between food and culture.

Regional Food: It is the local food available at a particular region. It may vary as per the geographical locations of the regions.

Chapter 7
Revenue Management in Portugal:
A Strategy to Compete

Marine Ferreira
Polytechnic Institute of Porto, Portugal

Teresa Dieguez
https://orcid.org/0000-0002-4886-1446
Polytechnic Institute of Cávado and Ave, Portugal

ABSTRACT

The world is constantly changing, and the hotel industry is one of the sectors where we can feel it the most. Nowadays, the hotel market is dynamic, diverse, difficult, and dangerous, which creates new challenges for hotel managers, particularly in terms of revenue, creation, and optimization. In order to overcome that, revenue management emerges as a crucial price management tool to face competitiveness and business growth. This study aims to analyze the practices and advantages of revenue management, as well as understand if its implementation in Portugal influenced the revenue and growth of the hotel industry. It also intends to analyze whether there is adequate revenue training or if further training should be required. A quantitative methodology was used, and 284 answers were collected. From those answers, 115 were validated, analyzed, and discussed. Conclusions were made, and finally some limitations were presented as well as suggestion for future research.

INTRODUCTION

Revenue Management (RM) known in the 1980s in USA initially was used on the airline industry. Only later it was applied on the hotel industry and the reasons are mostly related to the perishable characteristics inherent to its services: if a seat (airline sector) or a room (hotel sector) is not sold, the seat or room may never be resold (Littlewood, 2005). RM is a management tool that helps to sell the right product, to the right customer, at the right time, at the right price and through the right distribution channel (Kasavana

DOI: 10.4018/978-1-7998-4330-6.ch007

& Brooks, 1998). RM started to be a widely used practice in the hotel industry, not only to optimize the revenues of accommodation, but also the revenues of other departments, among which appears the F&B, the Conference Rooms, the SPA and the Golf. Its objective is to obtain the highest profitability of all services (Costa, 2017).

As the tourism market is a phenomenon that shows itself in a continuous growth, the evolution and maturity of the current market changes demands the implementation of new concepts and practices for business optimization (Costa, 2017), as well as a continuous improvement service coupled with customer expectations (KPMG, 2018) The rivalries in this industry are huge and hotels need to use price management tools to go along with the growth of tourism (Costa, Monte & Fernandes, 2013), and this is only possible through the use of a strategic tool: the RM (Altin, Schwartz & Uysal, 2017).

In Portugal, RM is a poorly developed concept and organizations have a small number of specialized Revenue Managers (RMr). It has slowly been incorporated on the hotel industry and only recently it is perceived as a price management tool with lots of advantages. Few departments are strictly dedicated to this area and for this reason the used practices and their advantages are still unknown to a large part of the population. The present study aims to understand which the incorporation of RM in hotels is. It begins by understanding what benefits its implementation brings to the economy as well as to hotels. Then, RM orients to understand the need that hoteliers have on specific training in this area, highlights the importance of RM' use in hotels and what are the used systems. This research is structured in three main chapters. In the first chapter, a theoretical framework is presented, through a literature review about RM. In the second chapter, a case study is presented, used methodology explained, analysis and discussion made. Finally, limitations and suggestions for futures research are presented, as well as conclusions.

REVENUE MANAGEMENT

1. Genesis

According to Huefner (2014), RM has its origins in the 1980s in USA, on the airline industry, more specifically in the "American Airline". In 1985, some low-cost airlines, for example People Express, flooded the market with lower fares compared to the ones presented by other companies, becoming a threat to the airlines that at time dominated the market. Since costs on a flight were almost all fixed, the loss of the sale of a seat represented an expense: make a discount and sell it was much better than not to sell it at all. American Airline was forced to combat the low fares presented by its competitors through the development of very selective pricing strategies, preserving its business with full prices whenever possible: in order words, sell all possible seats and implementing discounts on empty seats, in order to be sold. A seat on an airplane was considered a perishable inventory, since when the flight took-off, each empty seat could no longer be sold. The main challenge remained in the sale of seats at the highest fare, but also in filling all seats on the plane, implementing discount strategies. The deregulation of the airline sector created a highly competitive environment being the techniques of Yield Management (YM) and RM the savers of the problems (Kaswengi, 2011).

The earlier definition of YM evolved to RM, since both concepts are linked to the increase in revenues and the maximization of profits (Taliani, 2015). However, not being completely identical concepts, it is necessary to understand what each involves:

- Yield management is designed to predict and influence customer demand and manage a company's benefits (Taliani, 2015). In this technique, mainly used in situations of discount's application (Cross, Higbie & Cross, 2011), the discounts were applied to the empty places that would be lost in order to fight the competition and still optimize the revenue. In Hospitality, the YM concept is related to supply and demand and is used to maximize the results and profits from the sale of a product or service, taking into account three different factors, namely; availability, control of stays and discounts (Costa, 2017). It is essentially a concept based on forecasts and simulations of market segment trends.
- RM is an ancient but unknown concept, since all managers already had the same doubt and objective when they needed to answer questions such as "what is the ideal price to practice? ", "What offers to create and how?", "When to offer a lower price? (Talluri & Van Ryzin, 2004). RM is "the science that is in the historical past" and the current center of a company, in order to anticipate demand with more detail and precision. RM is designed to be able "to establish and appropriate prices according to the availability of the product, through the appropriate distribution channels" (Taliani, 2015, p.87). It also can be seen as a management tool that aims to maximize the revenues of a business, through the use of strategies and tactics that allow optimizing the product and price availability for each of the named segments (Gomes, 2004). Started to be widely used in the hotel business, RM not only optimizes and maximizes the accommodation revenue, but also departments as F&B, Spa, Golf, Conference Rooms, among others.

In the 90s, RM quickly started to be used not only in aviation, but also in hotels, mainly due to the perishable characteristics inherent to its services. On one hand, the large hotel chains invested in computer systems where they managed to track many sources of information. On the other hand, smaller hotel chains opted for manual systems and reports from their reservations department on price and availability management and all information on their property management system (PMS). According to Littlewood (2005), both hotel chains reached the same results: using RM and YM techniques, their revenues and profits significantly increased.

2. Concept

The author Kasavana (1998) proposed one of the first definitions of RM, in which he argues that RM is a commercial management technique that helps to sell the right product, to the right customer, at the right time, at the right price and through the right sales channel distribution. This definition represents its five inherent pillars: Product, Customer, Forecast and Pricing. Product, not being just a tangible good, it can also be an intangible good or a service, being perceived by consumers as a set of benefits that satisfy their needs (Santos & Fernandes, 2017). Thus, "customer satisfaction is derived from the proximity between the buyer's expectations and the perceived performance of the product... if it reaches expectations, the customer is satisfied, and if it exceeds expectations, he is dazzled. These feelings define whether the customer will buy the product again and whether he will approach favorably or unfavorably other people about the product" (Kotler, 2000, p.205). Customer is the most relevant factor for defining the price. All customers are different; each of them has their own motivation, their own expectation, their own interests and a willingness to pay different prices. In order to understand the environment in which each customer is inserted it is essential to segment the market (Taliani, 2015). In this sense, segmentation has become one of the most important aspects of the tariff as it helps to define the target audience and think the strategy of the entire company. The study of consumers and different interests and motives that will make them choose a hotel must be well planned. Authors such as Kotler, Armstrong, Brown & Chandler (2013) argue that in order to make an effective segmentation, it is necessary to meet certain criteria such as measurability, relevance, accessibility, differentiation and viability. The purpose of segmenting customers is to ensure that they can be actively targeted to the most profitable segments (Kimes, 2008). Moment of making the reservation is the key element in price management. It is a relevant moment for the customer, but also for the hotel. Forecasts based on historical data are the key to big decisions and good RM will always depend on forecasting demand, which represents a major challenge (Curtis & Zahrn, 2014). Making predictions essentially used to measure demand, react to low demand, define commercial strategies and apply tariffs. It is also useful to be able to restrict sales and control costs and commissions, helps with financial planning and human resources planning (Littlewood, 2005). Controlling the availability of services, i.e., efficiently managing an inventory to control "full price" sales is an important aspect (Curtis & Zahrn, 2014) and forecasting consists of estimating the KPIs of a hotel based on the defined segments for a given period, taking into account the micro and macroeconomic reality in which the hotel operates (Littlewood, 2005). Forecast helps hotels to organize themselves, since it allows managing the number of the needed employees for certain seasons or celebrations, investing in the necessary equipment, facilities and materials, depending on the demand forecast (Gilliland, 2010). Price is also one of the most important competencies of RM as it requires the ability to effectively evaluate products and services. According to Kotler (2000, p.476) "although other factors besides prices have become important in recent decades, price still remains one of the fundamental elements in determining market share and corporate profitability". As the RM is responsible for ensuring that the charged prices are in accordance with the customers' expectations and with the amount they are willing to pay, it is important to do it correctly, as the price is the element with the most important weight in the interaction of buy-sell action (Hayes & Miller, 2011). For an organization to increase the profitability of sales and consequently its profit it is necessary to know precisely at every moment what is the maximum price that the customer is willing to pay for the product or service (Taliani, 2015).

According to Kotler (2000) to establish a right price, Revenue Managers (RMr) are based on and take into account four essential points: the value of their product, the value perceived by the customer, the positioning and the competition. The authors Hayes & Miller (2011) add that "from the customer's point of view, price is easily the most visible, powerful and understandable part of the entire marketing mix" of a company and as Kotler argues (2000, p.475) "You don't sell for the price. Sell the price".

According to Tardieu (2004) and Kimes (2008), RM on Hospitality lays on five particular keys that define success, namely fixed capacity, ability to Segment the market, perishable inventory, products sold in advance and variable search

- Fixed capacity: changes in organizations for increasing capacity according to the level of demand are risky and require high costs, which makes these changes unfeasible (Kimes, 2008). So, once a hotel has been built, it is difficult and expensive to increase its capacity.
- Ability to segment the market: having the need to develop a distinct marketing plan for different types and segmentation of customers (Tardieu, 2004), selling a lower rate to customers who can give more value for this service would be a mistake (Kimes, 2008).
- Perishable inventory: unsold rooms will be lost forever and the hotel will not be able to put them back in inventory for use (Tardieu, 2004). Additionally having a fixed capacity, all the unused capacity cannot be stored for later use (Kimes, 2008).

- Products sold in advance: reservations can be made before the desired day and in the particular case of group sales reservations it can be made several years in advance. For the marginal price of a product to be considered acceptable and good for sale, its value must be higher than its variable cost, in order to combat fixed costs and thus generates profit. In this perspective, RM techniques make it possible to recognize which are the appropriate prices to determine in order to activate sales (Kimes, 2008).
- Variable demand: the objective is to be able to match occupation with demand (Kimes, 2008). RM can be used to help moderate fluctuations and demand, which varies according to the season, month or day of the week. Aiming to increase occupancy during low season and increase revenue during high season (Tardieu, 2004), variable demand is one of the most crucial points of the use of MR.

In order to implement RM in an organization, identify the following eight distinct steps is crucial: customer knowledge, segmentation and market selection, internal evaluation, competition analysis, demand forecasting, analysis and selection of distribution channels, dynamic value-based prices and inventory management (Tranter, Stuart-Hill & Parker, 2008). In a hotel, the RM system should include four structural elements (data and information, hotel RM centers, software RM hotel and RM tools) not only in the accommodation department but also in all services that generate revenue (Ivanov, 2014). RM at this moment has proclaimed, in many sectors and industries, a common objective: "the management of a space of time" (Yeoman, 2011, p.1-2).

3. Role of a Revenue Manager

The Revenue Manager (RMr) function appears around the 1990s and the mission of the yield (since the term revenue was not yet known) was simply to exercise the functions delegated to a current marketeer or sales manager (Legohérel, Poutier & Fyall, 2013). Nowadays RMr is a fully identified function and its positioning is already starting to be integrated in several corporate organization charts. According to Costa (2017) the RMr on the hospitality sector is responsible for implementing strategies and procedures to manage revenues. This means that through various practices and techniques, the RMr carries out a market study that involves his hotel unit, analyzing the competition and the demand, optimizing distribution and sales. Through analyzing historical data and forecast data (forecast), creates price and price management strategies as well as implementation of segmentation strategies. The role of a RM team is organized according to a division of five frequency levels (Ivanov, 2014): daily rates, weekly, monthly, quarterly and annual.

1. Daily – daily rates is the first step and consists on the analysis of the pick-up: the analysis of a report that provides an indication of the reservations made on the previous day compared to the present; analyzes the prices charged by the competition through a rate shopper for the next 24 days and review the rates of the hotel itself; additionally, analyzes of the hotel's performance through KPI's (ADR, OCC, RevPar, TrevPar, GOPPAR and RevPASH) and adjust them if necessary; reviews of the tariff restrictions implemented for the next 3 months and manages the social media platforms.
2. Weekly – weekly is the next step and consists on reviewing the pick-up from the last 24 hours for the next 3 months; analyzes of the charged price by the competitive set; adjusts tariffs based on the aforementioned reports; close of the restrictions and discounts applied on the dates with the

greatest demand and, therefore, management of the tariffs placed on the distribution channels; check of overbooking and update all CMS and OTA's content.

3. Monthly – it is important to review the pick-up of the last 24 hours for the end of the inventory to manage it; adjust all rates and restrictions/discounts if necessary; revving of the competitive set regarding the end of the inventory; analyzes of the overbooking level, if it is the case; revenues studies by distribution channel (associated costs and profitability); review MR reports against history and forecasts to make sure they are on schedule; prepare a meeting with the RM team, department directors and the general director to discuss the applied strategies.
4. Quarterly - analyze in detail the OTA's, that is, compare their production and costs to verify their profitability and review the entire sales strategy.
5. Annual - in this last level, annual forecasts are prepared; return to preparing the competitive set; define a budget and new strategies to be implemented according to market trends; renegotiate contracts with companies, OTA's, etc., taking into account hotel revenue, production and costs.

Since the RM was implemented in the hotel industry, the organizational structure of the companies has changed, giving a prominent place to the position of the RMr who, from a simple analyst which defined hotel prices, came to have power in the final decisions (Aubke, Wöber, Scott & Baggio, 2014). Furthermore and according to the organization chart presented by Legohérel, Poutier & Fyall (2013), in general an RMr should report directly to the top manager and must be aligned with the other directors of the various departments, as Figure 1 below represents.

Figure 1. Organization chart
Source: Legohérel, Poutier & Fyall (2013)

Since the RM had its origin in the airline sector, the second figure (Figure 2) depicts the organization chart used in that sector, prepared by Julie Adam (2013) (Corsairfly), which can easily be adapted to the Hotel sector. Thus, in the hotel sector, the economic analyze would be the financial department, the strategic marketing would be marketing & e-Commerce and the flights program and pricing would be just one sales & yield department. In this way, the function of the RM department is central to the other departments and can be seen as an essential and strategic function for a hotel organization, having to directly report to business intelligence management. Currently, the organization chart most used is presented in Figure 2 (Tranter, Stuart-Hill & Parker, 2014).

Figure 2. Corsairfly organization chart
Source: Adapted from Legohérel, Poutier & Fyall (2013)

The role of an RMr is not isolated and it is not just about converting day-to-day revenues, but essentially be able to inform, manage and train its entire team. RMr must be a good leadership and have persuasion skills (Ivanov, 2014). Also according to the authors and according to Van der Rest (2016, p.149), in RM, "knowledge is a resource" that must be shared among the entire work team, in order to facilitate internal communication. Communication, leadership and management become fundamental in the role of a RMr and the authors Aubke et al (2014) further argue that these skills are the *ex-libris* for the success of RM. RMr are strategic players and their mission is to anticipate activities and manage the capacity of the entire hotel. They must have a broad knowledge of the area in order to make sure that none night at the hotel is considered a lost sale (Legohérel et al, 2013). In addition, an RM team must collaboratively work and establish clear communication with the marketing, sales, reservations, commercial and/or front-office (reception) departments in order to be in constant synchronization.

RMr have a fundamental role in the management of hotel establishments because they are responsible for ensuring that the launched price of the product corresponds to the customer's expectation and disposition, taking into account the market competition and the optimization of revenues (Cetin, Demirciftci, & Bilgihan, 2016), A it is crucial to have an in-depth knowledge about price, supply and demand, depending on market trends. In addition to the techniques and specific training in RM, RMr must also have specialized techniques in marketing, strategic management and operational management, as well as know how to manage people, how to motivate, lead and influence its team (Cetin et al, 2016). They also must have two fundamental characteristics for their position as RM: immense attention to detail and an enormous capacity for analysis (Cetin et al, 2016).

The RMr function came to combat the idea that the more rooms sold, the higher the hotel's occupancy rate, regardless of the stipulated price. RMr came to show that this concept was and is wrong, since 100% occupancy in one hotel is not necessarily better than 65% occupancy in another hotel, as it all depends on the price per room at the time of sale and all extra services that a sale can purchase. "RM has had a huge impact in several sectors and is now an essential function" (Yeoman & McMahon-Beattie, 2017, p.70).

4. Computer Systems to Support Revenue Management

The first challenge in RM is to acquire the computerized RM systems for a large number of forecasts, given that in a hotel these RM systems must identify the various categories of the required forecasting (Curtis & Zahrn, 2014). Nowadays we are in a technological era where purchase on line is recurrently increasing. Consumers have easier acess to price comparison and the option of the best deal, triggered for the growth of several agents of online travel, is a major transformation (Yeoman, 2011). Furthermore, in RM it is important to correctly select the distribution channel in which it is desired to insert a hotel, because it is necessary to analyze whether it is worthwhile to be in a distribution channel with a high sales commission rate, for the simple reason of being known. Also it is necessary to understand which RM systems are helping and facilitating the entire price management process. Table 1 illustrates the advantages of the different support systems, which are all interconnected.

Table 1. Revenue management support systems

CRM	PMS	RMS
- Customer is the main focus of business processes in order to understand and anticipate their needs - it uses existing customer information from CRS and PMS to determine consumption patterns or type of customer	- It manages reservations, availability and occupancy, check-in/out, images, guest profiles, reports, etc - it momentarily receives reservations from all online and offline distribution channels	- Tool that serves to apply the principles of RM: price recommendation, forecast revenue, compet set analysis, databse analysis - it may or may not be integrated with CM and/or RS
CRS	**CM**	**RS**
- the main source of information about customer (through its reservation) - makes it possible to quickly optimize the hotel's reservation system, which in turn improves the occupancy rate - it retains hotel information, inventory and room rates to manage the booking and process	- it quickly updates all prices of different types of rooms, in the different associated online distribution channels - it safeguards overbooking, since when the room is sold, it will be automatically removed from all other online channels	- makes it possible to know the practiced rates by competing hotels - it provides data on the daily market demand - it is a tool that allows the positioning strategy to be resolved at any time

Source: own elaboration

4.1 Customer Relationship Management

In an organization, long-term price management needs to realize that regular customers occupy a more important position for the company than current customers (Taliani, 2015). In order to be able to offer better prices, it is necessary to evaluate the profitability and the period of stay of the customer within the company. CRM (customer relationship management) emerges to maximize, protect, retain and retain relationships with existing customers, in the long and short term, in order to target offers to customers who are loyal to their products and services and those who will become customers. I fact, it is more expensive to capture a new customer than to retain a current customer (Oliveira, 2000) and CRM "should be seen as a concept to obtain the best relationship with customers and not as a product or technology" (Oliveira, 2000, p.12-13). The duration of customer relationships increases due to the simplicity and speed of doing business with a company (Oliveira, 2000). It should also be noted that the basic point of action is not only to obtain customer feedback, but to act on it, since the improvement of the customer-company relationship is constituted by solving problems according to the "customer's point of view". This concept serves as a basis to be able to segment the market, because through a deep perception of the customers' needs it is easier to identify where advantageous and lasting connections can be built (Beck, Knutson, Cha & Kim, 2011). CRM is a "cycle" that operates in three fundamental stages: i) getting to know the client, ii) planning the marketing campaigns and interacting with the client, based on the first moment, and iii) carry out the necessary marketing and sales actions.

4.2 Property Management System

One of the basic systems of RM is the PMS designated as the property management system, considered as an infrastructure of the central data of a hotel (Kokaz Pucciani & Murphy, 2011). It is a hotel operating system used in the management of basic objectives, such as reservations, availability and occupancy management, check-in and check-out, images, guest profiles, inventories and report management, among others. It is used internally to control the activities of the hotel and is generally linked to other applications, mainly to the hotel's point of sale (POS). This management software is shaped according to the specific requirements of each hotel, and it can be used for only one independent hotel or for a complete hotel chain, with a central database set (Legohérel et al., 2013).

4.3 Revenue Management System

The revenue management system (RMS) has as main objective to maximize the business volume and the optimization of the company's margin, that is, the profit, reaching maximum occupancy, with the highest rate possible throughout the year (Yeoman & McMahon-Beattie, 2017). The RMS essentially helps to calculate the dynamic price to be executed, taking into account the customer's information and the organization's strategy, giving price alternatives in real time, according to the current state of the market (Yeoman & McMahon-Beattie, 2017). The authors Legohérel et al. (2013) add that, in general, its purpose is, above all, to analyze customer behavior and market changes. In order to make good use of the RMS, RMr perform an analysis through three specific divisions: "customer segment, price levels and RM levels for the hotel sector". This tool allows hotels to manage their revenue in the most effective way in their daily reservation system. The RMS thus provides four main benefits: "knowing the measures to be used every day; know the value of the customer segment to be communicated; adjust the strategy; and having the main portfolios of reservations and the geographical area for the hotel industry " (Legohérel et al, 2013, p.75).

4.4 Central Reservations System

The central reservations system (CRS), which is computer reservation software, allows storing and retrieving hotel information, room and fee inventories, managing reservations and processes. This computer system is connected with the website of the hotel and with its mobile booking engines. In the hotel chains, it centralizes the reservations of all its properties in a single system. A CRS offers hotel room rates and availability for many different distribution channels, namely: i) the GDS (global distribution system) - booking tool that passes on hotel inventory and rates to travel agents and allows them to make reservations, ii) OTA (online travel agency - online agencies whose websites allow consumers to book various travel-related services directly over the internet), iii) the hotel website and iv) third party websites, among others (Hayes & Miller, 2011).

4.5 Channel Manager

A channel manager (CM) is a computer system that ensures the management of rooms and prices, reservations and availability in the various online distribution channels and is usually connected to a PMS. According to Hayes & Miller (2011), this system emerged essentially due to the appearance of many online distribution channels that generated at least three huge needs: i) to be able to control the sales price to the end customer; ii) to be able, in a short time, to change, simultaneously, the tariffs of the different distribution channels and iii) to be able to control the existing (dis) parity. With a channel manager, it became possible to update at once, in a fast and intuitive way, all prices, availability and reservations on the different online platforms (OTAs). In addition to this functionality, a channel manager also has an advantage in terms of overbooking, as it avoids the risks of overbooking, since when a room is sold, the CM automatically blocks its availability in all other online distribution channels (Tranter, Stuart-Hill & Parker, 2014). Other advantages of this RM support system include: Control tariff parity; Manage confirmations via online; Manage reports and also their historical changes; Increase the possibility of hiring a variety of distribution channels. The probability of selling room increases, once it is associated with several online distribution channels, and in low season it will be easier to sell them at a good price.

4.6 Rate Shopper

To perform a good price management in a hotel, it is necessary to have data of a safe and reliable character, both in terms of internal and external variables, more specifically about the practiced prices by competitors considered direct. In order for this management to be viable, the rate shopper was introduced, which is a computer system that allows hotel units to analyze the charged prices by competing hotels, realizing their positionally in relation to the charged price before them. It also allows to view competitors' prices for 3, 6 or 12 months in a few minutes, which leads to more effective management in making decisions about the prices to be charged on different dates of the year, for each type of room and market, giving the possibility to apply restrictions. In RM, the main objective of obtaining a rate shopper is to increase the hotel's RevPar and consequently maximize total revenue (Faria, 2015).

The following figure (Figure 3) illustrates the relationship between the various computer systems to support RM, mentioned in this section. The rate shopper is a external tool to the information flow circuit.

Figure 3. Information flow circuit between computer systems to support RM
Source: own elaboration

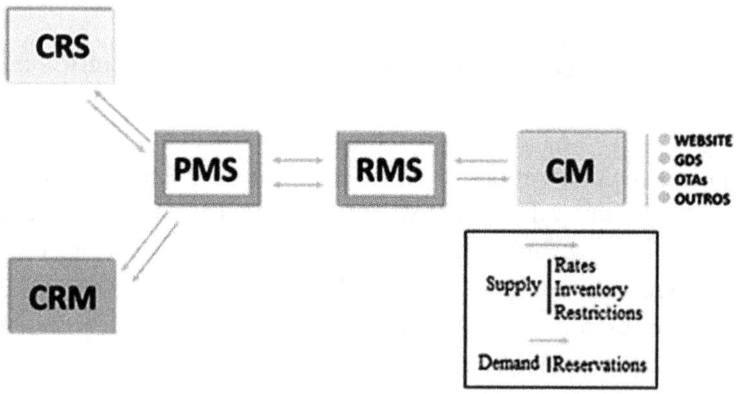

Depending on the systems in each hotel, the flow of the offer may be integrated into the PMS or RMS. In the presented scheme (Figure 3), the PMS and the RMS are interconnected, as being the "heart" of the communication flow transmission. The flow of information begins through a reservation either through an online (CM) or offline (CRS) distribution channel. As such, the PMS must communicate to the CRS the flow of the offer, that is, prices, inventory and restrictions, and in the same way it must communicate to the CM. Analyzing the demand flow, the extremities (CRS and CM) being responsible for the supply, are the ones who will receive the reservations, and as such they must communicate to the PMS. CRM is a computer system that allows working on the relationship or data with customers, where the main agent is the front-office manager or guest relation, with unlimited access to the PMS in order to absorb all customer information, and use that information to give a better experience to a customer. Generally, CRM can be integrated in a part of the PMS or in a separate system.

EMPIRICAL EVIDENCE AND RESULTS

1. Methodology and Research Question

1.1 Methodology

The study was carried out during 2019, the methodology is quantitative methodology and it used a questionnaire. The starting point of this investigation was: "Does RM have an impact on the growth of the hotel industry?".

In addition to the research question, it was essential to outline general and specific objectives. The present study has as main objective to understand which the incorporation of RM in hotel units is. It starts by understanding what benefits the implementation of RM brings to the economy as well as assessing the impact that the use of RM techniques has on hotel units. With the general objectives it intends to:

- Identify the need and the importance of the required RM training;
- Identify the used RM tools and systems in a hotel.

With the specific objectives it is additionally intend to:

- Identify the respondent, which department he/she belongs to and what function he/she performs;
- Know the respondent's level of training and knowledge about the culture of RM;
- Try to understand what the respondents perceive about the functions that an RM should perform;
- Identify the typology, the hotel establishment, its location, the existence (or not) of an RM department and its average occupancy rate;
- Identify the RM tools/systems that the hotel units use, in order to understand the answer to the question "did the use of RM techniques help increase revenues?";
- Identify whether RM helped to increase (or not) revenues as soon as implemented.

1.2 Questionnaire Description

According to the conducted above literature review, a script was prepared to start the questionnaire. The script was mostly composed of closed questions (twenty-three), and three open questions were also included, in order to enrich the information to be obtained. The guide was designed according to four major groups of questions: (i) demographic data; ii) Perception about the concept of RM; iii) data on the responding hotel units and iv) importance of applying the RM in the hotel unit in question. The script was validated by two professionals in the sector and two academic professionals and expertise on the research area. After validation, an online questionnaire was built, consisting of open and closed questions, using the LimeSurvey tool. In order to reach a large number of people and increase the level of responses, the questionnaires were distributed through a variety of online distribution channels, namely LinkedIn, Facebook, Instagram and via e- mail. The questionnaires were carried out only in the Portuguese language, as the study only addressed Portugal.

1.3 Data Collection

The criteria for data collection were established, within the following universe:

- Type of Accommodation: the universe was constituted by all accommodation in Portugal, within the hotel sector. With this it was gathered contacts for Hotels / Resorts, Hostel and Local Accommodation, Rural Hotel, Tourist Village, Tourist Apartment and Country Houses.
- Department: the sample it was outlined through the department in which each respondent was located. It was defined for the sample the departments of RM, Management/Administration, Commercial, Reservations and Sales, and Front-Office.
- Database Procedure: After outlining the type of accommodation and the department, it was necessary to start an online search on the different accommodations in Portugal (Mainland and Islands) and related to the hotel sector. It was collected emails from each of the accommodations made available through its websites, emails and telephone calls, in order to obtain the email from the RMr of each establishment, or from one of the five departments related to this study, since on their website they only provided the "general" or "email" info ". A database with 2,819 contacts was built and an email was sent to everyone with the questionnaire. The email was sent with the clear awareness that we would hardly get many responses or, even within the responses to receive, we would find many that would not be considered valid.
- Duration of the questionnaires: The collection of all questionnaires was carried out between March and May 2019.

The collected data, after being filtered and properly analyzed, were treated using the software IBM SPSS Statistics (IBM Corporation, 2010) and Microsoft Excel (Microsoft, 2010).

2. Descriptive Analysis

2.1 Sample Characterization

During this study, 284 questionnaires were collected. Only 121 were completed, corresponding to a rate of 43%. Thus, it was only validated the responses that contained the departments of RM, management / administration, sales, reservations /sales and reception / front office. In this way, the sample of the present study consists of 115 valid questionnaires, as it can be seen on Figure 4.

Figure 4. Total questionnaires vs Valid questionaires
Source: own elaboration

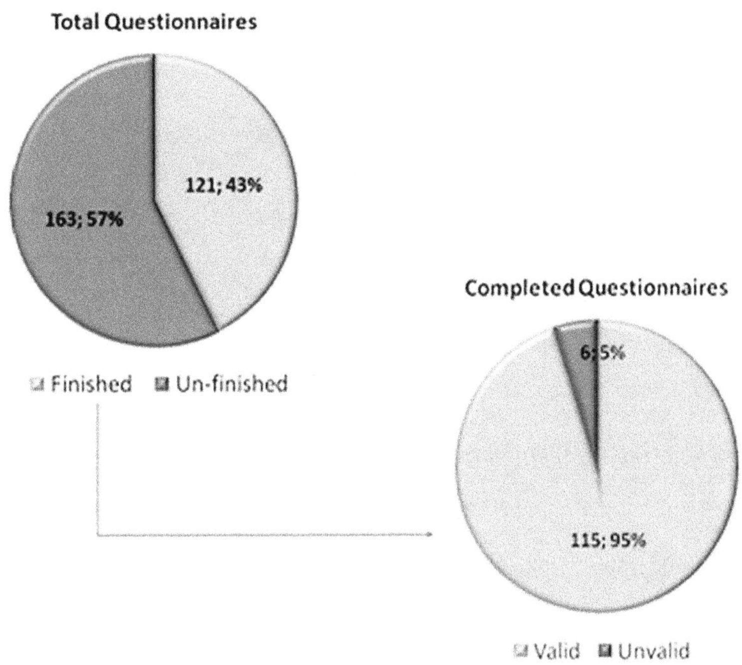

2.1.1 Demographic Data

The respondents were mainly "male", representing 59% of the respondents. Mostly (50%) are "between 21 and 34 years old", 43% has a "Degree", 42% belongs to the "Direction / Administration" department and 19% are "Managing Director", as it can be seen on Figure 5.

2.1.2 Perception of the Concept About RM

47% of the respondents have "regular training" on RM, 43% showed great need for RM training, 83% considered the RM function "very important", 83% pointed out "Total knowledge about the Area" as main skill/competence and 84% presented as the most important function of an RM the "rate management", as it can be seen on Figure 6.

Figure 5. Demographic data
Source: own elaboration

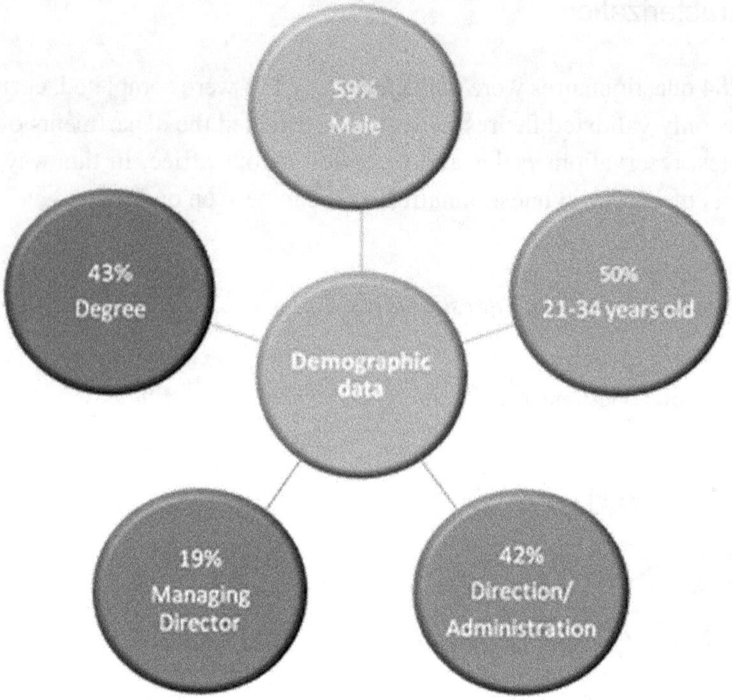

Figure 6. Perception about the Concept about Revenue Manager
Source: own elaboration

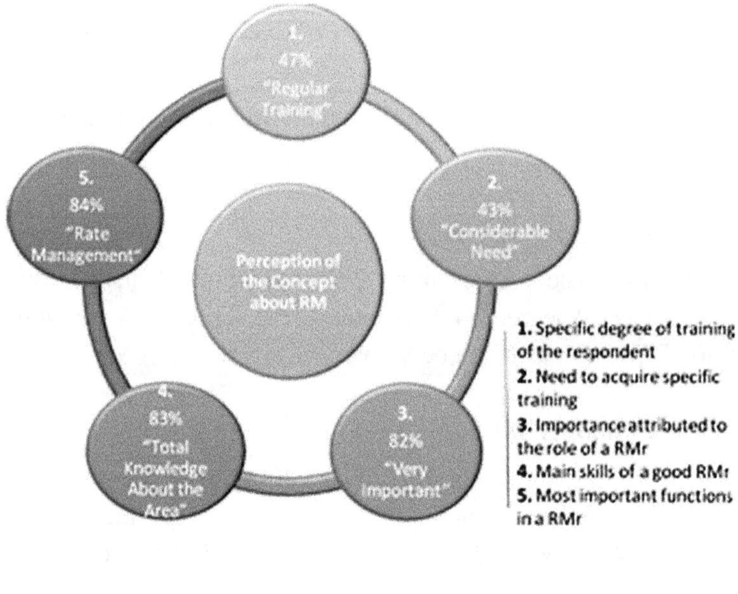

2.1.3 Hotel Unit Information

35% of the respondents belong to units from the Center of Portugal, 74% were Hotels. 44% were 4 stars Hotel and 43% has an average occupancy rate between 52 and 69%. Most hotel units (70%) where respondents work have an internal RM department. 22% of the respondents do not have the department installed in their unit and 8%, despite not having it directly on the company's organizational chart, outsource an RM team to implement the RM techniques (Table 2).

Table 2. Representation of the RM department in the Hotel Units

In the hotel unit the Revenue department is:	%
Internal	70
Outsourcing	8
We don´t have that department	22

Source: own elaboration

When trying to understand if the hotel units where the respondents work had a RMr, with specific training in the area, 51% of the respondents reveal that they have a RMr with specific training and 49% the opposite. It should be noted that this result comes from the total obtained in the previous question, where it was obtained a total of 22% of respondents without the RM department, which influenced the percentages in this answer. In this way, it was assessed the 70% of respondents with an internal RM department, and it was observed that 21% of them have no specific training in the area and 49% have specific training. It also was verified that if they did not have a RMr, it was expected to hire 5% in the next 12 months, 83% had no plans to hire one and 11% had any knowledge.

With regard to the areas of application of RM in the hotel units, it was understood that in addition to the accommodation, which evidently stands out 100%, there would be some more areas where it was applied. It was observed that 37% of the respondents use RM techniques in F&B, 10% in the SPA area, 4% in the Health Club and 5% in Golf (Table 3)

Table 3. Representation of used Tools to increase revenue

Application Areas to RM	%
Accommodation	100
F&B	37
SPA	10
Health Club	4
Golf	5
Others	0

Source: own elaboration

Once the percentage of RMr with specific training regarding the responses to the 115 questionnaires was analyzed, it was asked a question regarding the used tools to increase revenues and to measure the level of training of respondents. For 90% of the respondents, the "competition analysis" tool is the most important. The "creation of dynamic tariff", with 83% of the responses, is presented as being of a relevant nature. In addition, 6% of respondents use the "increase in distribution channels" to maximize revenues, 55% "segmentation by levels" and finally, "stay restrictions", with 47% of responses (Table 4).

Table 4. Representation of the application areas to the RM

Used Tools to Increase Revenue	%
Increase in distribution channels	66
Creation of dynamic tariff	83
Competition analysis	4
Stay restrictions	7
Segmentation by levels	55

Source: own elaboration

After analyzing the hotels and the tools they use to increase their revenues, it was also investigated the opinion of respondents regarding the help that the implementation of RM techniques had in relation to increasing revenues. Most of the respondents had a positive opinion, representing 57% of the statements that "completely agree", followed by 39% of those that put the option "agree". The most negative responses, "disagree" or "strongly disagree", together accounted for only 3% of all responses, which reflects a general agreement on this question.

Table 5. Representation of the Implementation of RM techniques facing Hospitality Growth

The Implementation of RM Techniques in the Hotel Unit Helped to Increase Revenues	%
Completely disagree	3
Disagree	1
Agree	39
Completely agree	57

Source: own elaboration

Regarding the computer systems to support RM, it was investigated whether the hotel units where the respondents work had a PMS or not and if so, which one. Answers showed that only 4% had no PMS. 32% use the Host, which is the highest percentage compared to other systems. With a percentage difference of 8 appears New Hotel with 24% of the answers, Opera with 15%, Protel with 6% and finally Maestro with only 2% of the answers. 17% of the respondents selected the option "others", with the PMS in use being Sihot (9%), Mews (2%), Alidata (1%), Cloudbeds (1%) and Own (4%) (Figure 7)

Figure 7. Representation of the used PMS
Source: own elaboration

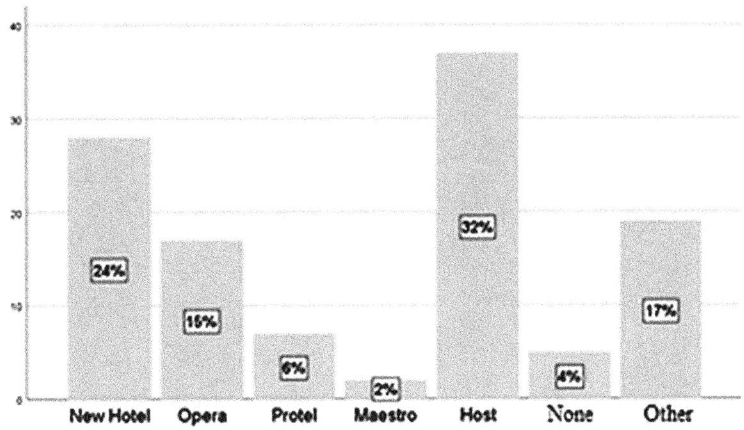

In what concerns the revenue management system (RMS), most hotels (47%) do not use any RMS. As for the remaining management systems,17% use Climber as RMS, 8% uses BookingSuite and 7% Duetto. In relation to Xotels RMS, the computer system obtained 3% of the answers, with Opera only holding 2% and Rate Wise 1%. It was also analyzed that 17% represents the "other" option, with the RMS being customized (5%), IdeaS (5%), Hotusa (2%), Cloudbeds (2%), E-GDS (2%) and Alidata (1%) (Figure 8).

Figure 8. Representation of the used RMS
Source: own elaboration

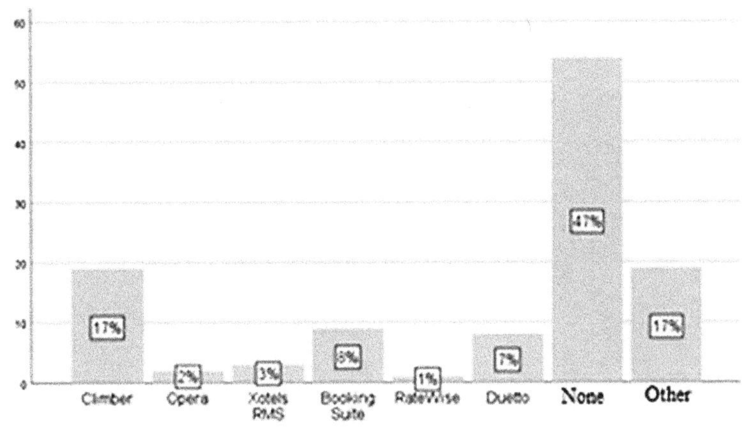

With regard to the Booking Engine (BE) used or not used by the hotel units in this study, the most used system is the Guest Centric, represented in 26% of units. Booking.com leads the second position, with 24% of the responses, followed by D-Edge (AvailPro) which obtained 11%. The least representative BEs are E-Gds with 7% and Paraty with 3%. It was also noticed that 6% of respondents do not use any BE. 22% represents the option "others", with BE being used Synexis (7%), customized (4%), Sinergy (4%), Host (3%), Omnibees (2%), SiteMinder (1%) and Saber (1%) (Figure 9).

Figure 9. Representation of the used RMS
Source: own elaboration

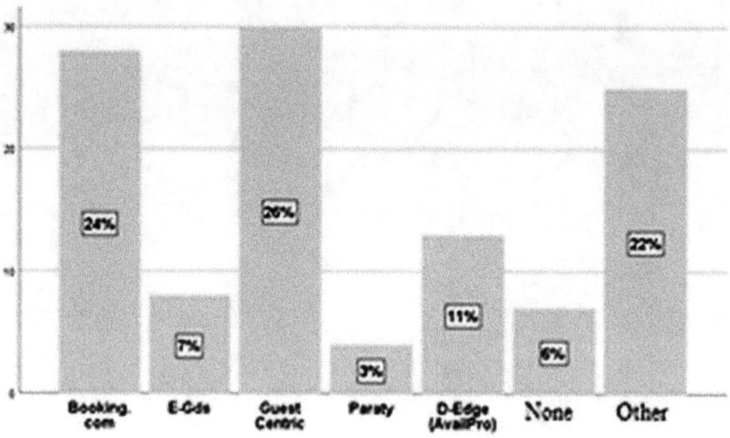

Regarding the Channel Manager, 9% of the hotel units in this study do not use any channel manager. In view of the responses given by the respondents, 31% of them use AvailPro as a channel manager in their hotel unit and 28% the Site Minder. 10% of the respondents use the Guest Centric and 7% the E-GDS. Although not mentioned in Figure 10, the questionnaire also included RateGain, which obtained 0% of responses as an option. It was also analyzed that 16% represents the "other" option, with the CH of use being Sinergy (4%), Synexis (4%), Omnibees (3%), Sihot (2%), SiteMinder (1%), customized (1%) and Host (1%) (Figure 10).

Figure 10. Representation of the used Channel Manager
Source: own elaboration

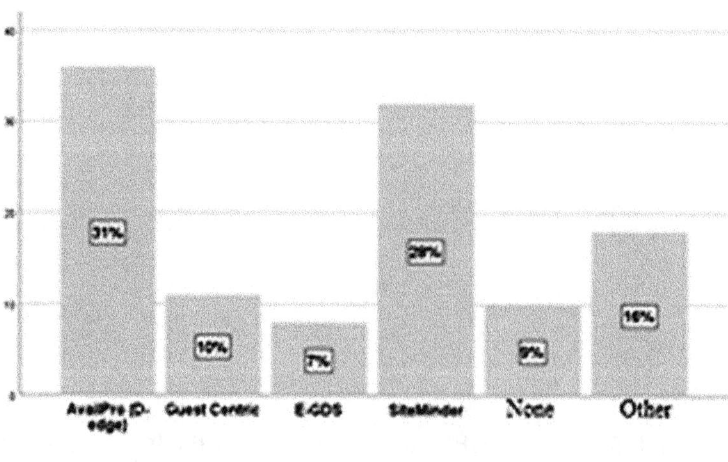

In view of the above and trying to summarize the main conclusions drawn from the sample from a demographic point of view, the following figure (Figure 11) represents the most significant indicators for the issues under analysis.

Figure 11. Summary Table of sample Hotel units Information
Source: own elaboration

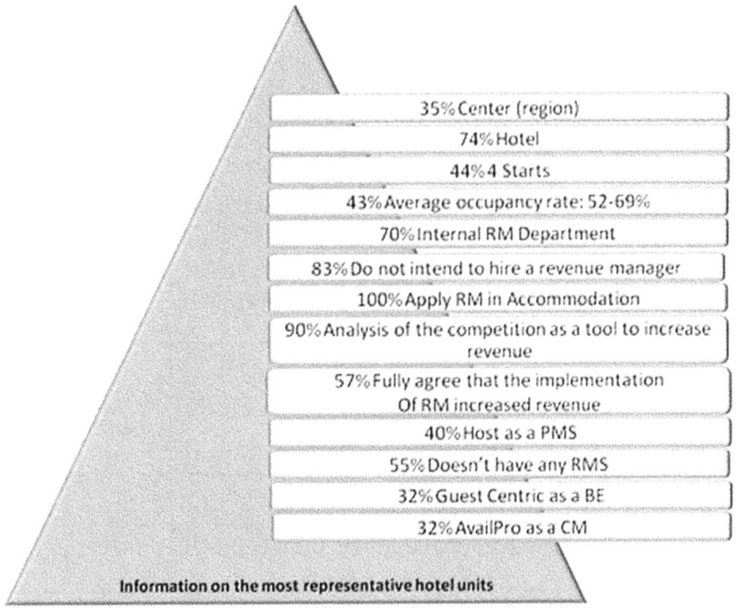

2.4 Importance of Applying RM in the Hotel

The last group of questions investigated the opinion of each respondent to the question "In your opinion, was the applicability of RM in the growth of the hotel industry important? Please justify. " Being an open question, not mandatory, 57 answers were obtained, corresponding to 49%. With regard to the 46% of respondents who agree with the asked question, they claim that the application of RM techniques was important in the growth of the hotel industry, as it enabled a more focused management and adapted to the reality and market circumstances. In this way, it allowed to increase the RevPar and the annual revenue of each establishment, which, in turn, helped to increase the growth of hospitality in Portugal. The same respondents point out that the application of RM strategies allowed, with the same resources, maximize the return of each stay to the maximum, and thus increasing profitability for each time of the year. Respondents consider that RM systems provide quickly and effectively information on the definition of pricing and segmentation strategies, in the short and long term. They argue that RM optimizes revenue and costs and helps to obtain information and define strategies to prepare company for the future and current market. They further argue that the application of dynamic tariffs based on demand and supply, helped to maximize revenue.

Regarding still 46% of respondents, 12% of them, specifically defend that RM came to combat the ideal of selling all possible rooms until reaching 100% occupancy rate, since the important thing is the price to be charged for each bedroom. They also indicate that one of the biggest problems in the hotel industry is the seasonality and that RM has reinforced and streamlined the supply and demand situation, which in turn increased revenue in the high season and made prices more desirable in the low season, helping to alleviate the big problem of hospitality related to seasonality.

DISCUSSION

Respondents, who reveal specific training in the area ofRM, represent 64% of the responses. However, when faced with the question about the main competences that an RMr should have, the answers contradict the referred degree of training, since one of the competencies considered essential in an RMr according to our literature review (leadership, persuasion and analysis), are considered unimportant for respondents. The following Table (Table 6), presents a summary picture of the convergences and divergences between the studied authors in the literature review and the answers obtained in the present study.

Table 6. Summary Table of Competencies for a Revenue Manager

Core Competencies of an RMr	Authors' Citations	Author		Respondent	
		Defends	Do Not Defend	Agree	Does Not Agree
Leadership ability	"(…) RMr must also know how to manage people, that is, have the power to negotiate, as well as being able to motivate, lead and influence their team" (Cetin et al, 2016).	X			X
Persuasiveness	"The function of an RMr is not isolated, it is not just about converting the day-to-day revenues, but essentially getting information, managing and training the entire team, being essential to have good leadership and persuasion skills" (Ivanov, 2014).	X			X
Communication skills	"Communication, leadership and management become fundamental in the role of a RMr and these skills are the *ex-libris* for the success of RM" (Aubke et al, 2014).	X		X	
Analysis ability	"(…) these professionals have immense attention to detail and an enormous capacity for analysis: two main characteristics for their positioning as RMr" (Cetin et al, 2016).	X			X
Total knowledge about the area	"(...)RMr are strategic players and their mission is to anticipate activities and manage the capacity of the entire hotel. They must have a broad knowledge of the area (…)" (Legohérel et al, 2013).	X			
Previous training in reserves				X	

Source: own elaboration

Although 64% of respondents claim to have specific training in RM, everything indicates that the importance due to the control of the disparity is not given, since only 50% of the respondents highlight this function. It should be noted that the researcher Ivanov (2014) considers this function as an essential function for the achievement of good RM. Respondents with regular or high training in RM (64%) were the same ones who felt the greatest need for training in the RM area (84%), perhaps due to the fact that they perceive that this tool can help them keep up with the competition and the growing evolution of the hotel sector. It was also assessed that respondents with little or no training in RM (36%), despite not feeling such an extreme need for training, begin to feel their need slowly. These data seem to indicate that RM is growing and people are beginning to feel the need to effectively graduate in this area. Most of them use almost all the RM exposed tools, which seems to indicate that its use has enabled better management and, therefore, an increase in the revenues of its hotel units.

Respondents use the Property Management System (95%), the Channel Manager (89%) and the Booking Engine (92%). Only the RMS (45%) stands out when compared to the others systems, maybe because of its high cost and answers seem to indicate that the RM support systems are essential for increasing revenues in a hotel.

Only 3% of the respondents affirmed not considered important the function of a RMr, because they didn't have specific training on RM and maybe didn´t understood its relevance. Only

22% of respondents had t no RM department and don´t consider contracting a RMr for the next 12 months. This analysis seems to indicate that the RM is not yet very present in the hotel units nor is it developed in the hotel sector, since the respondents who will not hire are the same ones who answered, in general, that the function of an RMr is in fact very important, and who feel the need for training in the area (except for two answers). This information may additionally indicate that the hotel units are not yet ready to train their employees in this area, even if they feel this need.

Respondents, who said that RM techniques did not help increase revenues, do not use half of the techniques they should, probably due to the lack of their knowledge or importance. Regarding support systems, it was observed that almost no respondents use an RMS and two do not use a channel manager, which makes it very difficult to manage prices in the various distribution channels. It could also means that they are not associated with almost any distribution channel. . It was also noticed that one of the respondents does not use any booking enginer, which being the booking engine may mean that the non-respondent does not have a website. Two of the respondents use Booking.com, which can be detrimental in terms of revenue, since is the platform with the highest commission percentage.

LIMITATIONS

Respondents were limited to answering closed questions and only one open question or none, and these were fundamental questions for the authors to analyze. Respondents were not representative, maybe because in Portugal there is some resistance to answering questionnaires, even if for academic purposes. No geographic area was delimited and the obtained information obtained was not as specific in character. However, it is relevant to note that the built database had 2,819 contacts and authors contacted them all.

FUTURE RESEARCH

For future research if only one area of Portugal was elected, the result will be more representative. Another suggestion would be to study two hotels or two hotel chains, depending on the size of the study, in which one hotel would use RM techniques and the other would not use them, thus being able to compare its revenues and determine whether RM effectively revolutionized management of a hotel and whether it really contributed to the increase of the hotel and tourism sector in Portugal. Use more open questions and qualitative methodology, as well as fewer questions in the questionnaire. As public policies, it is suggested that training programs / actions must be created at national level in these areas, as well as organized Workshops and Conferences where Good Practices would be disseminated and the quantitative and qualitative results presented.

CONCLUSION

In Portugal the concept of RM is recent and little developed? Organizations have a reduced number of specialized RMr, which has not allowed exploring all the advantages of its use. The number of departments strictly dedicated to this area is small and for this reason the used practices and their advantages are still unknown to a large part of the population. With a fundamental role in the management of hotel establishments, RMr are responsible for ensuring that the launched price of the product corresponds to the customer's expectation and disposition, taking into account the market competition and the optimization of revenues. RMr must have in-depth knowledge of price, supply and demand, depending on market trends. They must be specialist in marketing techniques, strategic management and operational management. They also must know how to manage people, how to negotiate, how to motivate and how to lead and influence their team, having attention to detail and an enormous capacity for analysis.

The first challenge in RM is to acquire computerized RM systems for a large number of forecasts, given that in a hotel these RM systems must identify the various categories of the required forecasting and help the entire price management process. In the present study, CRM (Customer Relationship Management), PMS (Property Management System), RMS (Revenue Management System), CRS (Central Reservations System), CM (Channel Manager) and RS (Rate Shopper) were addressed.

Most of the respondents consider themselves to have specific training in RM on a regular and high degree, however when analyzing their opinion regarding the skills of an RMr, it was found that respondents do not have solid information about the skills that an RMr must perform. In what concerns the need for training on RM, most respondents feel this need. There was a clear need for respondents with specific training to feel more in need of training. However, half of respondents with little or no specific training felt the same need. Regarding the existence of RM department of the hotel units, it was realized that although the RM is still under development, many units already have the RM department inserted in their organization chart. With the exception of accommodation, very few hotels have expanded RM techniques to other areas such as F&B, SPA, Golf and Health Club, among others. Despite the respondents' extreme need for RM training, the hotel units are not yet willing to train their employees in these areas, since most units that do not have an RM department or RMr do not plan to hire a professional with this training in the next 12 months.

Most hotel units do not use the RMS, which allows them to maximize profits, reaching maximum occupancy at the highest possible rate during the year. However, most units have already joined the price management support systems. The implementation of RM in hotel units was necessary due to the influx of tourism and in turn direct competition, and therefore it was essential to resort to new techniques.

As the starting question was "Does RM have an impact on hotel growth?", everything leads to conclude that it does. Respondents feel the need for training and consider that RM was a factor that not only helped companies to survive, but forced and obliges them to be up-to-date, have adequate software and to try to have their professional staff with specific training in this area. The application of RM strategies was highlighted by the growth of the hotel sector, since it allowed the increase of the RevPar as well as making it possible to maximize the return of each stay to the maximum. RM increased profitability and helped to optimize and maximize annual revenues, having significantly contributed to the weight of the hotel sector in Portugal. In addition to boosting revenue growth in the sector, companies are aware that RM is important, although, curiously, they do not plan to hire qualified Human Resources in this area.

REFERENCES

Altin, M., Schwartz, Z., & Uysal, M. (2017). Where you do it matters: The impact of hotels' revenue-management implementation strategies on performance. *International Journal of Hospitality Management*, *67*, 46–52. doi:10.1016/j.ijhm.2017.08.001

Aubke, F., Wöber, K., Scott, N., & Baggio, R. (2014). Knowledge sharing in Revenue management teams: Antecedents and consequences of group cohesion. *International Journal of Hospitality Management*, *41*, 149–157. doi:10.1016/j.ijhm.2014.05.010

Beck, J., Knutson, B., Cha, J., & Kim, S. (2011). Developing Revenue managers for the Lodging Industry. *Journal of Human Resources in Hospitality & Tourism*, *10*(2), 182–194. doi:10.1080/15332845.2011.536941

Cetin, G., Demirciftci, T., & Bilgihan, A. (2016). Meeting Revenue management challenges: Knowledge, skills and abilities. *International Journal of Hospitality Management*, 57.

Costa, M. J. (2017). *Total Hotel Revenue Management em Portugal* (Master Thesis). Escola Superior de Hotelaria e Turismo, IPP, Porto, Portugal.

Costa, V. F., Monte, A. P., & Fernandes, P. O. (2013). Tourism revenue for the north region of Portugal: an econometric analysis. *III Congresso International de Turismo, ESG/IPCA*, 241-259.

Cross, R. G., Higbie, J. A., & Cross, Z. N. (2011). Milestones in the application of analytical pricing and revenue management. *Journal of Revenue and Pricing Management*, *10*(1), 8–18. doi:10.1057/rpm.2010.39

Curtis, M., & Zahrn, F. (2014). Forecasting for Revenue Management: An Introduction. *The International Journal of Applied Forecasting*, *34*, 32–38.

Faria, A. G. (2015). *O Departamento de Revenue Management de um grupo hoteleiro: Análise dos Tivoli Hotels & Resorts*. Escola Superior de Turismo e Tecnologia do Mar, Instituto Politécnico de Leiria.

Gilliland, M. (2010). Defining "demand" for demand forecasting. *Foresight: The International Journal of Applied Forecasting*.

Gomes, J. (2004). *Como aumentar as receitas de um negócio: o conceito de gestão de receitas*. Revista Hotéis de Portugal.

Hayes, D. K., & Miller, A. A. (2011). *Revenue management for the Hospitality Industry*. John Wiley & Sons, Inc.

Huefner, R. J. (2014). An introduction to revenue management: Exploring common techniques and the CPA's role. *The CPA Journal*, *84*(6), 17–21.

Ivanov, S. (2014). *Hotel Revenue Management: From Theory to Practice*. Zangador. Zangador Ltd.

Kasavana, M., & Brooks, R. (1998). *Front Office Procedures* (5th ed.). Educational Institute of the American Hotel and Motel Association.

Kaswengi, J. (2011). Revenue management: fermeture d'allotements et coût d'opportunité dans le tourisme B to B. *Revue Management et Avenir*, (43), 57-76.

Kimes, S. E. (2008). Hotel Revenue Management: Today and Tomorrow. Cornell University School of Hotel Administration, 8(14).

Kokaz Pucciani, K., & Murphy, H. C. (2011). An investigation of data management and property management systems in hotels. *Tourism and Hospitality Management, 17*(1), 101–104.

Kotler, P. (2000). *Marketing Management, Millenium* (10th ed.). Prentice Hall.

Kotler, P., Armstrong, G., Brown, L., & Chandler, S. A. (1998). *Marketing* (4th ed.). Prentice Hall.

KPMG. (2018). *Tomorrow's experience, today Harnessing a customer first approach in a changing world. KPMG Global Customer Experience Excellence report*. KPMG International Cooperative.

Legohérel, P., Poutier, E., & Fyall, A. (2013). Revenue Management for Hospitality & Tourism. Goodfellow Publishers Limited.

Littlewood, K. (2005). Special Issue Papers: Forecasting and control of passenger bookings. *Journal of Revenue and Pricing Management, 4*(2), 111–123. doi:10.1057/palgrave.rpm.5170134

Oliveira, W. (2000). *CRM e e-business*. Matosinhos, Edições Centro Atlântico.

Santos, N., & Fernandes, P. (2010). Análise e Previsão da Procura Turística na Região Norte de Portugal. *Proceedings of APDR*.

Taliani, E. C. (2015). Impacto del Revenue management en los Sistemas de Control de Gestión. *Revista Facultad de Ciencias Económicas, 24*(1).

Talluri, K. T., & Van Ryzin, G. (2004). *The theory and practice of revenue management*. Kluwer Academic Publishers. doi:10.1007/b139000

Tardieu, J. M. (2004). *Marketing et gestion des services*. Chiron.

Tranter, K. A., Stuart-Hill, T., & Parker, J. (2014). *An Introduction to Revenue management for the Hospitality Industry: Principales and Pratices for the Real World*. Pearson.

Yeoman, I. (2011). Ten years of revenue management. *Journal of Revenue and Pricing Management, 10*(1), 1–7. doi:10.1057/rpm.2010.53

Yeoman, I. S., & McMahon-Beattie, U. (2017). The turning points of revenue management: A brief history of future evolution. *Journal of Tourism Futures, 3*(1), 66–72. doi:10.1108/JTF-11-2016-0040

ADDITIONAL READING

Abrate, G., & Viglia, G. (2016). Strategic and tactical price decisions in hotel revenue management. *Tourism Management*, *55*, 123–132. doi:10.1016/j.tourman.2016.02.006

Carlsen, J. (1999). A systems approach to island tourism destination management. *Systems Research and Behavioral Science: The Official Journal of the International Federation for Systems Research*, *16*(4), 321–327. doi:10.1002/(SICI)1099-1743(199907/08)16:4<321::AID-SRES255>3.0.CO;2-5

Hovinen, G. R. (2002). Revisiting the destination lifecycle model. *Annals of Tourism Research*, *29*(1), 209–230. doi:10.1016/S0160-7383(01)00036-6

Law, R., Bai, B., Ritchie, J. B., Tung, V. W. S., & Ritchie, R. J. (2011). Tourism experience management research. *International Journal of Contemporary Hospitality Management*.

Legohérel, P., Fyall, A., & Poutier, E. (Eds.). (2013). *Revenue management for hospitality and tourism. Woodeaton*. Goodfellow Publishers.

Martin, B. S., & Uysal, M. (1990). An examination of the relationship between carrying capacity and the tourism lifecycle: Management and policy implications. *Journal of Environmental Management*, *31*(4), 327–333. doi:10.1016/S0301-4797(05)80061-1

Russell, R., & Faulkner, B. (2004). Entrepreneurship, chaos and the tourism area lifecycle. *Annals of Tourism Research*, *31*(3), 556–579. doi:10.1016/j.annals.2004.01.008

Timothy, D. J. (1999). Cross-border partnership in tourism resource management: International parks along the US-Canada border. *Journal of Sustainable Tourism*, *7*(3-4), 182–205. doi:10.1080/09669589908667336

Wang, X. L. (2012). Relationship or revenue: Potential management conflicts between customer relationship management and hotel revenue management. *International Journal of Hospitality Management*, *31*(3), 864–874. doi:10.1016/j.ijhm.2011.10.005

KEY TERMS AND DEFINITIONS

Entrepreneurship: It is understood as a process for designing and running new business which may be initially or not small.

Management Revenue: This is a strategy in management oriented to develop an adequate product to be sold in the market.

Market Change: The different shifts and challenges the market faces in its history.

Pricing Management: Is the process of amalgamating all perspectives necessary to make decision about prices products should be exchanged in the market.

Tourism Management: The discipline aimed at protecting the tourist destination as well as overcoming all obstacles that lead to its well-functioning.

Chapter 8
The Engineering of Territorial Tourism Study Case of Algeria

Mohammed El Amine Abdelli
https://orcid.org/0000-0002-5784-4306
University of Zaragoza, Spain

Nadia Mansour
https://orcid.org/0000-0001-7690-7096
University of Sousse, Tunisia

Smaili Nabila
University of Tizi-Ouzou, Algeria

ABSTRACT

Tourism is a vital sector that presents one of the ways of economic diversification of a country. It has become a major economic activity that plays an important role in bringing people, countries, and regions together. In addition, it is an area promoted as a strategic sector since it provides a high percentage of foreign exchange earnings and represents a major share of GDP. Also, it provides a large number of jobs, especially for the most vulnerable groups in the society, including women, young people, and the low-skilled. Boosting this sector would result in remedying the inadequacy of the budgets allocated to the sector to allow maximum investments and to improve the business climate.

INTRODUCTION

Economic diversification has been considered since the 1930s as an important issue for national trade and industrial policies. Having said this, it is safe to say that many economic contributions have shown the benefits of diversification in terms of risk reduction and the contribution of diversification to the development process. The external trade in Algeria has been strictly controlled by the state during the period of 1970 to 1986. After this date, and the sharp fall of the international oil prices, the Algerian state started to think about opening the economy to the external world; the state opened its market progressively,

DOI: 10.4018/978-1-7998-4330-6.ch008

rather than suddenly. This openness coupled with the political circumstances that existed in Algeria at the time has led to strong economic and social instability. In 1999 and 2000, Algeria fought against the social strife and escaped the black decade. Consequently, the government reconsidered the trade policy and the openness of the market. From 2000-2008, the external trade was characterized by freedom and free exchange zones. In 2009, Algeria reassessed the adhesion to the free exchange zones, and worked on the limitation of imports as one of the tools to boost domestic production" (Elhannani and al, 2018).

Furthermore, Algeria suffers from recession investments and they are looking for ways to revitalize. Also, with regard to the sector of hydrocarbons, the business tourism is a vital sector that presents one of the means to break with the dependence of the Algerian economy, as one of the most important oil-exporting countries. It is, therefore, one of the ways of diversifying our country's economy.

Today and in view of a favourable global context, the potential for supporting the tourism industry is significant in Algeria. Geography and history give together to Algeria significant assets. The diversity of its landscapes and the vestiges of its history offer it the opportunity to develop several types of tourism. Indeed, one of Algeria's assets -in terms of tourism- is the vast Sahara and huge Saharan areas which are classified as world cultural heritage (Tassili, Hoggars, M'zab Valley). The diversity of the landscapes found here vast areas covered to infinity, dunes, immense rocky hills, and mountains with majestic shapes and dotted with oases, allow the development of different forms of tourism.

Figure 1. Geographical map of the wilayas of Algeria
Source: www.wikipedia.org/Geographical map of the wilayas of Algeria (2018)

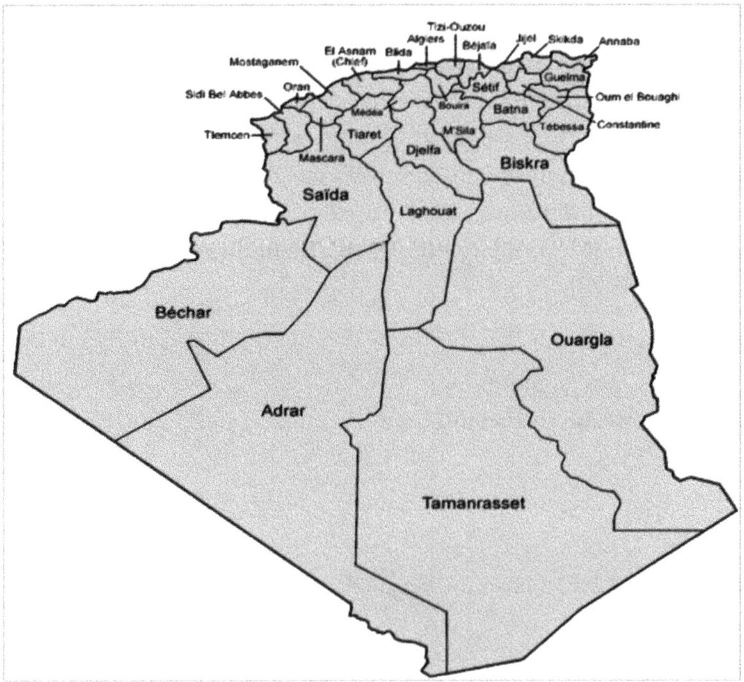

Our reflection began with a simple observation and this chapter aims to identify the specificities of the tourism sector in Algeria. Examining the strengths and weaknesses of this sector, its opportunities and threats will help to understand the current strategic directions. Therefore, we asked about the possibility of diversifying the economy in Algeria, by giving more interest to the tourism industry. The goal of this research is particularly adjusted to this problem. Since, on the one hand, tourism in Algeria has the resources to become a major destination on a regional or even global scale, today. The Mediterranean area represents the largest share of tourist flows worldwide, but Algeria is largely excluded. And, on the other hand, tourism development is obligatory and no longer a choice. The State, through the National Tourism Development Agency, claims to be the great prime contractor for tourism development as an essential factor in supporting the country's economic growth.

BACKGROUND

"Until a quarter-century ago, tourism was perceived as a privileged activity of a minor elite group, but today it became an ordinary activity of large communities. The development of tourism industry has been accelerated by the latest innovations created in transportation and technology, ubiquitous e-communications, disappeared geographical borders between the countries, curiosity of people with different cultures, and the people's increased expenditures on consumption as they prosper and have more leisure as a result of extended life expectancies, emanation of the healthy living concept, proliferation of the paid annual leave right, and decreasing weekly working hours, as well as increased number of new entrepreneurs participated to this industry in each of the mentioned areas" (Dincera and all., 2015).

At present, the record of international tourism actions is approved by European and North American countries. But, now, Asian and African regions have been increasing their parts in tourism.

In 2000 World Tourism grew by 7.4%, its highest growth rate in the decade and almost double in 1999. Total international arrivals reached a record 697 million. In the same year, revenues from international tourism reached 476 billion Euros, an increase of 4.5% compared to 1999. All regions of the world welcomed more tourists in 2000. Growth was particularly rapid in East Asia-Pacific (up 14.5% and 14 million more than in 1999).

Figure 2. World tourism towards 2030 (%)
Source: Association of The Mediterranean Tourism, 2012, p.3

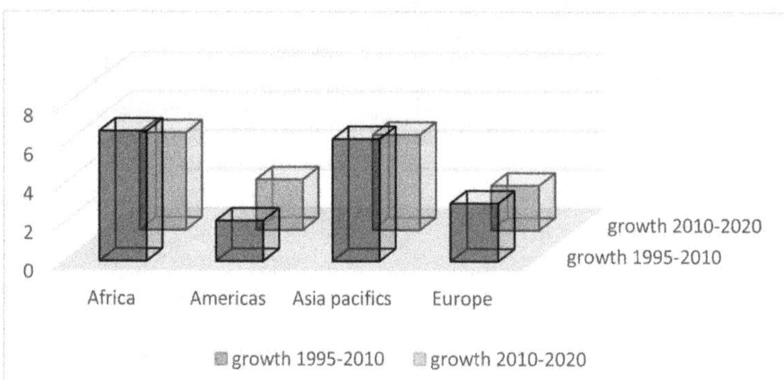

In 2001, the growth of international tourism, which seemed to be sustained, slowed down and international arrivals fell by 1,3% as a result of the terrorist attacks of 11 September 2001 and the economic downturn in the major tourism emitting markets. International tourist arrivals reached a total of 689 million in 2001 compared to 697 million in 2000. The last four months of 2001 saw a drop of more than 11% in arrivals worldwide. The last four months of 2001 saw a drop of more than 11% in arrivals in the whole world and significant declines in all regions: Africa (-3.5%), Americas (-24%), East Asia - Pacific (-10%), South Asia (-24%), Europe (-6%) and the Middle East (-30%).

After three years of stagnation, tourism recorded a spectacular rebound from 2004. This year the number of international tourist arrivals increased by 10% compared to 2003, which was a particularly difficult year due to the S.R.A.S, the war in Iraq and the economic slump. In Africa the year 2004 was particularly favourable for destinations located in the northern part of the continent with an increase of 17%, while destinations in the sub-Saharan part recorded only a 1% increase.

The year 2007 exceeded expectations with international tourist arrivals reaching 898 million, an increase of 6% in 2007 compared to 2006. The results confirmed both the sustained growth of recent years and the sector's elasticity to external factors. This growth was supported by the robust global economy, which for more than two decades has enjoyed its longest period of steady growth. In 2007, North Africa (8%) appears to have performed slightly better than Sub-Saharan Africa (7%), mainly due to the 14% increase in Morocco.

In 2009, global international tourist arrivals fell 4.3% to 880 million. "The global economic crisis, combined with the uncertainty of the H1N1 pandemic, has made 2009 one of the most challenging years for the tourism sector," said Taleb Rifai, UNWTO Secretary-General and he added: "2013 was a great year for international tourism. The tourism sector has demonstrated its ability to adapt to market developments, boost growth and create jobs around the world, despite persistent economic and geopolitical challenges. Indeed, for many economies, tourism has been one of the few industries to bring positive news".

As stated, "Tourism is one of the most remarkable economic and social phenomena of the century that is coming to an end. There is no doubt that it will retain this place in the coming century. (...) In most countries, tourism is now one of the most dynamic and fastest-growing sectors of the economy"(World Tourism Organization (O.M.T), 2000).

So, what is the definition of tourism?

TOURISM: WHAT DEFINITION?

For those who want to write about tourism, the most difficult thing is to define it (Boyer, 1982). To facilitate better reading and understanding of this chapter, we have considered it important to clarify certain terms that may confuse. For Leiper (1979) "Three approaches to the topic: economic, technical, and holistic are identified and analysed; it is argued that its multi-facets require a holistic definition. A systems methodology is used to develop a new definition of tourism. Five elements are isolated: tourists, three geographical elements (generating region, transit route, and destination region), and tourist industry. The process of tourism is dissected to show that it is inherently a partially-industrialized one, and the tourist industry is shown to contain several sectors with functional and spatial connections across the system".

Hence, the tourism industry includes activities undertaken by persons during their travels and stays in places outside their usual environment for a consecutive period not exceeding one year, for leisure, business and other purposes. On a more technical level, tourism can be defined as the set of relationships resulting from the temporary travel and stay of people who travel for entertainment or relaxation. For the majority, tourism is "synonymous with holidays" (Lozato-Giotart, 1993). Moreover, "Leisure is only a part of free time (...), tourism is an important part of leisure"(Dewailly, 2006). For Cooper and Hall (2008), for understanding the complexities of the contemporary tourism experience, we should utilize the concept of "the tourism system". Also, "the tourism system consists of consumption and production and the experiences that are generated…This approach is usually termed a geographical system of tourism and consists of four basic elements as follows:

- A generating: which is the permanent residence of the tourist and the place where the journey begins and ends
- A transit route: which is the path through the region across which the tourist must travel to reach his or her destination
- A destination region: which the tourist chooses to visit and which is a core element of tourism
- The environment: that surrounds the other three regions.

For Virgil and Popsa (2014), "Business tourism corresponds to very different domains, its field enrolling alike: global diplomatic meetings and meetings between representatives of different countries, economic and trade negotiations meetings, meetings for the exchange of scientific and technical information and communication of the results of certain research and development projects, cultural and educational meetings, as well as events specific to various forms of cultural manifestation, sports competitions, etc. Business tourism is a form of tourism for commercial, governmental or educational purposes, with the recreational (leisure) part as a secondary motivation. There are many kinds of business tourism: individual trips, group trips, and displacements at events (Meetings, Incentives, Conventions, Exhibitions (MICE), team building and training trips. Business tourism operates with a wide range of business corporations: hotel chains, tour operator travel agencies, organizers of trade fairs and exhibitions, business centres, which are grouped in most countries in professional structures or associations bearing the name Convention Bureau.

For Rodica (2003), Business tourism includes broadly human travelling for purposes related to their work, more specifically business and commercial trips or another kind of trips, participation in events organized by various economic enterprises or administrative bodies for their representatives. What is more important, "one of the main directions of modernization and increase of competitiveness of any country's tourism offer consists of the diversification of additional services, by encouraging business tourism. If the spatial extent of international tourism activity in each country depends to a decisive extent on the distribution and quality of natural factors and anthropogenic elements that attract foreign tourists, business tourism development, especially viewed through the prism of economic effects, lies in a direct correlation with the volume, diversity and quality of additional benefits" (stanciulescu, Tigu, 2005).

In Algeria, statistical definitions are used. The statistics published by the CNES (2000) tourist flows and hotel infrastructure comes from the Ministry of Tourism. The main concepts used are as follows:

- **Visitor:** A visitor is any person entering Algeria, and not engaged in any paid profession.

This definition covers two categories of visitors: tourists and excursionists.

- **Traveller:** He is any person entering Algeria, whatever the reason for his travel; the place of his residence and his nationality, except for excursionists on a sea cruise.
- **Tourist:** A tourist is a temporary visitor; staying at least 24 hours in Algeria and whose reasons for travel can be grouped into leisure (holidays, health, studies, religion, sports, pleasure...), business, family, mission.
- **Excursionist:** An excursionist is a temporary visitor, whose stay in Algeria does not exceed 24 hours, including maritime cruise passengers, excluding travellers who do not legally enter the national territory, as well as border workers working in Algeria.

Figure 3. Tourism characteristics in relation to time, distance, boundaries and description of purpose of travel
Source: *Cooper and Hall (2008)*

Time Dimension				
Years				Study/ working abroad / educational travel/student exchanges
Months			Travel to vacation homes	Seasonal travel for holiday work
Weeks				International Vacations
Weekends		Travel to second Homes		International short-breaks
Days	Visits	Excursion	Business travel	
Hours	Commuting Shopping			
	Local	Regional	National	International
Spatial and boundary dimensions				

ECONOMIC REWARDS OF TOURISM ACTIVITY

Tourism has important economic proportions because it is a collection of industries, companies, resources, attractions... (Sellal, 1980). So, it would be difficult to identify the boundaries of the tourism segment because many services are used jointly by tourists and residents. Of the same, goods and services are intended for both tourists and residents (Leiper, 1993).

Moreover, it is possible to consider that tourism is simply a service sector, in the sense that it does not transform raw materials such as water, air, snow... As it does not ensure the transport of these products to the consumer; it is rather, the tourist consumer who comes to the product. The term tourism industry seems inappropriate, in the sense that the tourism product is intangible and therefore a service. In other words, tourism activity remains closely linked to service activity because tourism is an important factor in the proliferation of service activity (Boyer, 2005).

Besides, the economic analysis of tourism activity highlights the importance of the activities associated with it, hence the overall ripple effects that this activity can have on the rest of the economy; although these effects remain difficult to measure and quantify. In other words, tourism is a productive investment that revitalizes many other sectors of the economy. Also, tourism has an impact, particularly on the economies of developing countries, through several mechanisms: According to the International Labour Office (ILO), the travel and tourism industry represents more than 4% of the world's gross domestic product and more than 3% of the workforce. Taking into account the share of the economic contribution of associated activities, this industry represents 11% of the gross domestic product and 8% of employees worldwide in value, according to the same source. Let us focus, in the following, on some purely economic elements (Kadri, 2007)) that make tourism one of the levers for breaking with our country's dependence on hydrocarbon revenues, as present in the figure 4.

Figure 4. The effects of tourism in different areas
Source: *CHERIF K., (2017)*

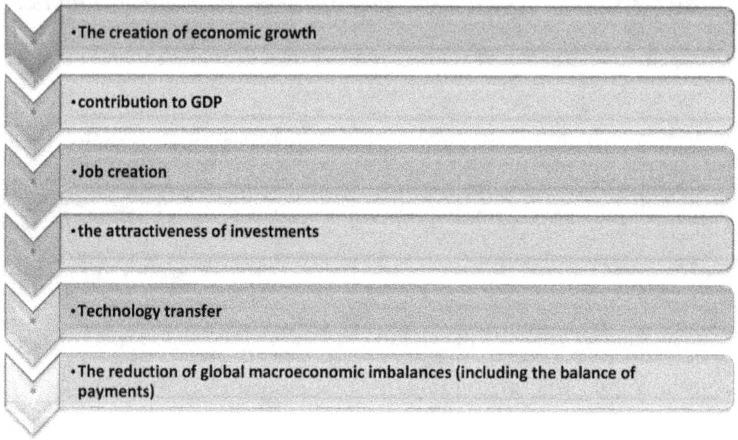

It should also be noted that tourism is a source of foreign exchange earnings, which makes it possible to vary the sources of income earnings and economic diversification, particularly in countries dependent on natural resource exports, such as Algeria. In the same vein, it should be noted that tourist spending constitutes a potential market for local products, particularly handicrafts.

TOURISM IN ALGERIA: STATE OF PLAY

Tourism is currently one of the most significant socio-economic phenomena. And according to the indicators, that it will maintain this position in the future. At the regional level, the Mediterranean basin, with its 30 countries, is the world's leading destination with 292 million international tourist arrivals according to the WTO (33% of the world total) in 2009, while forecasts predict the arrival of 400 million tourists by 2020 with$248 billion (179 billion euros) in international tourism revenues (29% of the world total). Algeria's thousand-year history, as evidenced by the impressive historical remains scattered throughout its vast territory, can help the development of a particular type of tourist by re-discovering the different moments that have marked the history of their country, and for foreigners to discover the prestigious past of a country that has often been presented to them as lacking historical depth. Algeria's natural assets and cultural, historical and heritage allow the practice of many tourist activities. Algeria has been a member of the WTO (World Tourism Organisation) since 1976, but tourism in Algeria is still limited. The country ranked 147th out of 174 countries in the world in 2008. The tourism sector in Algeria represents 9.3% of the volume of exports, 5.9% of the rate of productive investments and only 1.8% of the GDP.

The Algerian tourism industry is currently characterized by:

- Low contribution to the constitution of national wealth;
- A low tourist offer with a very poor quality of service delivery;
- Lack of well-defined tourism policy and strategy;
- The almost total absence of training programs in tourism occupations, etc.

Over the past decade, Algeria has made remarkable progress in catching up economically. Algeria's challenge now is to quickly find its way to growth. In the medium and long term, it will be necessary to create additional wealth beyond oil revenues, raise living standards and distribute them while avoiding factors that would block growth. Another challenge for the state is to deal with the informal economy. Trade liberalization policies implemented under the World Trade Organization had contributed to the expansion of the informal sector because liberalization is synonymous with increased competition for companies in the formal sector. To remain competitive, companies are striving to reduce costs by benefits to employees, by replacing permanent workers with part-time or by subcontracting certain jobs to the informal sector, and in particular to people working from home (crafts for example). Unfortunately, business tourism led to the creation of activities easy to exercise in clandestine workshops (guides, crafts activities…). To this end, Algeria has adopted a reform strategy that is part of a proactive macroeconomic policy oriented towards tourism for sustained growth and job creation. Algeria is the largest country in Africa and the 9th in the world, which gives it a variety of natural, historical and cultural resources. Through its geographical diversity, its triptych sea-mountain-desert, the country offers varied natural landscapes. Statistics indicate that Algeria's share receives no more than 1% of world tourist flows, ranking 147th out of 174 in the world ranking. It should also be noted that just over 2.6 million tourists arrived in 2012and the majority of them have nationalities from neighbouring countries among the 1.035 billion tourists in the world (UNWTO).

Economically, the tourism sector has only 7.3% of the value of total exports, a rate of 9.5% of the share of productive investment and 2.3% of GDP for 2010. From the early 2000s to 2015, the rise in hydrocarbon prices meant that the country did not consider it necessary to invest in areas other than the oil and gas sector. However, the non-hydrocarbon economy has long been compromised: decline in the industrial, agricultural and tourism sectors, too (it is also in this sense that much academic work has focused on the curse on natural resources in our country). It should also be noted that tourism investments are being delayed due to the slow administrative procedures (it turns out that out of 500 hotel projects in 2015, 340 are awaiting building permits). Rightly, it seems that there is an official willingness to do so, texts exist but the results remain limited. In Algeria, the delay in implementation is not necessarily due to studies, nor to the actual completion times, nor the duration of training; rather, it is the administrative processing times that determine the duration of implementation. And at this level, it should be emphasized that better management of deadlines would require improvement of institutional quality in general. Because of this, complementary tax measures should be implemented through the Finance Laws, actors in tourism demand a reduction in the rate of Value Added Tax in line with what has been done in other countries (Turkey). More generally, it should be noted that the development of the tourism sector in Algeria does not present a choice; rather, it seems to be an imperative for the Algerian economy, which is seeking a way out of its dependence on a single sector. It is rather an alternative to the crisis. Indeed, the tourism sector presents a privileged target for the country's economic policies. For this reason, the public authorities will set up a Master Plan for Tourism Development (SDAT) in 2008, based on the National Spatial Development Plan (SNAT) 2025, to promote "Algeria as a destination".

The Tourist Area: A Continent in a Single Country

At the end of 2017, the Algerian Minister of Tourism announced that 10,000 foreign tourists had visited southern Algeria since the end of September (the start of the tourist season in this region). The figure seems very low, but it shows a significant increase compared to 2016, which recorded only 7,500 visits.

The weakness of these figures shows the scale of the task required to bring tourism in Algeria up to the level of that of its main neighbours: Tunisia and Morocco. For proof, tourism would represent only 1.4% of Algerian GDP. But today, oil rents are no longer enough and economic priority is the diversity of production. In this context, tourism becomes a key sector. Hence we note new ambitions of the Algerian State.

4,4 million international tourists are expected in Algeria in 2027, against 2.4 million today, according to the WTTC (World Council for Tourism and Travel). Still quite, low targets for a country of this size. By comparison, Morocco recorded 11 million visitors for 2017. Algeria has invested only modestly in this industry. Today, public discourse seems to change, even if the objectives remain modest. Thus, the Minister of Tourism and Crafts, announced that 1812 projects to realize new hotel infrastructure had been approved by the ministry. This series of projects should increase the capacity to 240,000 beds, compared with 100,000 at present, and create 99,000 jobs. The number of nights spent by foreign tourists in hotels in Algiers reached 220,720 nights (from the beginning of the year 2018 until last June), which bodes well for the good foreign currency revenues earned by these hotels. The head of wilaya of Alger (2018) claimed that 581 visa applications were filed by foreigners wishing to visit the capital as a tourist. However, there is no shortage of tourist resources. Algeria extends over 1,622 km of Mediterranean coastline, and extends for more than 2000 km into the African continent, in the heart of the Sahara. Three major physical entities characterize the country:

- The Tell, 4% of the territory,
- The Highlands, 9% of the territory,
- The Saharan domain, 87% of the territory.

This physiographic variety, and the great diversity of landscapes and ecosystems, partly explains the richness of natural resources. The understanding and the acceptance of the principles of biodiversity and sustainability by a growing portion of the population was also involved in the development of ecotourism. It becomes synonymous to the benefit of nature, landscapes and species while respecting the ecosystem balance and antonyms to mass tourism (Grimes and al, 2017).

Also, Algeria is an immense country which gives it many types of climates and reliefs: coastline, cliffs, mountains, rivers, forests, steppes, endless desert expanses with oases. Besides, the country has 21 protected areas and natural parks spread throughout the country and covering more than 53 million hectares, most of which are located in desert areas. For example, The Tassili National Park (Illizi) covers 8 million hectares and is home to several protected animal and plant species such as dorcas gazelles,

Duperez cypress, acacia, etc. Algeria also enjoys a Mediterranean climate in the north in the coastal areas and a continental climate in the inland regions. In the south, the climate is dry and tropical with very wide temperature differences between day and night. The hot and humid summer months, with temperatures between 30° and 38° C, are ideal for swimming by the sea. In winter, the climate is mild and suitable for winter sports in the northern mountains where it snows. It is also the ideal time to visit the Sahara regions.

In the following photos, we can see a set of the main premises to visit in Algeria:

Figure 5. Top 25 tourist sites in Algeria
Source: https://www.cityzeum.com/tourisme/a-voir-a-faire/algerie.

1- Kasbah of Algiers
2- The Great Post Office of Algiers
3- The Algerian Sahara
4- Test Garden
5- Dardiaf: Ghardaïa M'Zab
6- The Saharan Atlas
7- Timgad
8- Babors
9- The Tadrart
10- Djemila
11- Ghardaïa
12- M'Sila
13- The Mzabregion
14- The dunes of Tin Merzouga
15- Tindouf
16- Soummam Abane-Ramdane de Béjaia Airport
17- Kabylia
18- Vgayeth Bejaia Candle
19- The National Museum of the Bardo
20- Martyr's Sanctuary
21- Cathedral of Our Lady o fAfrica
22- City of Annaba
23- City of Oran
24- Tamanrasset Region
25- Village of Tipaza

PROMOTION OF THE TOURISM SECTOR IN ALGERIA

The promotion of the tourism sector with regard to the National Tourism Development Agency (SDAT) is based on a strategy oriented towards tourism expansion areas. It should be noted that the SDAT is established by Act 02- 01 of 12 December 2001 on spatial planning and sustainable development. It is important to underline that the main lines of the SDAT take into account the lessons learned from neighboring countries as Tunisia, which have achieved remarkable successes in this area. The Tunisian tourism sector is well structured. In fact, despite the Arab Spring Storm and the different repercussions, this sector has made evolutions these 3 last years.

The tourism in Tunisia also presents the first competitor to Algerian tourism and especially of Sahara. Tourist tours are considered as the main attraction that the Sahara region offers. "This form of tourism concerns the Gafsa, Tozeur tourist zone. The Djerba, Zaris spa pole plays an important role in organized sightseeing tours in the Sahara regions of Tunisia. An analysis of the Saharan tourism image, given by foreign tour operators in their publicity brochures, shows that the organized trips and excursions represent the subject the most approached by travel agencies" (Souissi, 2011).

Figure 6.

The South of Tunisia is a strategically important region for sustainable tourism that allows the passage of seaside tourism, based on a single product, to the Saharan tourism developing multiple sub-producing and enhancing other natural and heritage resources. The setting in tourism of the Saharan natural and cultural heritage offers it a chance of conservation, the move towards a sustainable role that balances economic profitability with social equity and contributes to the creation of a rapprochement between all actors. The economic and social integration of circuit tourism in the Saharan region presupposes that it coexists with the traditional sectors to promote them. It is supposed to benefit the majority of the local population to create more jobs and power (Souissi, 2011). At this level of innovation, and development of this type of tourism that Tunisia has advanced compared to Algeria, given the services offered for tourists, hotels and the role of travel agencies. Despite the non-compliance of the prices proposed in Algeria with those proposed in Tunisia or other countries (competitive services with low prices).

Figure 7.
Source: WTO (2000,2007, 2012), OMT (2004, 2009, 2014)

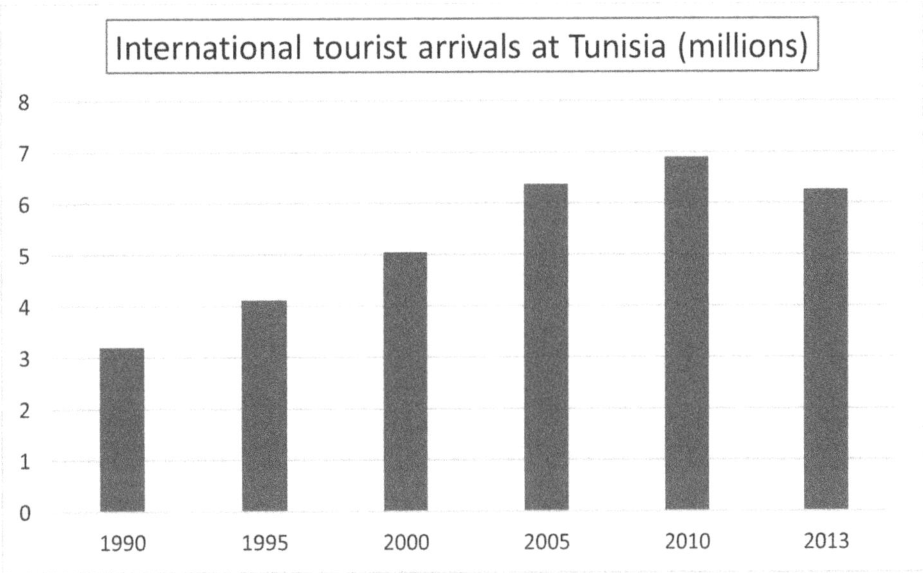

However, tourism is gradually becoming a pillar in bilateral economic relations between Algeria and Tunisia. In this context, and with all the commonalities between Tunisia and Algeria as some natural factors, the deserts…Algerian Minister of Tourism and Crafts, told the press, on the sidelines of a national meeting of tourism professionals, that 1.7 million Tunisian tourists had visited Algeria in 2019. While more than one million Algerians visited their Tunisian neighbours during the same period.

In 2018, Algerians had already established themselves as the first origin of tourists, with more than 2.7 million admissions, an increase of +9.2% compared to 2017. The trend has been noticeable since 2011, as Algerians come to the rescue of Tunisian tourism that has been ignored by Europeans and whose tour operators have been restricted by their governments. For this reason, for meeting Algerian demand which has supplanted traditional European markets, the Tunisian tourism sector has adapted. Tariff barriers have been revised downwards to become more accessible to Algerian purchasing power, with special attention paid to family offers. In sum up, to improve the reputation of Algeria and attract tourists, we must start by a territorial marketing approach:

- Websites focus on promoting the desert and cultural discovery;
- Few tour guides exist on Algeria (some very recent editions), so it is necessary to think about training schools in the field of tourism;

- Tourist brochures produced by the Tourism Directorates are sometimes high quality, but not sufficiently disseminated;
- The tools to monitor tourism activity are insufficient today;
- Information is scattered and impractical for both the consumer and the prescriber; Algeria's huge pool of natural sites. This plan is based on five basic elements:
- Make Algeria a competitive tourist destination;
- Develop tourist centres of excellence;
- Introduce quality controls;
- Encourage public-private partnerships; -Attract investment in general and FDI (Foreign direct investment) in particular to create added value and employment.
- More generally, the overall objectives assigned to the SDAT can be summarized as follows:
- Make tourism attractiveness one of the drivers of economic growth, as an alternative to the vulnerability of the hydrocarbon sector;
- Develop the other related sectors of economic activity;
- Enhance and preserve the historical and cultural heritage by adopting a sustainable local development approach;
- The most appropriate tool to achieve the objectives set out in the field of tourism seems to be expansion areas and tourist sites;
- Combine tourism promotion with the environment and thus develop ecotourism

On this last point, Grimes and al (2017) presented the method of biodiversity as a solution for the problem of pollution and human activities on the environment result of tourism. Indeed, "The lack of sensibility towards the environment is revelled throw the lack of good citizenship and the gesture which consist of getting rid of waste anywhere on the site after use. The fact is so serious since that goes till leaving all the strewn on the sand of the beach after departure. On the other hand, the relationship between tourism and the natural environment is ambivalent and confrontational by the fact that the existing potential for a relationship based on reciprocal benefits destroyed the cooperative relationship that is supposed to exist through ecotourism".

Algeria has 205 Expansion Area and Tourist Sites, divided by region (Master Plan for Tourism Development 2025, (2008)):

- On the coast, there are 160 Expansion Area and Tourist Sites, covering an area of 3706.5 hectares.
- In the Highlands, there are 22 Expansion Area and Tourist Sites, covering an area of 6464.83 hectares.
- In the south (Ghardaïa, Ouargla, Biskra, El Oued, Béchar), there are 15 Expansion Area and Tourist Sites, covering an area of 786.2868 hectares.
- In the Far South, (Tamanrasset, Adrar, Illizi), it is possible to count 8Expansion Area and Tourist Sites on a surface of 8941.92 hectares.

However, to conclude this discussion, it is worth mentioning some obstacles to the development of tourism activity, namely the existence of a real political will that can commit the necessary means to implement the projects defined, in theory. In concrete terms, this means addressing credit problems, while at the same time fighting bureaucratic delays and corruption.

State should invest in human resources, as well as in infrastructure, particularly urban infrastructure (transport, developed cities, etc.), by encouraging private investment. It should also increase the number of tourism schools throughout the country. Civil will also impose a constraint in this area; hence the need to take charge of human resources in general. The aim is to train the workforce in tourism professions in detail and to raise citizens' awareness of how to be civic in order to promote their city and enhance its potential. For Dincer and al, (2017: 17), "The most important factors which play significant roles in gaining an economic competitive advantage can be summarized as lower-costs in the production process, higher quality, more profitable customers, and a faster service provided to the customers. In these factors, the higher amount of human capital - both as entrepreneur and employee - in tourism is very crucial".

CONCLUSION

International demand is highly competitive as it is marked by variety and the diversity of existing products and offers (call products, consumer products, discovery and opportunity) which lean on the seaside product. It is finally modified by structural changes such as the reduction of working time which induces the multiplication of short stays and hence a staggering of the tourist season. Now, international demand is based on new articulated tourism Drivers on the creation of a new generation of quality products for different audiences (SDAT, 2025). It appears that one of the levers of Algerian economic diversification is the development of the tourism sector. This sector would be one of the pillars of diversification and economic dynamism. This is dictated in particular by the potential spillover effects that the development of tourism activity could have on the rest of the economy in terms of job creation and the attraction of external income. In this token, Algeria can make tourism practice in all the year, thus escaping the seasonality that affects the tourism sector in general. In fact, Algeria does not suffer in any way from the problem of seasonality, but the major challenge is to adapt each tourist sector in the best way to the demand of the domestic and international markets and to target a more promising and advantageous tourist product. For example, The Casbah is the soul of the capital and is the privileged destination for all tourists. Also, it is very important to prepare the tourist circuit of the Casbah-Alger with the rehabilitation of craft activities, which will increase the number of foreign visitors, according to the director of tourism of Wileya (2018). The opening of three hotels in 2018 in the municipalities of Chéraga, DelyBrahim and Bab Ezaouar for improving the services offered to tourists, adding that 7 other hotels (from 3 to 5 stars), will be opened soon in Bab Ezaouar, Ben Aknoun, Ain Beniane, KrimBelkacem and Dar El Beida with a capacity of 3300 beds. Moreover, the significant delay in meeting and demand requires a multifaceted use of available resources to diversify and improve the Algerian tourist offer. Considering the country's assets, there are many possible forms of tourism: seaside, archaeological, cultural, religious, sporting, hunting, climatic, rally. We believe that this approach to tourism can and should be generalized. Each type of tourism must constitute an upstream response to a tourist demand and a downstream response to a given situation. Thus, the implementation of the projects outlined in this field would undoubtedly require political and civil will, without which the country's geographical and cultural potential could not be transformed into tourism products. These products would be able to satisfy local demand quantitatively and qualitatively, and to make Algeria a competitive and attractive destination for foreigners. Last but not least, the Algerian Minister of Tourism and crafts, the responsible stressed the need to assemble all the means and all the necessary conditions for the success of the Saharan tourism season 2019-2020. Thus, he highlighted the "unique natural potential in the world" that the Algerian Sahara has, added to the availability of experienced professionals and to the public and private tourism infrastructure realized in the south of the country.

REFERENCES

Bouchair, A. (2015). Vernacular architecture: Towards a sustainable future. CRC Press, Taylor and Francis Group.

Boyer, M. (1982). *Le tourisme*. Editions du Seuil.

Boyer, M. (2003). *Le tourisme en France*. Editions EMS.

Cherif, K. (2017). *Economie du tourisme - un investissement d'avenir pour l'Algérie*. Casbah Editions, Alger.

Cooper, C., & Hall, C. M. (2008). *Contemporary tourism: an international approach*. Elsevier.

Dabour, N. (2003). Problems and prospects of sustainable tourism development in the OIC countries: Ecotourism. *Journal of Economic Cooperation, 24*, 25-62.

Davidson, R., & Cope, B. (2003). Business Travel. Prentice Hall.

Demen-Meyer, C. (2005). Le tourisme: Essai de definition. *Management & Avenir*, *3*(1), 7–25.

Dewailly, J. M. (2006). *Tourism & Geography: between peregrinity and chaos*. Harmattan.

Dincera, F. I., Yimaza, S., & Dincera, M. Z. (2015). The Economic Contribution of Turkish Tourism Entrepreneurship on the Development of Tourism Movements In Islamic Countries. *Procedia: Social and Behavioral Sciences*, *195*, 413–422. doi:10.1016/j.sbspro.2015.06.483

Elhannani, F. E., Bousalem, A. B., & Benbouziane, M. (2018). Economic Diversification and Trade Openness in Algeria Empirical Investigation. *Topics in Middle Eastern and African Economies Proceedings of Middle East Economic Association*, *20*(1), 1–09.

Grimes, S., Bouchair, A., & Tebouche, H. (2017). Sustainability of the Expansion Areas for Coastal Touristic Sites "E.A.C.T.S" Such as the case of El-Aouana in Algeria: Indicators for considering biodiversity. *Energy Procedia*.

Kadri, B. (2007). La ville et le tourisme: Relation ancienne, complexité nouvelle et défi conceptuel. *Téoros*, *12*(1), 76–79.

Leiper, N. (1979). The Framework of Tourism: Towards a Definition of Tourism, Tourist, and the Tourist Industry. *Annals of Tourism Research*, *6*(4), 390–407. doi:10.1016/0160-7383(79)90003-3

Leiper, N. (1993). Defining tourism and related concepts: tourist, market, industry and tourism system. In *VNR'S Encyclopedia of hospitality and Tourism*. Springer.

Lozato-Giotart, J.-P. (1993). *Géographie du tourisme: de l'espace regardé à l'espace consommé*. Masson.

Ministry of Regional Planning. (2008). *Environment and Tourism, "Schéma Directeurd' Aménagement Touristique 2025*. Official Document.

Ministry of Tourism and Handicrafts. (n.d.). *Synthesis of tourist flows in Algeria in 2012*. Author.

Minuciu, R. (2003). *Tourism economy*. Ed. Uranus.

Muhana, E. (2006). Sustainable Tourism Development and Environmental Management for Developing Countries. *Problems and Perspectives in Management*, *4*(2), 14–34.

Report of the National Economic and Social Council (CNES). (2000). *Contribution to the redefinition of national tourism policy*. Author.

Sellal, M. (1980). *Production touristique*. Société Nationale d'Edition et de Diffusion.

Sharpley, H. (2002). The challenges of economic diversification through tourism: The case of Abu Dhabi. *International Journal of Tourism Research*, *4*(2), 221–235. doi:10.1002/jtr.378

Sorsa, P. (1999). *Algeria- the real exchange rate, export diversification and trade protection*. International Monetary Fund, Working paper N° 99/49.

Souissi, M. (2011). Le tourisme de circuit dans le Sahara tunisien: réalités et perspectives. *Insaniyat. Revue algérienne d'anthropologie et de sciences sociales*, (51-52), 271-287.

Stnciulescu, G., & Tigu, G. (2005). Business strategies for tourism development in Romania. *Journal of Tourism, 2*.

Swarbrooke, J. (2000). Sustainable tourism management. CAB international.

Temmar, H. (2015). The Algerian economy: 1970-2014. In The public economic governance, University Publications Office.

Temmar. (2015). The Algerian economy: 1970-2014. In *The development strategies*. University Publications Office.

UNWTO. (2013). *World Tourism Barometer* (vol. 11). UNWTO.

Vellinga, R., & Abdelgalil, L. (2007). Trade liberalization and economic diversification: A dynamic CGE Model for the economy of Qatar. *Studies in Business and Economics.*, *13*(02), 1–17. doi:10.29117be.2007.0034

Virgil, N., & Popsa, R. E. (2014). Business Tourism Market Developments. *Procedia Economics and Finance*, *16*(1), 703–712.

World Tourism Organization. (2000). Highlights 2000 (2nd ed.). Author.

World Tourism Organization Guidebook. (n.d.). *Sustainable tourism for development*. Author.

Zhang, J. J., Joglekar, N., & Rohit, V. (2010). Developing measures for environmental sustainability in hotels: An exploratory study. *Cornell Center for Hospitality Research*, *8*(1), 1–17.

ADDITIONAL READING

Buckley, R. (2012). Sustainable tourism: Research and reality. *Annals of Tourism Research*, *39*(2), 528–546. doi:10.1016/j.annals.2012.02.003

Butler, R. W. (1999). Sustainable tourism: A state-of-the-art review. *Tourism Geographies*, *1*(1), 7–25. doi:10.1080/14616689908721291

Clarke, J. (1997). A framework of approaches to sustainable tourism. *Journal of Sustainable Tourism*, *5*(3), 224–233. doi:10.1080/09669589708667287

Farrell, B. H., & Twining-Ward, L. (2004). Reconceptualizing tourism. *Annals of Tourism Research*, *31*(2), 274–295. doi:10.1016/j.annals.2003.12.002

Hunter, C. (1997). Sustainable tourism as an adaptive paradigm. *Annals of Tourism Research*, *24*(4), 850–867. doi:10.1016/S0160-7383(97)00036-4

Korstanje, M. E., & George, B. (2015). The Imposible Sustainability: Applications Of Relational Perspective In Tourism Fields. *Ecoforum Journal*, *4*(1), 2.

Korstanje, M. E., George, B. P., & George, B. (2012). Sustainable Tourism and Global Warming: Panacea, excuse, or just an accidental connection? *Worldwide Hospitality and Tourism Themes*, *4*(4), 383–394. doi:10.1108/17554211211255710

Lane, B. (2005). Sustainable rural tourism strategies: A tool for development and conservation. *Revista Interamericana de Ambiente y Turismo-RIAT*, *1*(1), 12–18.

Liu, Z. (2003). Sustainable tourism development: A critique. *Journal of Sustainable Tourism*, *11*(6), 459–475. doi:10.1080/09669580308667216

Nedelea, A. M. Korstanje M, George B (Ed.). (2016). Strategic tools and methods for promoting hospitality and tourism services. Hershey, IGI Global.

Pearce, P. L. (2005). *Tourist behaviour: Themes and conceptual schemes*. Channel View Publications. doi:10.21832/9781845410247

Ritchie, J. B., & Crouch, G. I. (2003). *The competitive destination: A sustainable tourism perspective*. Walingford, Cabi. doi:10.1079/9780851996646.0000

Saarinen, J. (2006). Traditions of sustainability in tourism studies. *Annals of Tourism Research*, *33*(4), 1121–1140. doi:10.1016/j.annals.2006.06.007

Weaver, D. B. (2006). *Sustainable tourism: Theory and practice*. Routledge.

KEY TERMS AND DEFINITIONS

Development Theory: It is understood as a set of theories oriented to boost and recycle local economies.

Sustainable Tourism: An academic discipline emerged to give fresh and efficient responses to the problems of pollutions and contaminations generated by the tourism industry.

Tourism Consumption: It comprises the consumption of tourist experiences, products inserted in a specific territory or country.

Tourism Development: It is a net of techniques, theories and studies oriented to develop tourism industries in poor economies or countries.

Tourism Market: It is the encounter between tourist demand and offering.

Chapter 9
A Futuristic Approach to Sustainable Tourism Development:
Lessons From Kandovan Village

Hamed Rezapouraghdam
https://orcid.org/0000-0001-8190-6565
Bahçeşehir Cyprus University, Cyprus

Habib Alipour
https://orcid.org/0000-0001-5517-3118
Eastern Mediterranean University, Turkey

Arash Akhshik
https://orcid.org/0000-0001-6743-4725
Eastern Mediterranean University, Turkey

ABSTRACT

In spite of the enormous efforts and implementation of the masterplans for the development of rural areas in Iran, the majority face economic, social, and ecological challenges that jeopardize their future, particularly where tourism has been prescribed as a panacea. Among those, Kandovan, known as the only inhabited rocky village in the world that attracts half a million visitors, annually entails excessive consideration and as a case embraces imperative contributions to the sustainability-future research domain. This study aims to contribute to an enhanced understanding of the nexus between "sustainable development" and the "future" in tourist destinations. Through the analysis of the secondary data, this study takes a scenario-planning approach and discusses the current social, environmental, and economic themes: the foundation of four potential future scenarios. The underpinned framework in the context of Kandovan offers noteworthy implications both in theory and practice of tourism by proposing an area-based planning model according to the emerged alternative scenario.

DOI: 10.4018/978-1-7998-4330-6.ch009

INTRODUCTION

Tourism is considered as a potential future industry for particular communities, and it can serve different purposes in the contradictions between the use and conservation of natural resources (Moss, 2006). There has been tremendous global growth in nature-based, culture-based and heritage-based tourism (Balmford et al., 2009), particularly in small-scale rural communities where it is prescribed as a remedy to overcome the problem of centralization and regionalization and to achieve sustainable development (Perlik, 2006). The future is inherently medial in sustainability theory (Cameron & Potvin, 2016) as it refers to the long-term viability and integrity of human societies and natural systems to meet both the needs of the present and future generations (Brundtland, 1987).

However, as emphasized by Swart et al. (2004), the fundamental challenge in pursuing this sustainability is the search for potential future pathways that are viable under the influential circumstances of uncertainty, complexity and human choices. This is a process that can be explored via the scenario analysis approach, which refers to plausible alternative futures that result from the integration of trends in stimulating the relevant factors and policies (Yeoman et al., 2015). Although various development programs have been implemented at local level, majority of the rural areas in Iran are facing different ecological, physical and socio-economic challenges (Imani et al., 2017) where the issues of poverty, inequality, population migration, educational deficits and vulnerability against natural risks are evident in such areas (Maleki & Bigdeli Rad, 2017). This paper aims to present alternative future scenarios for one of the world's most unique and novel rural destinations, the Kandovan village, a national heritage archaeological tourism site in Eastern Azerbaijan province, Iran (Yahyavi & Shaghaghi, 2012) (Figure 1). Nowadays, this authentic stone village is a very popular tourist destination and attracts thousands of visitors annually (420,000 tourists in 2012 (Mohammadi, 2012), to experience a mixture of cultural and natural heritage that has a history of human habitation dating back more than 800 years (Fam & Keshavarz, 2015). In 5th of May 1997, Kandovan village was inscribed on the National Heritage List (reference number: 1857) (Irna, 2017). This destination is also endeavoring to share its fascination with a greater number of people worldwide, and to be included on the UNESCO World Heritage List; however, there are some obstacles which the major one is to reclaim its authenticity. Studies of this tourist destination, however, have reported serious environmental hazards and conservation issues (Eftekhari et al., 2011; Fam & Keshavarz, 2015; Razani et al., 2013), which indicate a lack of attention to these topics. As declared in "Our Common Future" document (WCED, 1987) sustainable development refers to "satisfying the needs of the present generation without compromising the ability of future generations to meet their needs" (WCED, 1987, ch 2, sec. 1, para. 1). Considering the unique archaeological characteristics of the study case at one hand and the various ecological risks this destination experiences on the other which endangers the sustainable development and the future of the site, the current study attempts to appraise the key trends of sustainability disposition in Kandovan village and to provide potential future visions through scenario planning approach. To better grasp the complexity of the future the scenario planning was recommended as a suitable approach (Yeoman et al., 2105). This is a timely discussion because the number of visitors to this village has been increasing that this has exacerbated the concerns of the local indigenous' socio-economic challenges and the environmental conservation difficulties. It also contributes to the extension of scenario planning approach (Yeoman, 2012a) to the study of heritage tourism and sustainable development in rural destinations in developing countries. Potential future pathways that are viable under the influential circumstances of uncertainty, complexity and human choices can be explored via the scenario analysis approach, which refers to plausible alternative futures that result from the integration of trends in stimulating the relevant factors and policies (Yeoman et al., 2015).

CONCEPTUAL BACKGROUND

Tourism and Sustainable Development

Nowadays, sustainability and sustainable development either within the tourism organizations (Rezapouraghdam, Alipour, & Arasli, 2019; Karatepe, Rezapouraghdam, & Hassannia, 2020) or tourism destinations (Rezapouraghdam, Alipour, & Darvishmotevali, 2018; Alipour et al., 2018) are the main source of focus among the scholars the entire world. Particularly, in the destinations where tourism is the main source of economy, the sustainability issues gain much more significance (Rezapouraghdam, Behravesh, Ari, & Lema Doh, 2018). The UNESCO World Heritage Convention postulates the link between a heritage site and the sustainable development of the area (UNESCO, 2016) which the current study uses this notion and integrates that with sustainable livelihood concept (Scoones, 1998) which concerns with promoting the essential characteristics of sustainability such as social equity, economic efficiency, ecological integrity, and resilience in local communities (Tao & Wall, 2009). The UNESCO underlines the significance of heritage in the mainstream sustainable development debate (WHC, 2015) and indicates that the nexus between the heritage and sustainability has a pivotal role in societies' achievement to social, economic and environmental goals. The utilization of this approach was recommended in the post-2015 development agenda of the UN that supports the relationship between heritages and the three pillars of sustainability; that is, the conservation of natural resources and outstanding heritage sites has a direct and principal contribution to the issue of environmental sustainability. Moreover, a 'well-protected' heritage can contribute to the alleviation of poverty and job creation, and can also provide services to the destinations that contribute to the economic and social aspects of sustainability discussion (UNESCO, 2015). This perspective also focuses on the advantages of a heritage (world heritage and heritage in general) concerning the well-being of present and future generations of people from different locations. For example, the conservation of a heritage and equitable access to that heritage, together with the production of benefits for all of the stakeholders, results in the development of a sense of attachment to the place and the creation of social cohesion (UNESCO, 2016). The scope of this approach, while indicating the diverse functionalities and contributions of heritages to the sustainable development of communities and their inhabitants (Alipour et al., 2020), very explicitly expresses the striking value of 'conservation'. Concerning heritage sites, the matter of conservation carries such importance that the 26th Session of the World Heritage Committee (Budapest, 2002, p.4) stressed that there should be an appropriate 'balance between conservation, sustainability and development'. Even the Operational Guidelines of the Sustainable Development Notions stated in 2005 that 'the protection and conservation of the natural and cultural heritage are a significant contribution to sustainable development' (paragraph 6). Sustainable tourism, in the long run, must be environmentally stable, economically executable, yet besides, it should be socially and morally ethical in connection to the indigenous populace. The development of such tourism fulfils the necessities of present-day stakeholders and secures any future conceivable outcomes. Sustainable tourism guarantees that the local community lifestyles and habits are significant and the locals should proportionally benefit from the positive economic impacts of tourism without exploiting the natural and constructed environment and socio-cultural assets of the society (David, 2011). On another side, the participation of the local communities in tourism has been described as a significant element for the sustainable development of the destinations which creates them better opportunities to gain more and balanced benefits from tourism (Tosun, 2000) and builds positive attitudes toward tourism development and conservation of resources (Inskeep, 1994). Tourism, if managed sustainably, can enhance the local livelihood and give diversification to the sources of income for the community otherwise it can interrupt the conventional economic activities of locals and reduce livelihood diversity (Tao & Wall, 2009). Referring to Chambers and Conway (1992), Scoones, (1998, p. 5) delineated that "a livelihood comprises the capabilities, assets (including both material and social resources) and activities required for a means of living. A livelihood is sustainable when it can cope with and recover from stresses and shocks maintain or enhance its capabilities and assets, while not undermining the natural resource base". The tourism activities are preferred to be incorporated into a community's current mix of livelihood strategies to enrich rather than to replace people's traditions (Tao & Wall, 2009). A sustainable livelihood approach intrinsically uncovers the multi-sectoral character of real-life, integrating ecological, social and economic issues into a holistic frame, which is a chance to advance the kind of cross-sectoral and cross-topical approach that ought to be the sign of sustainability (Helmore and Singh, 2001). Accordingly, the contribution of heritage-tourism in a context to the sustainability can be regarded positive when it integrates into the local's livelihood as an alternative and supports the conservation of the heritage and environmental assets of the society, promotes the social well-being of the local indigenous and the community and sustains the economic prosperity of the destinations as well (Toker & Rezapouraghdam, 2019). Regarding the aforementioned frame, in the next section, the authors based on the shreds of evidence collected from the literature, analyze the current circumstances in the study setting.

The Study Area

Located in the remote north-west of Iran (674 km distance from Tehran), 60 kilometers from Tabriz, the capital city of Eastern Azerbaijan province (see Figure 1), Kandovan village exemplifies manmade, cone-shaped cliff dwellings that are still inhabited. These cones are the result of volcanic eruptions of the Sahand Mountains; the local indigenous population dug hideouts in the volcanic rock and changed them into permanent houses 850 years ago (Yahyavi & Shaghaghi, 2012). These astonishing troglodyte homes have between two and four floors, and, according to the latest census reports, 700 people currently live in Kandovan (Razani et al., 2013).

KANDOVAN: THE PAST AND PRESENT

It is no more than four decades since Kandovan was first introduced as a tourist destination, to alleviate the poverty and underdevelopment of the area and to accelerate the economy of the indigenous locals. Before then, the lack of accessibility was the main reason why people from great distances away did not know of the existence of Kandovan. In 1997, this archaeological site was included in the list of national treasures of Iran, and the area has gradually become popular ever since, such that thousands of people now annually travel to Kandovan; according to statistics, around 420,000 tourists visited this village in 2012 (Mohammadi, 2012).

Figure 1. The location of Kandovan village.
Source: adapted from google maps.

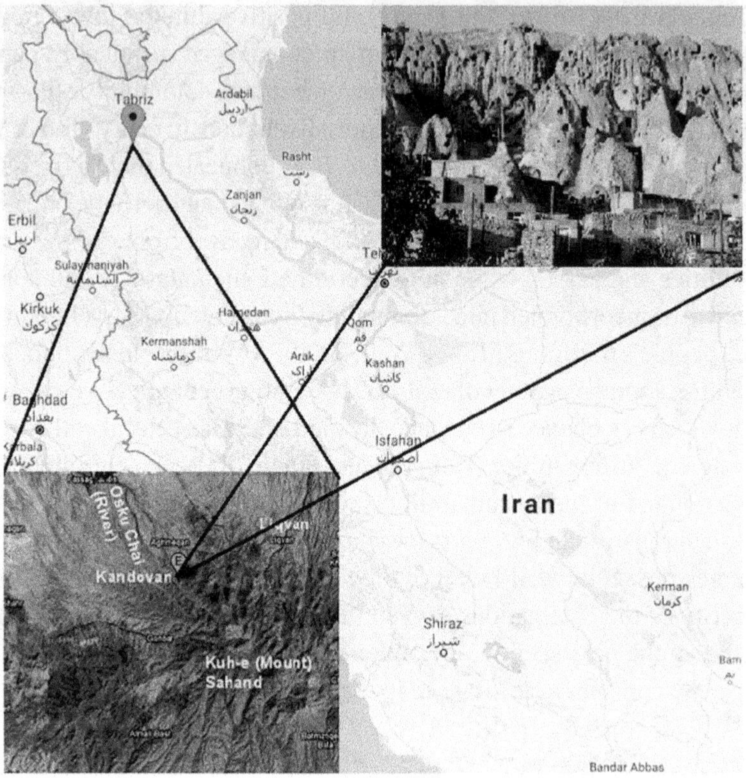

However, Kandovan at present struggles with several issues that their identification can succour visioning the future, such as migration of the indigenous locals, as a result of population growth and lack of space, disagreement with the relevant authorities about expansion of the locals' residential homes and the building of new establishments, a shift in the locals' lifestyles from husbandry, handcrafting and agriculture to seasonal shopping and, most importantly, the destruction of the caves (Razani et al., 2013).

Aien et al. (2012) stated that population growth has meant the indigenous locals of Kandovan have no choice other than to build new settlements (Figure. 2) in which they can reside in the long-term. Zargar (1994) stated that the stone homes of Kandovan are very humid and that the locals are unable to efficiently equip them with suitable facilities to allow them to live comfortably. Also, Ashrafi (2013) mentioned some difficulties that locals are experiencing as a result of unsustainable development and uncontrolled construction in the past:

'The haphazard road infrastructure in the village has led to the creation of streets and alleyway too long and narrow to be effective. All major public amenities such as the village clinic, commercial centers, and the telecommunication center are concentrated in one location, the entrance to the village, depriving the majority of the inhabitants of easy access. Furthermore, no domestic natural gas is available. In absence of a proper sewage system, households resort to digging shallow wells to deposit toilet sewage in and disposing of their wastewater in the village streets and alleyways. A major factor in spreading epidemics, such unsanitary practices have resulted in the contamination of underground water supplies' (Ashrafi, 2013, p. 75).

In a comprehensive review of the sustainability issues regarding seven historical and tourist destinations in Iran, including Kandovan, Eftekhari et al. (2011) reported that these places had unfortunately not been sustainably developed and that the indigenous populations were experiencing many problems as a result. They also stated that, in some of the destinations they assessed, the tourism managers forced the locals to remain in their uncomfortable homes to avoid the destruction of the authenticity of those sites. Moreover, they also added that the environment and entity of these historical settlements are at genuine ecological risk.

Figure 2. Alteration of the historical texture
Source: Authors.

Methodology

To illustrate the future of tourism, scenario planning has been widely used in past studies (Yeoman, 2012a, b). Analysis of the alternative futures is the major discourse in the futuristic studies which is opposed to making a single prediction for the future and is along with the assumption of the complexity and uncertainty of the phenomena in the globe (Son, 2013). The use of scenarios in addressing the concept of sustainable development has been widely corroborated in the literature (Rotmans et al., 2000; Walz et al., 2007). Sustainable development functions over the long run, into a future whose details are difficult to prediction and per se, thinking about that is ofttimes done by developing scenarios (Rotmans et al., 2000).

The strategy-driven and inquiry-driven scenarios are the two broad strands of the alternative future analysis where the purpose of the former is "to explore and promote radical imagination and raise curiosity to develop future-related questions" while in the latter it aims "to enable decision-makers to develop different possibilities for establishing the organization's strategic planning" (Son, 2013, p. 28-29).

This study adopts a scenario-planning approach which is "a process of predicting multiple, plausible and uncertain futures" (Heijden, Bradfield, Burt, Cairns, & Wright, 2002) and offers a "tool for ordering one's perception about alternative future environments" (Yeoman, Robertson and Smith, 2011, p. 509). The study is expert-oriented (vs participatory) in nature and qualitative-descriptive, forward-directed scenario method that seeks the "future consequences of a sequence of assumptions… (and) state an ordered set of possible events irrespective of their (un)desirability" (Rotmans et al., 2000, p.812). Scenarios are defined as "archetypal descriptions of alternative images of the future, created from mental maps or models that reflect different perspectives on past, present and future development" (Rotmans et al., 2000, p. 810). The eminence and magnitude of the scenarios do not refer to their ability to forecast the future but to provide insights into the present (Rotmans et al., 2000). Discerning the weak signals that are the seeds of the future is so momentous that could be dominant in the future (Rotmans et al., 2000). In this study, the environmental, economic and social aspects have been epitomized as key drivers of the developing the potential future scenarios. The aforementioned phases were included in this paper since they are according to Brundtland Report, the chief elements of sustainability (WCED, 1987) and the major focus of the debates in the secondary data drawn from reports, articles and academic knowledge. A thorough review of the published literature regarding the research site was conducted and an inconsiderable number of available English resources and also the materials published in Persian (the official language in Iran where the site is located), and a number of academic papers, books and magazine reports were analyzed. The results of various sustainability arguments about this tourism destination are presented in Table 1, 2 & 3. Four scenarios were developed based on the resources as follows:

Table 1. Environmental forces

Factors	Sources
1. Annihilation of architectural integrity.	(Abdollahfam & Keshavarz, 2015);
2. Loss of physical identity.	(Akbarpour et al., 2013);
3. Unplanned additions of commercial and residential structures.	(Ghasemzadeh, 2013);
4. Incompatible and contradictory buildings to mushroom in and around the village.	(Imani et al., 2017; Karami, E., & Sharifi, 2017);
5. The poor condition of roads.	(Khorasani et al., 2017);
6. Non-compliance of new constructions with the future development of the village.	(Kalajahi & Birami, 2015)
7. Improper condition of physical development in the village	
8. Absence of a proper waste collection system and landfill in the village	
9. Lack of attention toward the aesthetics of the environment in the village.	
10. Destruction of the structure due to low durability and atmospheric factors.	
11. Agricultural land loss.	
12. Unplanned construction growth in riverside.	
13. The area is prone to earthquakes happening.	
14. Dark inner space because of the thickness of the bounds.	
15. Destruction of orchards and farms.	

Note: Some factors were mentioned in more than one source repeatedly while some sources included only one of the factors

Table 2. The key social forces

1. Inadequate local participation in rural management	(Akbarpour et al., 2013);
2. A high rate of migration	(Abdollahfam & Keshavarz, 2015);
3. Low rate of local awareness	(Ghasemzadeh, 2013);
4. Population growth and the need for new housing.	(Ilgren et al., 2014);
5. Cultural conflicts.	(2014; Imani et al., 2017);
6. Congestion and traffic.	(Karami, E., & Sharifi, 2017);
7. Unequal distribution of advantages	(Khorasani et al., 2017)
8. Pollution and open swage in passages.	
9. Destruction of traditional and local culture.	
10. Change of occupation from farming to seasonal shoppers	
11. Destruction and the loss of historical and cultural legacy.	
12. Potential health hazards of the rocky houses.	

Note: Some factors were mentioned in more than one source repeatedly while some sources included only one of the factors

Table 3. The economic forces

1. The unsustainability of the local economy.	(Imani et al., 2017;
2. The high rate of unemployment.	(Akbarpour et al., 2013)
3. Less welfare of local indigenous.	
4. Revenue leakage.	
5. Profits of tourism get out of the village to the authorities and investors.	
6. Inequality in the distribution of tourism income.	
7. Majority of business owners are outsiders.	

Note: Some factors were mentioned in more than one source repeatedly while some sources included only one of the factors

The First Scenario: "Winning the Battle, Losing the War"

The evidence (Table 1) exhibit that there has been a battle with no winner between inhabitants and the environment in this tourism destination since long ago. The question which arises here is that: Whether the rocks once used to be a (suitable?) shelter for the human escaping from various social and environmental threats and risks whose security (physical safety) was the primary concern in the past centuries, can still be considered a (suitable) home for them, regarding the advances (e.g. technological, social, …) achieved by human and experienced (even the indigenous locals of this village) in 21 century?

The villagers are combating to smooth the harsh conditions of living by manipulating the nature which now does not provide them with the facilities they demand and need for a comfortable living in the current era. In this scenario, the struggle will go on for a long till the surrendering of one of the sides in which the outcome is either the total/partial destruction of the site or departure of the villagers who during an overwhelming battle have ruined the authenticity of their place where had been a treasure for attracting tourists. In the first phase of this man vs nature scenario, the dwellers will do their best to manipulate and alter the cone-shaped caves much closer to what they have been experienced in a new age. They change the shapes, attach trespassing, put new doors/windows, establish piping water system, and create modern-looking swage tools, transfer cooling/heating mechanisms, buy large screen TVs and satellite receivers, use high power projectors for illumination and any other change which may shape their homes like "homes" they wanted to live in. Even worse will be the construction of several-floored apartments so that on the site under their shadows the so-called the cone-shaped rocks will appear as small mushrooms. Notwithstanding all the efforts, in the second phase of this scenario, the habitants will not achieve what they have expected from changing their caves into homes in which they may feel comfortable. Nature will resist such changes and will be destroyed under that huge pressure. The sensitive and fragile and poor site will no longer bear the on-going strikes it receives from man. And now the man has no choice but to leave the damaged site. In this circumstance, the place is no longer a treasure attractive for tourists, an identity for indigenous.

Second Scenario: "The Stranger Left No Card"

The locals who get advantage from the mass flow of visitors to this small village are in favour of tourism which regardless of the carrying capacity of the site remains unfavourable effects to the community. Table 2 which is the result of studies conducted in the Kandovan village shows that the traditional livelihood sources of the locals are diminishing. Those who have a little share of the tourism benefits are only the bearers of negative social impacts of this phenomenon (e.g. congestion). Moreover, the majority of the business owners are not locals and high rate of unemployment is evident among locals. The conflicts between the tourism lovers (beneficiaries) and haters (not- beneficiaries) at one hand and the lack of space for the population growth will force the haters to prefer migration on the other. The future picture of this situation influenced by the inadequate local participation and awareness about tourism will create a destination characterized by marketization and overdependence to the tourism industry; an unsustainable stage which can no longer satisfy the remained locals who hate what they loved before. In this scenario, since there is no other source of income except than the cursed tourism, the remained locals (now haters) will tend to move gradually from the site. Then there will be only the shadows of an authentic village occupied by outsiders that has no meaning but a market product just in there to be sold which sooner or later will reach to the end of its life-cycle.

The tourism which was a strange phenomenon in Kandovan at first gradually penetrated to all over the community and affected every aspect of the community lifestyle. In this scenario, the tourism advocates who now comprise almost all of the society are the marketers and traders of the merchandise they import from outside. The traditional and natural products that once were being manufactured in locals' hands now are replaced with industrialized stuff. The visitors are buying handicrafts, dairy products, honey, herbs and even meat that is produced elsewhere and now labelled as local. The mineral water that in the past was freely available for locals and visitors now is owned by non-natives and the bottled water is sold to both visitors and so-called indigenous. The term tradition is now used very rarely and the fruit of unsustainable tourism modernity is evident in the museumized Knadovan village. The living of people is tightly fastened with tourism and that old stranger with its kind smile has eradicated the whole livelihood sources of income but itself. The community should now be afraid of the stranger who has murdered all the traditions of the people and may flee silently sooner or later.

The Third Scenario: "A Hard Day's Night"

The revenues from tourism leak out of the community and thus the indigenous villagers are still waiting to experience the welfare resulted from tourism. In this scenario, the disappointed locals will be frustrated as the time goes on and they will lose their trust to the promises given about the positive outcomes of tourism to the community. The support for further tourism activities in the area will fade. The tourism will be a nonsense term in such a context and the community will act against the industry that within their community enriches others but not themselves. The desperate and ignored indigenous, tired of poverty and hardship try to deliberately destruct the cone-shaped cliffs. They want to end up their hard days at any cost even by destroying the heritage they inherited. The decision they make is not right, it is not logical. The only aim they have now is to revenge what they believe is the reason for their misery; the tourism and what brought the tourism into their living place, the tourist attraction, their cliffy homes. They are against what is related to tourism, the tourism advocates, beneficiaries, outsiders and even the tourists. In this scenario, the messy situation gradually will stop tourism. It will destroy the unique heritage site. For them no matter what will happen next and how they will survive without their cliffy homes. They do not think about the future anymore. In this scenario they stop the days of suffering and penury in their perception by removing the tourism, however, the tourism has left its trace in every aspect of their lives during the past several years that without it achieving a better life seems to blur and vague in Kandovan.

Fourth Scenario: "Happy Ever After"

Finally, in this alternative and hopeful scenario, a conservation-based sustainable development is imagined that emphasizes the long-term benefits for all stakeholders and a well-maintained heritage for future generations. In this sustainable scenario, not only will future generations inherit Kandovan as a genuine example of human-environment harmony (not the battle of man vs nature) and a source of sustainable inspiration (not a story of love and hate), but a well-maintained heritage which has been providing welfare for generations (not waiting for Godot) which is known as a wonderland. This archaeological monument now is known as a symbol of steady source of socio-economic well-being for the local indigenous that has been warranted a sustainable development for the local area and all of the stakeholders including the visitors through astonishing them about how the Mother Nature in such a remote and harsh condition could have been providing mercifully a haven for their ancestors during the past centuries; a natural educational travel, it is too. In this scenario, locals have moved to a pre-planned area close to the main site of the heritage. The locals who have been educated, encouraged and empowered by authorities to participate in the management of tourism are enthusiastically providing services to visitors, producing their handicrafts and local diaries and foods prepared by themselves in their agricultural lands. The benefits are for all of the families and they are truly happy and motivated to preserve their treasure. The tourists who visit the site also enjoy experiencing other activities such as sightseeing and other nature-based recreations available in the area. There is carrying capacity for visitors and vehicles are no longer permitted to enter the main archaeological site to create pollution, noise and traffic to disturb man and nature. Various sources of income flourished by tourism that provides enough occupation for the inhabitants, the fair distribution of the benefits and the enhanced traditional and cultural assets of the locals leave no reason for the youth to migrate and for some to hate the tourism. At this end the locals are happy, travelers are happy and nature is happy too.

DISCUSSION

Sustainable development which refers to satisfying the needs of the present generation without compromising the ability of future generations to meet their needs is the key for ensuring the well-being of the community and the whole stakeholders involved in the case of Kandovan village. In other words, without the preservation of heritage which is the contemporary process through which human societies engage with, and make use of, their pasts (Harvey, 2001), there will be no tourism in the region. That literary means the livelihood of the future generations will be compromised. Heritage tourism is a major economic activity that attracts people to destination and generates livelihood capabilities to the locals. This highlights the linkage among future of the people in the region, the existence of a well-planned tourism, a preserved heritage and the sustainable development concept. In this research, it was tried to outline the current challenges and obstacles of sustainability in a rural heritage tourism destination. Each of the social, environmental and economic forces in this village was separately discussed and as a key driver, a scenario was constructed that illustrates what the future of this destination will look like. The three initial scenarios; "Winning the battle, losing the war", "The stranger left no card", "A hard day's night" show how unsustainability will make Knadovan. These scenarios are there to provide insights to the present stakeholders and managers and authorities to take action not to let the unpleasant future happen. The tourism in this destination has not been able to provide a sustainable livelihood for the locals and the integrity of the site is in danger. Unfortunately, previous studies of the environmental impacts of tourism in Kandovan indicate that serious ecological issues threaten this destination (Eftekhari et al., 2011; Fam & Keshavarz, 2015; Razani et al., 2013).

Bruner (1994) stated the context authenticity of heritage sites is often formed as a result of social constructs, and even the interpretations of the heritage managers and visitors. This means that the authentication of the Kandovan village can be reconstructed in the future as a mere authentic heritage site in which the indigenous locals used to live in the past. This also does not threaten the authentication of the site. As asserted by Park (2013), authentication of heritage tourist sites is negotiable and reflects a site's attributes and tourists' perceptions. Therefore, in the future, it can easily be claimed that the new form of Kandovan is an authentic tourism destination with qualities other than its current 'living village' attribute. All of this indicates that ensuring the authentication of Kandovan does not mean that its current inhabitants must live there forever. The majority of world heritage sites once used to be the shelter of a specific population, and although this is no longer the case, they are still considered authentic. Unfortunately, the Kandovan heritage site has remained embroiled in an indecisive approach to modelling/ planning a tourism system based on the needs and potential of the area and its locals. A lack of serious attention to this situation has taken its toll on the authenticity of the site, undermined the conservation attempts and prolonged the inhumane living conditions of the inhabitants. Furthermore, an unbridled tourist invasion, which has already inflicted sufficient damage to disqualify Kandovan from earning UNESCO World Heritage Site recognition, will continue.

In the alternative scenario, however, a comprehensive master plan is available for sustainable development of the area. As a result of continuous meetings between the authorities, experts and all of the stakeholders, with the locals and the owners of the dwellings at the top, alternative land close to the site is chosen. This sustainably well-planned area which also has sufficient expansion capacity to answer the needs of the current and future indigenous population of Kandovan is built by the budgets provided by the private or public organizations that have mutual shares in the project.

The main Kandovan area will be cleaned, and the additional and altered constructions will be carefully removed by archaeologists and experts, such that the site reverts to its original appearance and status. Next, using sustainable materials, a limited number of paths will be designed and constructed to allow visitors to walk through and explore the area without coming into direct contact with other sections. Information signs, to remind the visitors of the value of the treasure they are visiting, are distributed across the entire area. A number of the indigenous locals will provide the visitors with information about the history of the place and the experiences of the people living there. The site is regularly monitored and protective initiatives against environmental and human damage are established. The trained and educated locals serve the visitors in the resorts, and the visitors feel welcomed by the local dishes and souvenirs and the smiley faces of their hosts. The locals will be empowered and educated to be a part of the managing system in the village. The tourism will be a complementary livelihood source for the locals whose traditional lifestyles and ancient occupation have been revitalized by tourism The Kandovan site sustains its benefits for future generations of locals and remains a treasure for visitors. The process of the planning and outcomes of the plan is illustrated in Figure 3.

Figure 3. Kandovan sustainable livelihood model
Source: author, adopted from Scoones (1998).

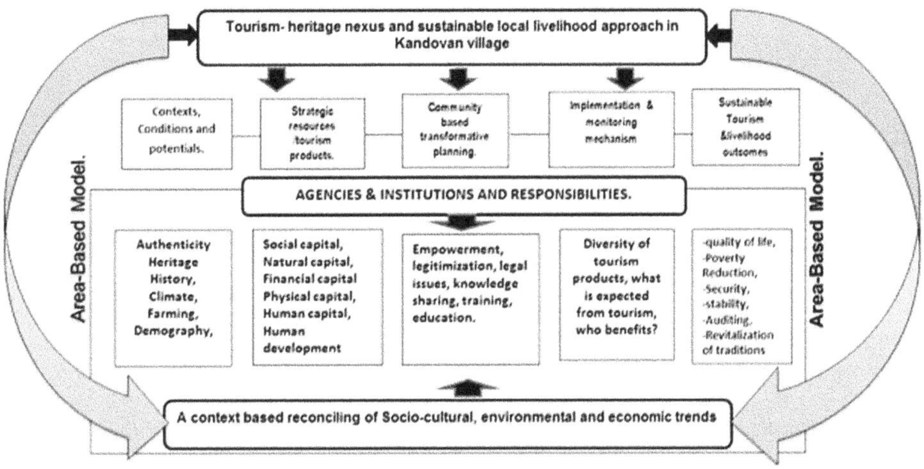

CONCLUSION

The concept of future is so salient in the initial definition of sustainability (WCED, 1987) in which beside emphasize on the needs of present generation the future ones also were given part. Sustainability theory tries to provide a mechanism for human being about the appropriate ways of living so that the resources can be protected for the future. Sustainability therefore can be described as engaging future generations today and now. Living in cliffs, in mountainous area is as difficult and harsh as it seems exciting, which makes inhabitants much vulnerable to anything. This brings extra difficulties to their daily life since for some of the locals renting traditional custom and turning their home to a photography studio for tourist is a source of income, can be troublesome for next-door neighbors. The only development path possible and future requires the satisfaction of local as a key priority and conservation of the site with better management that adopts new modes of governance and more cooperative and coordinating actions (Alipour et al., 2017; 2019).

In Kandovan village locals want to improve their living conditions and for that, they try to engage in commercial/tourism-oriented activities; on the other hand, they want to also keep their way of living which is important for authenticity keeping. However, balancing is very difficult and finding a solution is not easy. If the site is well-conserved and managed properly it not only can be an alternative and sustained source of livelihood for the locals it also can broaden and build positive social and environmental impacts. The highly important contribution of the Kandovan site to the sustainability phenomenon that exacerbates the vitality and priority of the conservation principle is the role it can play in familiarizing the mutual adaptation between humans and the environment. This site emphasizes how the interaction between humans and the environment and their effect on each other has made an outstanding contribution to the co-evolutionary existence process. Conservation of the Kandovan for future generations is not only a part of sustainable development principals, but it will also remain a source of motivation and inspiration for future generations to adopt, such that they may be influenced to behave in sustainable ways and treat the environment responsibly. Strengthening local's sense of ownership can be a potential strategy that can evoke their sense of responsibility (Abubakar et al., 2019) towards the conservation of the heritage site. More research still is recommended to be done for scenario planning for the future of this site through a participatory approach by including all of the stakeholders which were a limit for this research.

REFERENCES

Abdollahfam, R., & Keshavarz, P. (2015). Feasibility Study of Tourism Development Impacts on Stable Regional Development Dimensions: The Case Study. *Life Science Journal*, *12*(7), 88–96.

Abubakar, A. M., Behravesh, E., Rezapouraghdam, H., & Yildiz, S. B. (2019). Applying artificial intelligence technique to predict knowledge hiding behavior. *International Journal of Information Management*, *49*, 45–57. doi:10.1016/j.ijinfomgt.2019.02.006

Aien, S., Kordjamshidi, M., & Pourfallah, H. (2012). *An investigation on thermal performance of rocky architecture approaching thermal comfort with less energy load*. 28th international plea conference opportunities. Limits & needs towards an environmentally responsible architecture, Lima, Peru.

Akbarpour, M., Mosaviaraghi, M., Rahimi, E., Rahimi, M., & Yaqobi, R. (2013). The Approach of the Host Community to the Expansion of Tourism in the Rural Areas (Case Study: Kandovan village, Oskoo City, Iran). *Global Journal of Science. Engineering and Technology*, *13*(1), 15–29.

Alipour, H., Malazizi, N., & Rezapouraghdam, H. (2018, June). Complementing Sustainability through Green Marketing: From Tourism Operator's Perspective. In *8th Advances in Hospitality and Tourism Marketing And Management (Ahtmm) Conference* (p. 636). Academic Press.

Alipour, H., Olya, H. G., Hassanzadeh, B., & Rezapouraghdam, H. (2017). Second home tourism impact and governance: Evidence from the Caspian Sea region of Iran. *Ocean and Coastal Management*, *136*, 165–176. doi:10.1016/j.ocecoaman.2016.12.006

Alipour, H., Rezapouraghdam, H., & Esmaeili, B. (2020). Public Health and Tourism; A Personalist Approach to Community Well-Being: A Narrative Review. *Iranian Journal of Public Health*, *49*(3), 447–453.

Alipour, H., Rezapouraghdam, H., & Hasanzade, B. (2019). A Holistic Analysis of the Second-home Tourism Impacts in the Caspian Sea Region of Iran. *Experiencing Persian Heritage (Bridging Tourism. Theory into Practice*, *10*, 275–293.

Asghari Kaljahi, E., & Amini Birami, F. (2015). Engineering geological properties of the pyroclastic cone-shaped rocky houses of Kandovan, Iran. *Bulletin of Engineering Geology and the Environment*, *74*(3), 959–969. doi:10.100710064-014-0679-4

Ashrafi, N. (2013). Preservation of Kandovan, Based on Pathology of Population Decline. *International Journal of Architecture and Urban Development, 3*(4), 71–80.

Balmford, A., Beresford, J., Green, J., Naidoo, R., Walpole, M., & Manica, A. (2009). A global perspective on trends in nature-based tourism. *PLoS Biology, 7*(6), e1000144. doi:10.1371/journal.pbio.1000144 PMID:19564896

Boorstin, D. (1964). *The Image: A Guide to Pseudo-Events in America.* Atheneum.

Brenner, L., & Aguilar-Garcı'a, A. (2002). Luxury tourism and regional economic development in Mexico. *The Professional Geographer, 54*(4), 500–520. doi:10.1111/0033-0124.00346

Bruner, E. M. (1994). Abraham Lincoln as authentic reproduction: A critique of postmodernism. *American Anthropologist, 96*(2), 397–415. doi:10.1525/aa.1994.96.2.02a00070

Cameron, L., & Potvin, C. (2016). Characterizing desired futures of Canadian communities. *Futures, 82*, 37–51. doi:10.1016/j.futures.2016.05.003

Chambers, R., & Conway, G. (1992). *Sustainable rural livelihoods: Practical concepts for the 21st century.* IDS discussion paper 296. Brighton, UK: Institute of Development Studies, University of Sussex.

Dávid, L. (2011). Tourism ecology: Towards the responsible, sustainable tourism future. *Worldwide Hospitality and Tourism Themes, 3*(3), 210–216. doi:10.1108/17554211111142176

Eftekhari, A. R., Mahdavi, D., & Pourtaheri, M. (2011). *Evaluation of tourism sustainability in cultural and historical villages of Iran: The emphasis on sustainable tourism development paradigm.* Scientific Information Database. Retrieved from http://www.sid.ir

Fam, R. A., & Kesavarz, P. (2015). Feasibility Study of Tourism Development Impacts on Stable Regional Development Dimensions: The Case Study. *Life Science Journal, 12*(7), 88–96.

Ghasemzadeh, B. (2013). An opportunity for tourism development with troglodytic architecture. *Research Journal of Applied Sciences, Engineering and Technology, 5*(12), 3294–3297. doi:10.19026/rjaset.5.4570

GhouchiA. (2014). http://press.jamejamonline.ir/newspreview/1532554389176220663

Haji Ebrahim Zargar, A. (1994). *Assessing the rocky phase of Kandovan Village.* Shahid Beheshti University.

Halewood, C., & Hannam, K. (2001). Viking heritage tourism: Authenticity and commodification. *Annals of Tourism Research, 28*(3), 565–580. doi:10.1016/S0160-7383(00)00076-1

Harvey, D. C. (2001). Heritage pasts and heritage presents: Temporality, meaning and the scope of heritage studies. *International Journal of Heritage Studies, 7*(4), 319–338. doi:10.1080/13581650120105534

Heijden Van der, K., Bradfield, R., Burt, G., Cairns, G., & Wright, G. (2002). The sixth sense: accelerating organizational learning with scenarios. Chichester, UK: Wiley.

Helmore, K., & Singh, N. (2001). Sustainable livelihoods – Building on the wealth of the poor. Bloomfield: Kumarian Press.

Ilgren, E. B., Kazemain, H., & Hoskins, J. A. (2014). Kandovan the next 'Capadoccia'?–A potential public health issue for erionite related mesothelioma risk. *Epidemiology, Biostatistics, and Public Health, 12*(1).

Imani, B., Khalifeh, E., & Ebrahimi, K. (2017). Assessing the Impacts of Implementing a Guidance Plan in Rural Sustainable Development (Case Study: Kandovan Village, East Azerbaijan Province). *Journal of Research and Rural Planning, 6*(2), 57–71.

Inskeep, E. (1994). National and regional tourism planning. Routledge.

Irna. (2017). Retrieved from http://www.irna.ir/ eazarbaijan/fa/ News/ 82010759.

Karami, E., & Sharifi, E. S. (2017). Rocky Habitats Architecture (Kandowan Village as study sample). *Archipolis, 1*(1), 1–10.

Karatepe, O. M., Rezapouraghdam, H., & Hassannia, R. (2020). Job insecurity, work engagement and their effects on hotel employees' non-green and nonattendance behaviors. *International Journal of Hospitality Management, 87*, 102472. doi:10.1016/j.ijhm.2020.102472

Khan, S., Sara, L. M., Sydenstricker-Neto, J., Sutherland, C., & Hordijk, M. (2015). Scenario building as a process and tool in urban governance. In *Geographies of Urban Governance* (pp. 193–214). Springer. doi:10.1007/978-3-319-21272-2_10

Khorasani, M., Hatami, L., & Kiakojoori, D. (2017). Strategic Planning of Rural Tourism Development Using SOAR Model: A Case Study of Kandovan Village. *Journal of Sustainable Rural Development., 1*(2), 171–188. doi:10.29252/jsrd.01.02.171

Kosow, H., & Gaßner, R. (2008). *Methods of future and scenario analysis: Overview, assessment, and selection criteria. DIE Research Project Development Policy: Questions for the Future*. Deutsches.

Lominé, L., & Edmunds, J. (2007). *Key concepts in tourism*. Palgrave Macmillan. doi:10.1007/978-1-349-92040-2

MacCannell, D. (1976). *The Tourist*. Schocken Books.

Maleki, S., & Bigdeli Rad, V. (2017). Social Sustainability Measures for Rural Areas in Iran. *Space Ontology International Journal, 6*(1), 79–84.

Mohamadi, T. (2012). *Narrated by the Actuary of Eastern Azerbaijan Travel Facilitating Organization*. Retrieved from https://www.yjc.ir/fa/print/3670234

Moss, L. A. (Ed.). (2006). *The amenity migrants: Seeking and sustaining mountains and their cultures*. Cabi. doi:10.1079/9780851990842.0000

Nasirahmadi, A., Madarsara, F. A., & Aghdam, H. R. (2014). Cultural issues and teaching literature for language learning. *Procedia: Social and Behavioral Sciences, 98*, 1325–1330. doi:10.1016/j.sbspro.2014.03.549

Park, H. Y. (2013). *Heritage tourism*. Routledge. doi:10.4324/9781315882093

People Travel. (2015). Retrieved from https://financialtribune.com/articles/people-travel/16570/kandovan-village-may-be-nominated-for-unesco-list

Perlik, M. (2006). 15 The Specifics of Amenity Migration in the European Alps. *The amenity migrants: Seeking and sustaining mountains and their cultures*, 215.

Razani, M., Emami, S. M. A., & Baghbanan, A. (2013). *The Necessity of Applying Decay Mechanisms Investigation in the Rock Hewn Architecture of Kandovan; a Historical Village in Azerbaijan, Iran*. Iranian National Symposium on the Application of Scientific Analyses in Archaeometry & the Conservation of Cultural Heritage. Tabriz, Iran.

Razani, M., Emami, S. M. A., & Baghbanan, A. (n.d.). *National conference on the application of scientific analysis in the anthropology and restoration of cultural heritage*. Academic Press.

Rezapouraghdam, H., Alipour, H., & Arasli, H. (2018). Workplace spirituality and organization sustainability: A theoretical perspective on hospitality employees' sustainable behavior. *Environment, Development and Sustainability*, 1–19.

Rezapouraghdam, H., Alipour, H., & Darvishmotevali, M. (2018). Employee workplace spirituality and pro-environmental behavior in the hotel industry. *Journal of Sustainable Tourism*, *26*(5), 740–758. doi:10.1080/09669582.2017.1409229

Rezapouraghdam, H., Behravesh, E., Ari, E., & Doh, W. L. (2018). Cost-benefit analysis of educational tourism in North Cyprus: a qualitative study of the socio-economic impacts. *e-Review of Tourism Research, 15*(6), 457-479. https://www.researchgate.net/profile/Elaheh_Behravesh/publication/324601003_Cost-Benefit_Analysis_of_Educational_Tourism_in_North_Cyprus_A_Qualitative_Study_of_the_Socio-economic_Impacts/links/5b8cddf6299bf1d5a73a0de4/Cost-Benefit-Analysis-of-Educational-Tourism-in-North-Cyprus-A-Qualitative-Study-of-the-Socio-economic-Impacts.pdf

Rezapouraghdam, H., Behravesh, E., Ari, E., & Doh, W. L. (2018). Cost-Benefit Analysis of Educational Tourism in North Cyprus: A Qualitative Study of the Socio-economic Impacts. *Ereview of Tourism Research*, *15*(6).

Rotmans, J., van Asselt, M., Anastasi, C., Greeuw, S., Mellors, J., Peters, S., Rothman, D., & Rijkens, N. (2000). Visions for a sustainable Europe. *Futures*, *32*(9-10), 809–831. doi:10.1016/S0016-3287(00)00033-1

Samuel, R. (2012). *Theatres of memory: Past and present in contemporary culture*. Verso Books.

Scoones, I. (1998). *Sustainable rural livelihoods: A framework for analysis*. Institute of Development Studies, University of Sussex.

Son, H. (2013). Alternative future scenarios for South Korea in 2030. *Futures*, *52*, 27–41. doi:10.1016/j.futures.2013.06.005

Swart, R. J., Raskin, P., & Robinson, J. (2004). The problem of the future: Sustainability science and scenario analysis. *Global Environmental Change*, *14*(2), 137–146. doi:10.1016/j.gloenvcha.2003.10.002

Tao, T. C., & Wall, G. (2009). Tourism as a sustainable livelihood strategy. *Tourism Management*, *30*(1), 90–98. doi:10.1016/j.tourman.2008.03.009

Taylor, J. P. (2001). Authenticity and sincerity in tourism. *Annals of Tourism Research*, *28*(1), 7–26. doi:10.1016/S0160-7383(00)00004-9

Toker, B., & Rezapouraghdam, H. (n.d.). *Educational Tourists' Familiarity With the Traditional Foods in Their Host Community: The Case of a Mediterranean Island*. Academic Press.

Tosun, C. (2000). Limits to Community participation in the tourism development process in developing countries. *Tourism Management*, *21*(6), 613–633. doi:10.1016/S0261-5177(00)00009-1

Walz, A., Lardelli, C., Behrendt, H., Grêt-Regamey, A., Lundström, C., Kytzia, S., & Bebi, P. (2007). Participatory scenario analysis for integrated regional modelling. *Landscape and Urban Planning*, *81*(1-2), 114–131. doi:10.1016/j.landurbplan.2006.11.001

WHC. (2015). Retrieved from: https://whc.unesco.org/ en/ sustainable development

World Commission on Environment and Development. (1987). Our common future (Brundtland report). Oxford, UK: Oxford University Press.

World Heritage and Sustainable Development. (n.d.). Retrieved from UNESCO Centre: http://whc.unesco.org/en/sustainabledevelopment

Yahyavi, S., & Shaghaghi, S. G. (2012). Investigation of the Structural Characteristics of Kandovan Village from Sustainability Perspective. *Journal of Basic and Applied Scientific Research*, *2*(9), 9200–9204.

Yeoman, I. (2012a). *2050: Tomorrow's Tourism*. Channelview. doi:10.21832/9781845413033

Yeoman, I. (2012b). Tourism2050: Scenarios for New Zealand. Victoria University of Wellington.

Yeoman, I., Palomino-Schalscha, M., & McMahon-Beattie, U. (2015). Keeping it pure: Could New Zealand be an eco paradise? *Journal of Tourism Futures*, *1*(1), 19–35. doi:10.1108/JTF-12-2014-0017

Yeoman, I., Robertson, M., & Smith, K. (2012). A futurist's view on the future of events. The Routledge handbook of events, 507-525.

ADDITIONAL READING

Beekman, V. (2004). Sustainable development and future generations. *Journal of Agricultural & Environmental Ethics*, *17*(1), 3–22. doi:10.1023/B:JAGE.0000010842.88648.36

Butler, R. W. (1991). Tourism, environment, and sustainable development. *Environmental Conservation*, *18*(3), 201–209. doi:10.1017/S0376892900022104

Coccossis, H. (2016). Sustainable development and tourism: Opportunities and threats to cultural heritage from tourism. In *Cultural tourism and sustainable local development* (pp. 65–74). Routledge.

Correia, A., Kozak, M., & Rodrigues, A. I. (Eds.). (2019). *Experiencing Persian Heritage: Perspectives and Challenges*. Emerald Group Publishing.

Cronin, L. (1990). A strategy for tourism and sustainable developments. *World Leisure & Recreation*, *32*(3), 12–18. doi:10.1080/10261133.1990.10559117

Hardy, A., Beeton, R. J., & Pearson, L. (2002). Sustainable tourism: An overview of the concept and its position in relation to conceptualisations of tourism. *Journal of Sustainable Tourism, 10*(6), 475–496. doi:10.1080/09669580208667183

Hobson, K., & Essex, S. (2001). Sustainable tourism: A view from accommodation businesses. *Service Industries Journal, 21*(4), 133–146. doi:10.1080/714005050

Hunter, C. J. (1995). On the need to re-conceptualise sustainable tourism development. *Journal of Sustainable Tourism, 3*(3), 155–165. doi:10.1080/09669589509510720

Lane, B. (1994). Sustainable rural tourism strategies: A tool for development and conservation. *Journal of Sustainable Tourism, 2*(1-2), 102–111. doi:10.1080/09669589409510687

Murphy, P. E., & Price, G. G. (2005). Tourism and sustainable development. *Global Tourism, 3*, 167-193.

Rezapouraghdam, H., & Esmaeili, B. (2017). Sustainable desert-tourism development strategies in Khara, Iran: a swot analysis approach. In *Conference proceedings, 7th Advances in Hospitality & Tourism Marketing & Management (AHTMM) Conference, Famagusta, Cyprus, 10-15 July 2017* (pp. 504-510). Eastern Mediterranean University and Washington State University.

Seyfi, S., & Hall, C. M. (Eds.). (2018). *Tourism in Iran: Challenges, development and issues*. Routledge.

Shahgerdi, A., Rezapouraghdam, H., Ghaedi, A., & Safshekan, S. (2016). The Tourism-Environment Nexus; Challenges and Opportunities. *Journal of Sustainable Development Studies, 9*(1).

Williams, P. W., & Ponsford, I. F. (2009). Confronting tourism's environmental paradox: Transitioning for sustainable tourism. *Futures, 41*(6), 396–404. doi:10.1016/j.futures.2008.11.019

Yfantidou, G., & Matarazzo, M. (2017). The future of sustainable tourism in developing countries. *Sustainable Development, 25*(6), 459–466. doi:10.1002d.1655

Chapter 10
Monginis:
A Unique Innovation in the Franchise Model

Christo Fernandes
Independent Researcher, India

Babu George
https://orcid.org/0000-0002-2791-828X
Fort Hays State University, USA

Ajit Mishra
Independent Researcher, India

ABSTRACT

Monginis was started in 1970 at Chembur, Mumbai, India, with just one store. The franchise model that it pioneered in the industry embraced some real-time practices of human development that were carefully captured through the recruitment strategy, job engagement, training, positive work environment, along with strategic practice of embracing local taste in product that has summoned success. How will Monginis reach out to every Indian? While the franchise model has led to expansion and growth, the company does not wish to sit idly on past laurels, but how should it move ahead? This is an updated case study that touches upon two additional dimensions: 1) getting children to bake a cake program and 2) training intervention and buddy support to the franchisee team.

INTRODUCTION

Food habits of Indians have been changing for quite some time. The 'ready-cooked food to go' craze has become the new norm. This is a reflection not only about what is fashionable but also about changes in work culture (Kar & Kundu, 2014). The traditional bakeries and eateries have either adapted themselves to this or perished (Giri, 2018). There is very little room for food sector businesses to survive based on their past glories alone (Fernandes, George, & Mishra, 2016). It is important that these places maintain

DOI: 10.4018/978-1-7998-4330-6.ch010

brand personality congruence while innovating; in the absence of this balancing act, change would disrupt continuity and thereby blow off a key basis of competitive advantage. Surviving brand reorganization is harder especially when it comes with the need to reorganize business processes. This was what happened to food service businesses in India (Agrawal & Singh, 2016).

In this case study, the authors examine the success story of the Monginis brand, one of India's leading cake brands with close to a thousand retail outlets across the country. Having tasted success with the franchise model and banking upon rising middle class income, Monginis has determined to carve a major slice for itself from Indian market where cakes still constitute just about 20% of the total desserts and sweets market. Overseas expansion is no longer an aspiration but a reality, too.

BACKGROUND

Monginis traces its roots back to its humble beginning to a time when it was a favorite with the Europeans in Mumbai. Little excuse was needed for Englishman to pick up a Monginis cake whenever there was an occasion in his family. A birthday, and an anniversary, a wedding or even tea-time would not be complete without Monginis. Today, it is the nationwide headquarters to which all manufacturing and cake shop franchisees report.

Early in the 20th century, two Italian brothers of the Monginis family ran a catering service in Mumbai's Fort precinct, one of which was popular with the city's European residents at the time. In 1958, Monginis Catering was bought up by the Khorakiwala family, and became Monginis Foods Limited. In 1971, the company adopted the franchise model of business, with a stated emphasis on localized production for local tastes. It also models itself on the "food boutique" concept, focusing on quality, presentation and service. It has thereby expanded its brand and reaches across the country, hitting a total worth of about 950 million rupees by 2012.

Today, the 42,000-sq-ft HACCP certified headquarters of Monginis is the place from where all the yummy stuff emerges from. It is the nationwide headquarters, to which all manufacturing and cake shop franchisees report. It is the fountainhead of ideas that are good enough to eat. The Monginis headquarters is also a model bakery with state-of-the-art manufacturing facilities designed to produce a wide range of cakes, gateaux and pastries, tongue-tickling savouries and a variety of any-time snacks. Here quality assurance steps at each and every stage of the manufacturing process ensure world-class soft n fresh cakes, snacks and baker's ware. Monginis headquarters has a well equipped lab and has gained the HAACP certification since past 2 years for maintaining and assuring best of the Hygiene and quality standards of our products and for our customers.

Globally Monginis has its strong presence in Cairo, Egypt,and besides supplying to neighboring countries like Libya, Yemen, Sudan in the African continent. Monginis brand also caters to the prestigious UK market and will soon be expanding to other European countries.

Monginis, based in an emerging economy (India) and a part of a growing bakery market, always had a challenge to understand the local palette in order to become successful and to become a national brand. It was then in 1990 that Monginis decided to franchise even the production of the cakes. Hence the cake shop moved outside Mumbai for the first time, opening stores in Rajkot, Pune, Goa and Ahmedabad

Monginis now have tens of production facilities, some owned directly by it while several others are franchised. Today the franchisee manufacturers account for about one half of total production and the company monitors quality of the produced by deploying its own executives to be part of the franchise team. All Monginis products are made at these centralized bakeries and then sent to the various outlets, which greatly helps Monginis to optimize the retail space - the cake shops are only about 200 sq ft. Monginis charges the franchisee bakers a royalty of 4% on their total sales to franchisee outlets.

IDEAS AND INNOVATIONS

Ideas and innovations make it possible for the high-level excitement to be sustained in the brand.

- Pioneers of a fantastic cross delivery system which connects 11 manufacturing Franchisees through an intricate network to simplify the means so that your beloveds can get a cake delivered in any part of India ordered from anywhere in the world.
- A helpline number attending to customers to assist to make any celebration possible even at a short notice.
- An interactive website which allows online bookings of our products from all across the world to be delivered in parts of India, where we have a presence.
- Pioneers in the introducing corporate cakes with company structures, logos and products too.

Clients and Customers

Having top notch players as their clients is something Monginis has always aspired for.

Monginis follows a simple "doughnut principle" whereby the customer remains that valued creamy centre around whose satisfaction all activity revolves.

1. ***Customer orientation***: The company produces, keeping in mind the feelings of the end consumer in mind, be it son, daughter, father, mother and make the products with the same love and affection as it were made for a family member. They strongly believe that good intentions breed good products.
2. ***Value for money***: Monginis offers consistent value for money products, their prime USP (unique selling proposition).
3. ***Fairness***: They try to be fair in all dealings with the stake holders (shareholders, suppliers, employees, franchisees, dealers, consumers).
4. ***Relationships:*** Monginis constantly strive to build strong relationships based on understanding each other and mutual cooperation.
5. ***Excellence***: The Company constantly tries to innovate and maintain excellence in their day-to- day work and in the quality of the goods and services they provide.

Monginis is the 'Celebration King' with virtually no popular or special occasion remaining outside its range. Clients have now become habituated to celebrate with Monginis whether birthdays, anniversaries, weddings, engagements, Children's Day, etc.

Products

Monginis sells itself as "The Cake Shop ", producing ready-made as well as order-made cakes for catering and carry-out. Individual cake slices are also kept in Monginis stores for dine-in customers. Be it chocolate cakes or cakes in general, Monginis has mastered the art of making cakes over a period of time. A specialist in making cakes begins right from selecting right quality ingredients in precise quantities, blending them together to the best of knowledge and baking to the level of perfection. The soft and moist sponge so made is then sumptuously layered and coated with cream flavored with dark chocolate or milk chocolate or with various fruit flavours. The chain sells both Indian and Western savouries including samosas, cutlets, puffs and doughnuts. Apart from these, snack foods and breads are also sold at Monginis shops.

Monginis has a product line for diabetics, and offers themed products during Diwali, Christmas, Easter, EID and other festivals. Fast food Snacks include more than 50 items like Burgers, Pizzas, Patties, Cornato (bread cone with garlic chicken), Cornizza (veg version), Hotdog & so on .

Monginis produces more than 30 different gateaux primarily in round, square and heart shape both in egg and eggless category. It has cakes in regular, premium and super premium segments in which Black forest, truffle Dutch premium Shimmer and premium Zanzibar are most popular cake family among customers. Price of Monginis cakes is reasonable as compared to its competitors. The prices of cakes vary on their size. On an average a half kg cake would cost something around 200 and it goes up to a 1000 depending on the size and the flavor.

Monginis: Goa Relationship

People in Goa love food and food sector is one of the fastest growing sector here, in terms of the life style and purchasing power of the customers the growth rate of the bakery and confectionary industry is excellent. One of the most important pillar of food industry is baking industry.

Bread baking in Goa dates back to the time of European missionaries however over a period of time it got molded to goan flavor by addition of local goan ingredients. During the Portuguese rule bakers were one of the most respected persons in society. Goans truly love a great pastry chef and nearly every neighborhood in Goa has local favorites.

In the recent times due to heavy influx of multi-cultural and foreign tourists there is a paradigm shift in the bakery industry in Goa. New age products are preferred over the traditional baked products as a result of which new types of bakery have come up all around goa. One of this kind of bakery which has been in the Goan market for quite some time now is Monginis, its brand name originated as a single Italian bakery at Mumbai, Fort area, over a hundred years ago. The brand has grown remarkable in the form of highly successful Monginis cake shop franchise network. Goan culinary traditions are exactly as varied as Goan cultural traditions. Goa was ruled by the Hindu rulers and then by the Portuguese. It is this fusion of East and West in Goan cuisine that makes it irresistible to tourists who flock to the state throughout the year. Indeed, all Goans combine three things: rice, curry and fish. The daily ration of local residents includes these products as well as many vegetables and fruits. Now on the Goan sweet tooth, Bibinca, Godshe, Patoleo, Bolinhas sounds strange enough, but they are the famous Goan desserts. The Goan sweets may not be as famous as its seafood and feni, but anyone who have travelled to Goa must have heard of it, especially during Christmas season. They involve a lot of coconut, coconut milk, jaggery and Goan red rice and are usually baked in traditional stone oven.

Monginis follows a simple "doughnut principle" where the customer remains that valued creamy centre around whose satisfaction all activities revolve. This is by creating value-for-money products without compromising on quality in terms of taste or appearance. Monginis' promotion as 'The Cake shop' seems to go well with the cake tradition of Goa, well known for its celebrations during Christmas and other festivals. Monginis' workers put extra effort to cope up with the increase in demand during seasons. Monginis launched Pattam Pattie and a sponge cake especially for Goan market. Monginis also managed to become one of the most popular cake shops in Vasco-da-Gama, South Goa. Goa is known for Chocolate Cakes, Blackforest, Vanilla and Strawberry Pastry, Veg and Non Veg Puffs, Burgers and other bakery products. Monginis business diversification also helped it to build its customer base in Goa.

FRANCHISE MODEL: THE MONGINIS WAY

Franchising provides benefits for both seller and buyer. For franchisors, the primary benefit is the ability to use other people's money to expand the brand more rapidly than they could either on their own or through investors or lenders. The initial franchise fee and ongoing royalties they collect allow franchisors to build their brand without sacrificing control to outsiders or the pressure of repaying lenders. The fees and royalties are used to fund operations at corporate headquarters, train and support franchisees, market and advertise the brand, improve the quality of goods or services, and build the brand in the marketplace.

For franchisees, benefits include:

- Higher chance of success than in a sole proprietorship;
- Shorter time to opening; initial training and ongoing support;
- Assistance in finding an optimal site;
- The selling power of a known brand;
- lower costs through group purchasing;
- Use of an established business model; national and regional advertising campaigns.
- Customer lead generation through websites and centralized call centres.

As with any business opportunity, there is no guarantee of success, and there are trade-offs to be made. In some ways, franchising is like paying condo fees instead of owning a home. In a condo association, monthly fees are pooled for common external maintenance (mowing, snow removal, roof repairs, etc.) -- a tradeoff many are willing to make to free themselves to concentrate on their "core business" of living their lives (or business) within the walls of their condo (or franchise) unit. And unlike renters, who can be evicted (or corporate employees who can be fired or "downsized") franchisees have some power of their own: a franchisor cannot "fire" a franchisee that is operating in conformity with the franchise agreement. It is better to be prepared with complete knowledge before you take the plunge. Before you make the decision to buy into a franchise, make sure you have a thorough understanding of the total the franchise investment you need to make.

"Monginis" exclusive cake shop business operates on a franchisee model. It franchises out not only the retailing but manufacturing also. It strongly believes in inherent human entrepreneurship and strives to bring it out by training and by giving a lot of freedom in working so as to create "entrepreneurs". The "entrepreneur" franchisees then work in a team with the brand striking a perfect "win-win" situation for both. A Monginis Manufacturing Franchisee virtually owns the entire Monginis business of the city he has been appointed for. He is responsible for the procurement of raw materials, manufacturing, logistics and local marketing. He has to work hard to develop a set of reliable local vendors, get competent and skilled people to handle manufacturing, create a robust logistics and appoint strong retail franchisees to take care of the critical last mile. He has to invest not only in plant and machinery but in training manpower, efficient transportation of the products, creating and sustaining retail footfalls and all the market development activities.

A Monginis Retail Franchisee owns and operates a Monginis Cake Shop. To begin with he has to receive a proper training in "store management". He has to properly receive and store the products. He then sells the products by using his best of the selling abilities. He has to constantly strive to offer the best shopping experience to the Monginis customers so that they not only keep coming back again and again but spread a positive word of mouth for the brand and get additional footfalls also.

The success of Monginis may largely be attributed to its franchisee management skills -starting from selection to servicing and ensuring profitability of franchisee. Monginis has utilized its franchisees effectively as part of its marketing strategy. It is not only an outlet but a brand shop. The shop speaks of Monginis quality and creates its brand image. It is also a mode of its advertisement. The neon signs and the electrical board acts as advertisement for the company. Monginis is the only bakery chain which does not have any owned showroom and all their outlets are franchised. The cycle of managing its franchisees starts from the selection of the shop. The company looks into the location of the shop as the first criteria for selecting the shop since it is considered as a prime factor for driving sales. It insists on a minimum carpet area of 200 square feet and a frontage of 10 feet. It stresses upon the area and frontage norm in order to create an amicable ambience for the customers. The company believes that customers come to the store not just to buy bakery products, what matters is the overall experience during the buying process. To be a market leader, the stores need to provide the customers with that experience. A friendly atmosphere where the customer can enjoy their snacks is prime consideration for the company. The overall investment of a Monginis franchisee is approximately INR 0.5 million, which includes a security deposit of INR 0.2 million on, which an interest of 10% per annum is paid by the company. The remaining around INR 0.3 million is the estimated cost of decoration. The company has a standardized format regarding the décor, principal colors, infrastructure like display counter, air conditioners, refrigerators and microwave oven for its franchised shops. The company pays an average commission of 14% on the goods sold. The company provides the design for the interior decoration and it is up to the outlet owners either to employ their own work force or take the help from the company sources. The outlet is also allowed to keep soft drinks and ice creams of companies having tie ups with Monginis. There is a constant monitoring of the outlets every month, for which there is standardized format for evaluation of the shop on varied parameters.

The parameters for evaluation are given below:

1. Comparison of average daily sales (same months of present and past year)
2. Ambience
3. Product display
4. Profile behavior, knowledge of the store service personnel
5. Customer handling
6. Availability of items and beverages
7. Overall attitude of the shop

The parameters are judged on a three-point scale i.e. 'average' (0-4), 'good' (5-7) and 'excellent' (8-10).

The payment of bills is done on every alternative day by the franchisees as sales are mostly by cash. As such this allows a clean flow of cash throughout the year .As discussed earlier, managing perishability and ensuring delivery of fresh products to the customers is a key challenge for a bakery, especially for a craft bakery selling through multiple retail points. As such, the key challenge is to supply the shops fresh products on a timely basis, which requires a structured and efficient distribution system. The system needs to be reliable and responsive to the needs of the franchisees.

Distribution Process of Monginis in the Shops

The distribution system of Monginis catering to the franchised shops is such that the outlets must have product replenishments at least once and twice if the outlet sells well. Monginis provides for delivery twice daily to its franchised shops. The second delivery ensures that the shops remain well stocked throughout the day. The first delivery starts at6.30 a.m. and ends at 8.30 a.m. and the second delivery is done in the afternoon and this process is followed daily. The afternoon distribution enables it to collect empty crates and return the unsold stocks. Monginis assures its franchised outlets 100% return of unsold stocks. As such, it developed a forecasting system to ensure right amount of production and low amount of returns. The company feels that stock returns in the range of 3.5% to 5% is acceptable, since below 3.5% would indicate that the outlet is selling stale products and anything above 5% would result in additional loss to the company. Furthermore, the company has never given any sales target to the franchised shops. A monitoring system is in place to check that the outlets do not order for quantities that they cannot sell. The production department at Monginis produces based on the projected sales as forecasted. This helps in attaining lower level of return from the outlets. The production department works in three shifts a day i.e. general shift, afternoon shift and night shift. Night shift bridges the demand supply gap in the orders received from the outlets. The flow chart below shows the distribution process followed by Monginis with respect to its franchised shops.

Human Capital

The company lays huge focus on preserving the core value system which looks at the needs of its employees whether on payroll or through franchises.

1. ***Employee selection process:*** Monginis has more than 140 employees on its rolls and on contract. The employees are selected by a rigorous process which starts with prospect being referred to senior accountant and then to the HR manager and finally to respective department heads. This stringent process is followed for both skilled as well as for the unskilled employees since employees are considered to be the most important asset of the company. The employee trainings are conducted on department levels after the induction process.
2. ***Training:*** Resources are sent to Mumbai for training as and when required and the decision for the same is taken by the department heads. An employee needs to undergo trainings under various departments like sales and distribution, logistics, designs, flavouring and garnishing, cake composition and methods etc. Monginis invests heavily in their work force as they are the key components who are responsible in consistently delivering the high quality of products to the customers. Training

on the job extends even after unit starts operations with key company persons remaining there to correct any kind of recurring problems and assist new recruits in ensuring quality standards.

The staff at Monginis is cushioned with Life-long learning with year-round training programs with also a focus on the ISO 22000 standards and manufacturing practices.

Monginis has very effective motivational policies which make the employee work harder. The company encourages employees to acquire new or advanced skills, knowledge, and viewpoints, by providing learning and training facilities, and avenues where such new ideas can be applied.

Below are the ways through which Monginis inspires employees and increase satisfaction:

1. ***Employee Orientation*** – One of the best way Monginis follows for the employees is to make sure they're pleased from the get-go. The company offers a thorough orientation which ensure expectations are realistic and that new staff doesn't come in with rose-colored glasses that will quickly fade.
2. ***Positive Work Environment*** – At Monginis upbeat workplace is a necessity. The company ensures that employees are encouraging one another, avoiding micromanagement, giving positive feedback and ensuring criticism is constructive in all ways in order to keep the environment a place where employees can not only survive but they can also thrive!
3. ***Workforce Engagement*** – Employees that do not find their work interesting or do not feel they are contributing to the mission of the firm will not be engaged. The company believes that for employees to be satisfied, they must feel like they are part of something bigger than just what their individual work tasks are. The company ensures to include staff in goal setting and how they fit in the corporation fabric to increase engagement and satisfaction.
4. ***Developing Skills*** – The Company provides training opportunities, mentoring, online courses or external training to improve its employee's skills.
5. ***Job Enlargement*** – The Company follows a technique wherein there is an increase in the number of tasks associated with a certain job. This is critical for the functioning of the bakery industry because of the variable production needs. For the same reason, interested employees are facilitated with extensive training in the field of interests from the training centers at the Monginis Mumbai branch.
6. ***Initiative and recognition awards*** – Goa is renowned for its celebration of Christmas and other festivities. And since bakery is a critical industry where demand increases exponentially with celebrations, employees are expected to work twice as hard to rake in the orders. With no holidays and excessive work load at the time of festivities, the company by paying doubles the salaries; make sure that employees are never demotivated. With the help of these motivated employees the franchises aims at achieving recognition through various awards like:
    a. Highest percentage growth award
    b. Highest value wise growth award
    c. No cheques bounce award
    d. Most innovative idea award.

Standards and Moral Responsibility

Very high levels of hygiene are maintained at Monginis and employees are therefore supposed to follow a certain code of conduct to work in the company and under the same consumption of tobacco, alcohol, chewing gums, supari etc. is not allowed. Outside eating material is not allowed in the key production areas and the area is completely sacrosanct. Non-adherence to these norms results in termination of the employee. This dominant has over the years has resulted in developing a strong culture where employees adhere to all the codes of conduct without failure.

Monginis is a HACCP & ISO: 22000 certified company. It believes in ensuring the highest degree of sanitation and hygiene and follows Good Manufacturing practices levied by the AIB: The American Institute of Baking.

It upholds all the laws levied by the FDA. It believes in using the highest quality of raw materials and ingredients in the manufactured goods and takes pride in its processes which ensure consistency and ensure safety from hazardous mishaps.

As part of Corporate Social Responsibility Monginis takes up the following activities in Goa.

1. On fresher's day at GEC, it provides refreshments to all students as a token of love and appreciation.
2. Monginis puraskar to class 4 students of all major schools in Goa, which has garnered huge response and positive feedback. It is one of the most coveted trophies at this level.
3. Monginis also started the 'Monginis Quiz' in the year 2007 which was the biggest quiz event in Goa for schools.

CHALLENGES

"Companies that are breaking the mold are moving beyond corporate social responsibility to social innovation. These companies are the vanguard of the new paradigm. They view community needs as opportunities to develop ideas and demonstrate business technologies, to find and serve new markets, and to solve longstanding business problems." - Rosabeth Moss Kanter, Harvard Business Review

Monginis has been proactive in augmenting their time, effort and money in three areas namely Green and Sustainable Processes, Community Outreach and embracing their staff as their family.

- Their solar panels in their factories create a clean and reliable source of energy while interestingly their Air conditioning systems have been programmed to recycle heat to their boiler to reduce thermal waste. The use of gas and electric burners attempts to reduce carbon emission, while they enhance the absorption of carbon dioxide by planting trees around their factories.
- Monginis is the official partner of the National Skill Development Council of India- a facility within a Monginis factory that allows the promotion of baking sciences and training as part of college curriculums. Yet again Mongiland is an effort to educate the youth of India.

Some of the lingering challenges include the following:

1. Threat from competitors: Monginis has a threat from competitors like Birthdays, Hangout etc.

2. All the companies that position their product in terms of celebrations are also threat to Monginis. Examples: Cadbury, McDonalds, packaged cakes from Britannia and competition from local market
3. Online ventures: Various online ventures which facilitate home delivery of confessionary are a threat to Monginis (Monginis has an online portal where orders can be placed online. But the popularity is very low. It would need some aggressive marketing for the same)
4. One important challenge would be maintaining the quality and hygiene-inconsistency can lead to negative perception
5. The franchise system of business can lead to sometimes misuse of the brand name. It can also hamper the quality of the product and also the service offered by the employees at the store.
6. Diversification: Diversification and introduction of new products and flavours play an important role in the future on Monginis.
7. Rising Input costs: In the current economic scenario the costs of raw materials have been escalating. It has hiked prices every year since 2009 — it risks losing its 'affordable' tag.

CONCLUDING REMARKS

"We felt that partnering with people who knew their neighborhoods and communities would allow us to further our goal of being a part of their celebrations and memories," - Qusai Khorakiwala, Director, Monginis Foods Pvt Ltd.

Oftentimes, there is a tendency to discount the possibility of innovations in the bakery industry. In fact, a lot of innovations and change that happen here go unnoticed and underexamined by the scholarly communities (Ocampo & Barrientos, 2008). As bakery products become a regular food in our day to day menu, this cannot continue to be the case. Scholarly attention needs to fall on marketing strategies and branding of bakeries, especially those in the small and medium scale enterprises. Bakery marketing practice, likewise, needs to move beyond the current dominant focus on promotion and advertising (Kiumarsi, Jayaraman, Isa, & Varastegani, 2014). Increased attention on Creating Shared Value (CSV) is critical in this sector, observe Jeong et al. (2013). Monginis has already invested heavily into quality management practices and supply chain management. It has established reliable and lasting partnerships with others in the value chain, especially suppliers and buyers. It is a household name in India and marketing budget need not be spent any longer on creating an awareness or interest. At this stage, Monginis should spend more effort in gathering business intelligence hidden in customer transactions and this should inform further decisions regarding not just promotions but also in product design and management. Businesses that do this will have access to a mine of vital actionable information (Fitriana, Saragih, & Luthfiana, 2017). Understanding all the competitive forces and designing marketing strategies based on this has proven to be successful for bakeries, observe Renko, Sustic, & Butigan (2011).

Increasing internationalization initiatives of Monginis also mean that they need to adapt the offerings to national cultural differences. Changes in consumer tastes and differences in legal requirements need to factor in (Moslehpour, Aulia, & Masarie, 2015). While Monginis already has a world class production system, it also needs to keep eyes open for emerging baking technologies that allow mass customized manufacturing. Baking business is complicated by the fact that manufacturing and service business coexist here. The complex interactions and incompatible expectations of these two sides make managing this business very hard (Lee, 2012).

History has shown that the Monginis strategy has been successful so some consideration by providing high quality bakery products, connected customer service, delivered and served in a comfortable and a hygienic environment. Through decade of thought and service, the Monginis strategy is what it is today- successful, customer centric and business-oriented cake shop. Innovative interventions have led the teams to attract young children to join their journey to cake making baking and the know how that encapsulates an active educational program of hygiene and quality. This one of a kind program will not only delight the customer by will also allow them to gain hands-on experience of a business working and influences. Children attending this program receive Monginis tiffin and a certificate!

It was in the 1960's that this brand was bought over by the Khorakiwala family, who by 1971 fostered the luminous idea of a nationwide franchise network. One of the key element in this vision according to Huseini T. Khorakiwala's vision was reach out to every customer in the Indian neighborhood. The strategic advantage of building and sustaining a good franchisee model is by constantly conducting training for the franchisee teams. The journey today has led to over 1000 cake shops across pan India. Reaching out across to diverse age groups of people draws the true flavor of the secret sauce. The vision of the franchisee is continuously unfolding with cultural initiatives of drawing children into the baking room and then unleashing the power of influence on the other age group in the society.

REFERENCES

Agarwal, S., & Singh, R. (2016). 21 Corporate social responsibility practices in the Indian food industry. *A Stakeholder Approach to Managing Food: Local, National, and Global Issues*, 297-305.

Fernandes, C. F., George, B., & Mishra, A. (2016). Delivering Hospitality through Franchising: The Case of Monginis, India. In Strategic Tools and Methods for Promoting Hospitality and Tourism Services (pp. 139-148). IGI Global.

Fitriana, R., Saragih, J., & Luthfiana, N. (2017, December). Model business intelligence system design of quality products by using data mining in R Bakery Company. [). IOP Publishing.]. *IOP Conference Series. Materials Science and Engineering*, *277*(1), 012005.

Giri, A. V. (2018). *Hurdles and Barriers in the Development of Bakery Business*. Idea Publishing.

Jeong, J. S., Jeong, H. N., Ko, K. H., Moon, J. H., & Kang, H. G. (2013). Success Factors of Paris Baguettes Bakery Franchise Business: Industry Competition and Core Competence Analysis. *Asia Marketing Journal*, *15*(2), 1–27.

Kar, A., & Kundu, S. (2014). The changing perspective of food habits in Kolkata: A case on Monginis Kolkata. *International Journal of Business Performance Management*, *15*(1), 64–71.

Kiumarsi, S., Jayaraman, K., Isa, S. M., & Varastegani, A. (2014). Marketing strategies to improve the sales of bakery products of small-medium enterprise (SMEs) in Malaysia. *International Food Research Journal*, *21*(6), 18–24.

Lee, K. S. (2012). Development of baking technology and analysis of domestic and international bakery market and trends. *Food Science and Industry*, *45*(4), 16–20.

Moslehpour, M., Aulia, C. K., & Masarie, C. E. (2015). Bakery product perception and purchase intention of Indonesian consumers in Taiwan. *International Journal of Business and Information, 10*(1), 8–20.

Ocampo, A., & Barrientos, A. (2008). From the bakery to the brain business: Developing inducible yeast models of human neurodegenerative disorders. *BioTechniques, 45*(4), vii–xiv.

Renko, N., Sustic, I., & Butigan, R. (2011). Designing marketing strategy using the five competitive forces model by michael e. Porter-Case of Small Bakery in Croatia. *International Journal of Management Cases, 13*(3), 376–385.

ADDITIONAL READING

Board, N. P. C. S. (2013). *Confectionery Products Handbook (Chocolate, Toffees, Chewing Gum & Sugar Free Confectionery): Food Processing & Agro Based Profitable Projects, Food Processing Industry in India, bread manufacturing project, candy manufacturing process, Food Processing Projects, How to Start a Food Production Business, How to Start confectionery and bakery Processing Industry in India, How to Start Food Processing Industry in India.* ASIA PACIFIC BUSINESS PRESS Inc.

Brester, G. W. (1999). Vertical integration of production agriculture into value-added niche markets: The case of wheat montana farms & bakery. *Review of Agricultural Economics, 21*(1), 276–285.

Broderick, B. (2017). Food with Purpose: Dudley Dough and Haley House Bakery Café. In *Managing for Social Impact* (pp. 73–86). Cham: Springer.

Chládková, H. (2006). Internal environment of a bakery and its analysis. *ZEMEDELSKA EKONOMIKA-PRAHA, 52*(2), 83.

Iacob, C., & Ţaicu, M. (2012). Managerial Accounting And Environmental Performance Of Bakery Companies. *Scientific Bulletin–Economic Sciences, 11*(1), 16–24.

Kwak, S. H. (1998). A study on the current problems in the bakery business. *Journal of Food Service Management, 1*(1), 5–40.

Meshalkin, V. P., & Katerishchuk, M. Y. (2014). Effective business processes reengineering on the bakery enterprises. *International Journal of Advanced Studies, 4*(2), 3–8.

Mitreva, E., Gjurevska, S., & Gjorshevski, H. (2018). Optimization of the business processes in a company for manufacture of bread and bakery products in Macedonia. *Tem Journal, 7*(2), 439–445.

Nindiani, A., Hamsal, M., & Purba, H. H. (2018). Product and Service Quality Analysis: An Empirical Study of Customer Satisfaction in a Bakery. *Binus Business Review, 9*(2), 95–103.

Oh, M. C., Oh, C. K., & Yang, T. S. (2007). A Study of the Effect of Bottleneck in Bakery Management on Sales and Job Satisfaction-Focusing on Bakery Owners in Jeju. *Culinary science and hospitality research, 13*(1), 179-191.

Onwumere, J., Nwosu, A. C., & Nmesirionye, J. A. (2012). Assessment of the effect of external capital mobilization on the equipment investment of bakery enterprises in South Eastern Nigeria. *Management, 1*(9), 83–89.

Seiler, D. A. L. (1998). Bakery products. In *Principles and applications of modified atmosphere packaging of foods* (pp. 135–157). Boston, MA: Springer.

Seiler, D. A. L. (1998). Bakery products. In *Principles and applications of modified atmosphere packaging of foods* (pp. 135–157). Boston, MA: Springer.

Yoo, J. Y. (2016). The impact of job characteristics and job satisfaction of franchise bakery employees on turnover-Focused on Gwangju. *Culinary science and hospitality research, 22*(3), 22-32.

KEY TERMS AND DEFINITIONS

Bakery: A bakery is an establishment that produces and sells flour-based food baked in an oven such as bread, cookies, cakes, pastries, and pies. Some retail bakeries are also categorized as cafés, serving coffee and tea to customers who wish to consume the baked goods on the premises. Confectionery items are also made in most bakeries throughout the world.

Franchising: Franchising is based on a marketing concept which can be adopted by an organization as a strategy for business expansion. Where implemented, a franchisor licenses its know-how, procedures, intellectual property, use of its business model, brand, and rights to sell its branded products and services to a franchisee. In return the franchisee pays certain fees and agrees to comply with certain obligations, typically set out in a Franchise Agreement.

Innovation: It is a new idea, creative thoughts, new imaginations in form of device or method. Innovation is often also viewed as the application of better solutions that meet new requirements, unarticulated needs, or existing market needs.

Product Management: It is an organizational function within a company dealing with new product development, business justification, planning, verification, forecasting, pricing, product launch, and marketing of a product or products at all stages of the product lifecycle.

Chapter 11
Modeling Destination Competitiveness:
The Unfamiliar Shift for Destination Rebranding, Restructuring, and Repositioning With DMOs

Bindi Varghese
Christ University, India

Shazin Aboobacker
Christ University, India

ABSTRACT

Tourism is a tactical economic practice across the globe, but the urban and provincial transformations in the industry are strongly contemplated in the light of an unfamiliar shift in tourism business. This chapter discusses an integrated concept with a framework relating systematic approach of managing the destination and its competitiveness. An investigation on the impact on tourism and the recent narrative of national, regional, and local planning approach directs towards efficient destination management organizations (DMO) in practice for future development. This has proceeded by the formation of a competitive approach, emphasizing on the DMO roles and responsibilities helpful for a destination management during an unfamiliar business trend. Modeling destination competitiveness demands an absolute mechanism through destination rebranding, restructuring, and repositioning with DMOs for enabling competency.

INTRODUCTION AND BACKGROUND

One of the biggest challenges for the tourism industry is the need for travel destinations to enhance their competency holdings with an improvised competition and global market. This challenge is critical and crucial and is represented by various prominent complications. The increased competitiveness of the

DOI: 10.4018/978-1-7998-4330-6.ch011

destination among international destinations perceived the significance of brand value and destination quality as an important element to make the destinations attractive. Concurrent studies have indicated the destination product as a widespread experience provided by the destinations in the tourism industry. This experience is emulated not by a single stakeholder but by many significant players/actors who engage in the travel and hospitality business. This is subsequently impacting the tourist experiences; mainly tourism stakeholders and other organizations supporting the DMOs and public sector who is often considered as the facilitator or aggregator who take the lead role which improvise public infrastructure such as roads and other general developments as well as government departments, other local residents and community.

The multiple stakeholder perspective in the tourism industry involves various supply and delivery services required for the tourist experience which make the destination management comparatively complicated than management of other products developed by a single organization. For effective and efficient tourism, there should be proper coordination and cooperation among the various participants. A collaborative effort by the Government bodies, local community, Non- Governmental Organizations and other stakeholders result in tourism success thereby leading to the success of the destination (zuzic, 2012). Tourism, being one of the largest economic sectors, plays a very important role in regional development.

To emerge successfully as a competent destination, the tourism sector should be backed by the strong network of policies, laws and decision-making capabilities. This calls for systematic planning and research before making giant leaps in the market. The fragmented nature of tourism industry requires utmost coordination and collaboration among its players. Tourism in its vitality plays a very critical and crucial role not only in the overall economic development but also in globalization and maintaining cordial relations at the global front. It unites the whole world to come together with fewer border concerns and facilitates being global. Tourism has a multiple benefit in the economy if not appropriately managed can even lead to exploitation of destination resources. This leads to a disaster component starting from limited employment opportunities for the local community to environmental issues and social deterioration. An unmanaged destination can also cause irrevocable and irreparable damages to the destination and its resources.

For effectively managing a destination, the government and the other regulatory bodies formulates policies and plans to standardize the tourism practices in a destination (Marzano & Scott, 2009).It is prominent that a destination modeling is a result of factors based on the destination cycle. Several DMOs most importantly marked quality and value as a significant element to rejuvenate the products and as critical objective to revitalize the destination. Depending on different approaches of developing tourism, destination is defined as an amalgam of various products and services that can attract various people beyond its spatial confines. Quality is the positive distinguishing characteristics, which is an initiative for destination marketing and pricing of a quality experience when combined perfectly will lead to a visitor feedback as value they got (Murphy, Pritchard, & Smith, 1999).

Need for Destination Management

For a destination to be successful and attract tourist inflow, it needs to develop a full-fledged destination management organization which functions as an apex organization to manage the resources and at the same time ensure all the facilities and activities satisfies the needs of the tourists. The success of tourist destinations calls for a balanced tourism development and integrates understanding on research, and a co-operative effort to sustain the sought tourism initiatives for retaining competitiveness. Destination management ensures optimum allocation and utilization of resources for the best possible use and in the most sustainable manner. It provides a plinth for the marketing and promotional activities at the destination and ensures a smooth conduct of the tourism industry. Destination management essentially includes activities like visitor management and resource management.

(Magas, 2010) discusses on the tourism trends and latest difficulties like social, cultural and technological aspects with the integrated approach in tourism planning towards tourism experience. The article indicates the pitfalls of the actual system of Tourism Boards in their functionality especially on optimal management of a destination. The author considers DMO as a strategic method to address the challenges faced by destinations as, they could enable, coordinate, and mediate business co-operation and thereby support. DMOs demand co-operation from a number of organizations who are working with a common goal and work with a pragmatic approach with a coherent strategy. Rather dominating other stakeholders, it assembles and manages resources and expertise, and encourages certain levels of independence and objectivity in order to follow the right direction. Thereby, DMO can, therefore, be a systematic developer and coordinator who will simplify the stakeholder participation depending on the common goal of the destination.

For ensuring the quality of services and facilities at the destination, it is highly essential to develop a DMO that can monitor the quality aspects of the resources at the destination. Due to the exponential rate of increase in the tourism industry, controlling the impacts on the destinations is a necessity. The economic gain and the impacts on the environment have to be balanced. Here, DMOs play an important responsibility of instigating appropriate measures and practices to maintain and stabilize the growth of tourism and control its negative impacts.

At the point of growth in tourism destination competitiveness, the restrictions of the traditional concepts need to be transformed by latest techniques and proposes an integrated system for planning and management of destinations, debates (Carlsen, 1999). Similarly, (Vajcnerova & Ryglova, 2010)pressure of relying on the legislatively accommodated model wherein there is an organizational behavior is utilized by DMOs, the most significant factor being the quality in the services provided by tourism destinations, Thus DMOs are putting effort to bring in competitive advantages and at the same time concentrates of sustainable development also. There are many means of managing a destination, such as the cluster method as adopted by (Zupanovic, 2010)where stress is laid on the concept of cluster making, and forming a destination management organization for each cluster required to form a management structure, implemented through DMOs. On similar lines, (Magas, 2010) introspects on the trends of tourism which depicts new challenges and opportunities. The author indicates the ignorance of the real system of tourism associations based on proper destination management. The success of a destination demands a balanced tourism development and integrates research, planning and development. DMOs are envisaged in a strategic way to address the challenges faced by the destinations. DMOs call for stakeholder cooperation and coordinated efforts by the various players in the tourism industry. This process facilitates assembling resources and encourages a certain level of independence and objectivity in order to follow the right path. Thus, DMO can take the form of a systematic leadership and development that will ease the collaboration and cooperation based on the common interest and goal in overall development of the destination.

Therefore, to achieve competitive advantage, destinations have to provide outstanding value based experiences to the tourists. The complexity in the tourism business due to its fragmented nature from the tourist arrival to departure, the destination efficiency is perceived by various services and experiences; this vary from public and private service involvements, environmental and hospitality efficiency. Providing "excellent value" will be based on various organizations driving towards a common goal with unity. DMOs take the initiative of collaborative effort of different interests to drive towards a specific point to make sure the efficiency and effectiveness of their destination now and in long term. It is a common approach these days to have Destination Management Organizations or DMOs to direct the way. Thus, a destination is comprised of several organizations all functioning in unison; in brief, it is the governing bodies that administer the region, the business that are managed therein and the community that resides there.

The uniqueness of the tourism industry is that there needs to be harmony in the destination most importantly the destination should join hands in welcoming tourists to the destination; the tourist expects an "all-in-one experience" or integrated services. To provide a wholesome experience to the tourist, all stakeholders of the destination must impart value to the tourist. A minor setback in the total experience offered to the tourist can create a totally negative impression of the destination which in turn leads to bad publicity and a slack in the chance of re-visitation.

Furthermore, (Presenza, Sheehan, & Ritchie, 2005)contemplates on the managerial concerns of a destination, stakeholder coordination and destination audit. The primary functions of DMO include marketing and managing a destination in the most effective and systematic way. DMOs also contribute towards the framing of the destination's views, aim and future achievements. While introspecting on the functions and activities of a DMO, (Buhalis, 2000)discusses the tactical marketing and management of destinations as a prime function of DMOs. Literature substantiates that marketing a destination should be able to cover up the interest of all the stakeholders objectives with sustainable approach of local resources. A destination is categorized by a combination of brands (products, services, and experiences provided locally) and a political and legal framework, this complexity of destinations extends to the multifaceted relationships of stakeholders, thus, destinations are considered as a complex entity to the market. Hence, Destination Management Organizations (DMOs) are essential so as to plan and market a region as well as authority to control the resources to fulfill the strategic goal. The strategies and plans need to take into action the interests of the stakeholders and effectively manage conflicts.

Tourists overall experience from a destination is an amalgam of varied inputs like intermediaries such as taxi drivers, hoteliers, stewards, as well the components of local attractions like museums, theaters, beaches, theme parks etc. The criticality of resources develops an overall image of the destination post visitation. Therefore in order to effectively market a destination as a whole, each individual supplier at the region should also be considered as the participants in the system or network which are closely linked to each other. (Maccannell, 1999) notes that: "Taken together, tourist attractions and the behavior surrounding them are one of the most complex and orderly of the several universal codes that constitute modern society".

The present day tourists are very well informed and knowledgeable about various destinations, attractions, accommodation facilities, amenities available and accessibility. They plan their holiday very carefully as information is readily available through social forums. They also value the environmental resources which are the assets of a destination; therefore, if not managed sustainably it can often lead to irrevocable damages (Lehto, Morrison, & O'Leary, 2006) have aptly mentioned that "the internet has radically transformed the tourism and travel industry. (Luna-Nevarez & Hyman, 2012). Employing the World Wide Web can be both advantageous and disadvantageous to DMOs as they should use it effectively so as to attract visitors to their destination quickly and directly. As (Kim, Shaw, & Schneider, 2003) explains that "DMOs should refine their websites for two reasons: Primarily, as the main interface between a destination and potential tourists, web users believe such websites represent destinations, and secondly, such websites allow visitors to evaluate the products, services, and experiences (e.g. events, attractions, places to visit, culture) offered by a destination" (as cited in (Luna-Nevarez & Hyman, 2012)). Thereby, with the help of analyzing the contents of top rated destinations, it would understand the present conventional wisdom on the outstanding design. In such websites, hence the aim of the study evolve around identifying and evaluating the most significant model in tourism web design across the globe and to figure out and explain latent segment (clusters) of destination website depending on design and feature advantages. The literature review disclose that social forums are expected to create a favorable impression as people often develop an initial impression of a destination within two to seven seconds (Kim, Hwang, & Fesenmaier, 2005) and can assess a website's visual appeal within 5 minutes (Lindgaard, Fernandes, Dudek, & Brown, 2006) (as cited in (Luna-Nevarez & Hyman, 2012)) thus the importance of having an appealing website design is essential. As a poorly constructed website can foster negative beliefs that encourage website departure and alternative seeking. The visual aesthetics of the website is to be considered so as to raise the interest of the visitors.

According to WTO (Global Code of Ethics for tourism - Article 3) " All the stakeholders in tourism development should safeguard the natural environment with a view to achieving sound, continuous and sustainable economic growth geared to satisfying equitably the needs and aspirations of present and future generations. This specifies the need for controlling the environmental elements within a destination which is a responsibility of the stakeholders. Destination management is used so as to bring about increased destination competitiveness and at the same time to focus on sustainability aspect also." (Global Code of Ethics for Tourism, n.d)

Tourist destinations and the business of tourism depend entirely on the response of tourists to what a destination has to offer. In the face of the competition that exists between these destinations, destinations strive to provide a stay to their visitors that are based on an unforgettable experience and value, both economic and otherwise.

This excellent value, however, cannot be provided exclusively by any one agent that makes up the tourist experience, but rather, requires a coalition of various agencies working in unison towards a common goal. An amalgamated work process, therefore, becomes a pre requisite to achieve the common target. This in turn ensures the sustainable integrity of the tourist destination that is not short lived. (Prahalad & Ramaswamy, 2004) Destination Management Organizations are now spearheading this process, becoming the nerve center of these destinations. These DMOs are multifaceted in the functions they perform. These functions range from being the storehouse of market information of the concerned destination, using the available information to coordinate and keep a check on the tourist flow over the years, use the trends presented by the available data to tackle present and anticipated challenges and also provide a common platform for the various stakeholders and agencies that come into contact with current and potential tourists. DMOs can therefore be seen as a broad management system of a tourist destination. The functions of DMOs, in totality, lead to placing focus on the following arenas of a tourist destination:

- The overall management of a destination
- The destination's competency
- Management of stakeholders
- Development of infrastructure
- Quality control and governance
- Advertising and marketing the destination as a potential tourist spot

The DMOs, in fulfilling these functions, set indicatives of the quality of the services and products that a tourist may expect on visiting the destination and in turn, lead consumers; here they are the tourists, to achieve a satisfactory experience. In order to comprehend the nature of organizations related to the tourism process, it is imperative that one develops an understanding of the various factors that combine to create this complex system. Pollock (n.d) delves into the historical aspects of DMOs in order to explore the source of origin of the concept of DMOs. It was with the advent of mass tourism that the need for DMOs was felt. Their primary objective in the wake of mass tourist trends was to increase the number of tourists visiting a destination, thereby providing it business. DMOs, originally named Convention and Visitor Bureaus (CVBs), started as organizations that were city or community based with the aim to firstly, take care of the visitors visiting a tourist destination and secondly, increase the footfall in these destinations.

The five primary roles of a DMO (Morrison, Bruen, & Anderson, 1998)can be outlined as follows:

1. **Economic Driver:** A DMO generates income and employment, leading to tax generation. It helps in diversifying the economy of a destination or locality.
2. **Community Marketer:** Identifying the select visitor market, a DMO presents the most favorable image of a destination along with its facilities and attractions.
3. **Industry Coordinator:** By providing a clear focus of the economy, the DMO enables the industry to work as one whole and not in fragments. By doing so, it enables every individual agency of the economy to partake the benefits enjoyed by the destination.
4. **Quasi-Public Representative Body:** DMOs make the tourism more legitimate and provide a sense of security to tourists, individuals and groups alike.
5. **Building Community Pride:** DMOs enhance the quality of experience provided by a destination which makes both visitors and residents, come under a common head.

Successful destination management strategies that are affected largely by the quality of service provided, representing the function of DMOs are discussed by (Saftic & Rafajac, 2010) discusses the successful destination management strategies, which depends on the overall level of services quality, which represents the function of DMO and facilitates tourism firms to understand the rising significance of strategic management. Destination management undertakes long-term activities, and thereby, DMOs are essential to foster collaboration and integration and strive towards a system of indicators to monitor the effects of actions on destination competitiveness and the performance of stakeholders in the system. DMOs enhance destination efficiency and facilitate branding by introducing information systems in the promotion. Information and investment capital are essential for destination management and strategic planning (p.137).

Therefore, it is comprehensible to imply on the significance of the DMOs role in the administration formation of tourism, so as to have an authoritative body that supervise and control the tourism activities in the destination which results in customer satisfaction. A DMO acts as an epicenter of the tourism destinations where the important and confidential information regarding the market conditions are maintained, it takes the responsibility to bring in solution to all the management issues and acts as a coordinator to control and direct the visitor flow in the destination which includes dealing with the opportunities and difficulties and enabling a platform to connect the stakeholders with the target customers. The Destination Management Organization (DMOs) takes the call on serving several categories including: Destination Management System (DMS), maintain competitiveness, quality check, stakeholder collaboration, developing infrastructure, sales and marketing strategies eventually resulting in setting standardized quality and value of products and services offered to the customers and thereby improving the level of customer satisfaction and feedbacks (Varghese, 2013). Consequently, it is essential for a destination to adopt a proper management system and enhance its Destination Management Practices. Implementation of Destination Management Organizations (DMOs) could influence and vitalize the tourism activities in the destination and lead to DMOs in bringing an exclusive brand image and can create a competitive edge for destinations.

All these points make it clear why a destination should work on formulating a Tourism Management System and enhance the management practices of the destination. Implementation of these DMOs will definitely be valuable in influencing and bringing to life the tourist movement in and around the area. They will also bring the destination an edge when it is juxtaposed with other tourist destinations (Varghese, 2013).

FRAMEWORK OF DESTINATION MANAGEMENT ORGANIZATION

To bring in a paradigm shift in the destinations, it is important to adopt value based model in networked environment. Thereby to build professional competency; Destination Management Organizations are to foretell the future challenges. This chapter takes through, a conceptual framework building process where the focal point is stakeholders' attitudes, values and practices influencing the destination management process. Developing such a framework acknowledged wider perspectives, to establish relatable Destination Management Organization prototypes. As discussed by (Andrews, 2009), these inclusions contribute to the context and helped to shape specific research questions worthy of exploration. Thus, it was imperative to clearly reflect upon, the ontological and epistemological perspectives to reveal the ways in which the constituents contribute to the development of appropriate research paradigms.

A Priori

In the context of research, the Latin phrases "A Priori"; elucidates the construct which is said to be "from causes to the effect". This process explains the thought process with clarity and is adopted in the 20th century by the logical empiricists. This construct presents the perceived role of predominant stakeholders of tourism industry with strategic role played by DMOs based on Internal Destination Development (IDD) and External Destination Marketing (EDM).

The model is designed in such a manner that there exists correlation between each and every factors associated with the tourism sector. In the center of the model, the body of DMO is placed, which is essentially responsible for the External Destination Marketing and the Internal Destination Development functions (Ritchie & Crouch, 2003). The structure and composition of the DMO articulate the membership structures and other affiliations and alliances which can be incorporated by the state government. The roles and responsibilities to be performed by the DMOs are highlighted, which is the most integral aspect depicted as the central theme. (Stephan, 2009)studied the role of DMOs in cross national sites, and explains that the main motive of DMO'sis to manage and market the destination alongside two other important concepts ; External destination marketing (EDM) where the DMO has to cover all activities that tend to attract the tourist to the destination for example by use of web marketing, publication, multi-channel advertising, tradeshows etc. the next major concept is Internal destination development

(IDD) this focuses on the activities that the tourism destination develops and maintains for example by coordinating the deployment of resources, support human recourses development and sustainability. The various global factors revolving around the business environment are considered in the outer strata of the model which includes the major functions performed by a DMO. The EDM function includes activities like strategic destination marketing by building alliances with stakeholders and tourists, destination branding and image building, De marketing to control seasonality by promoting alternative tourism destinations and developing destination alliances with collaborative marketing and inter organizational relationship building.

The IDD functions typically includes development of a destination information system to educate the tourists on the various activities and services at the destination, collaborating the various stakeholders involved in tourism, performing destination audit periodically to improve and strengthen the operational standards and destination governance through adequate visitor management techniques. (Bornhost, Ritchie, & Sheehan, 2010) states that a DMO acts as a leader in tourism destination management. The DMO has a role of discernible body that brings in focus towards tourism, so inhabitants of the product recognize the importance of the visitor industry and thereby ensuring the construction of outstanding tourism attractions with entertainments and programs that will fulfill the position and uphold the competitive advantage in the experiences provided; Supporting and being helpful with the provision of customer services including advance information and on arrival information. Most importantly DMO also takes the responsibility of assisting external organizations, such as conducting meetings with travel agents and mass bookings who is pro- actively working to bring customers to the destination (p.573).

Destination management involves long-term activities, and thereby, DMOs are effectively taking a call on fostering collaborative integration and strive towards a technical analysis to monitor the influence of action on competitive advantage and stakeholder performance in the model(Saftic & Rafajac, 2010). (Castelltort & Mader, 2010)Explains that media image is also prominent to DMOs as the important motive of the organization is promotion of a recognizable destination and focusing on competitive and sustainable environment, also have to assess a ROI of their management efforts. According to (Pike & Ryan), " most prominent feature of a DMO should be to be able to monitor the efficiency of their effort, not only for further promotional improvements but also to have a control over their activities which is essential for their survival" (as cited by (Castelltort & Mader, 2010)). (Min, Martin, & Jung, 2013) in their seminal work indicatethe challenges to the destination managers, especially when the tourists have mixed images (that is positive and negative),DMOs and other bodies can develop a compelling tourism advertisement by utilizing campaign goal progress frame as well as progress rate to motivate committed travelers.

(Stepchenkova & Zhan, 2013)Substantiate the role of DMOs to control destination imagery and produce marketing material such as brochures, videos etc. the tourists also use and create materials to showcase their perception of destinations, especially through the usage of the Web 2.0 technologies such as Facebook, Flickr, various blogs and so on. The study portrays "non-promotional" material gains more attention than the marketing efforts taken by the DMOs, so to counter act this, the DMO must be aware of the images that are depicted in the internet so as to "reinforce positive images or counter unfavorable images, if necessary" (p.590). (Pestana, Laurent, Nicolas, Elisabeth, Bernardin, & Assaf)mentions that "DMOs must now make the tourist central to their decision-making" and that the DMOs must "manage their attractions as efficiently as possible in order to maximize the number of tourists it receives" (p.145).

DMOs try to meet the increased challenges of today. In order to achieve possible benefits for DMOs and their stakeholders, it becomes important for DMOs to establish organizational settings that sustain the development of inter-organizational affiliation to build long-lasting partnerships that create value for both the DMO and the State. Furthermore, a structured approach towards innovation, reckons development of DMO in the system for efficacy in destination management. Therefore, all the paradigms of functioning of a DMO are incorporated in the strategic model for destination management. Such a model can contribute towards improving the overall competency of the destination thereby leading to socio economic prosperity of the destination which is the ultimate aim of every destination.

Figure 1. Archetype of destination management organizations

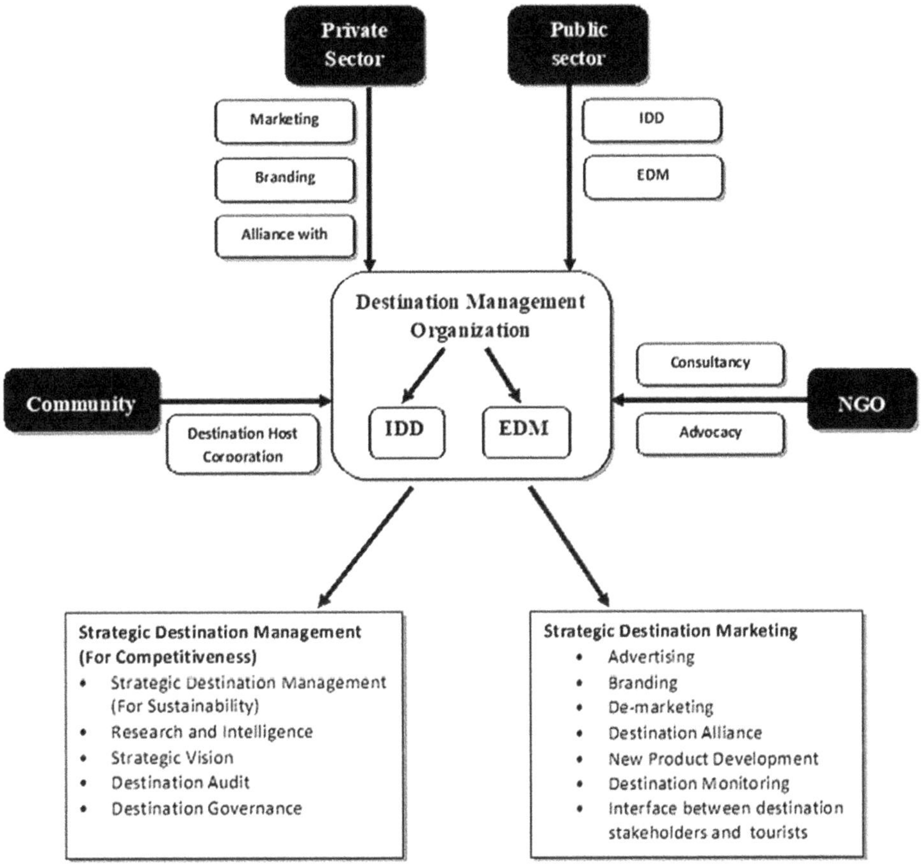

Implications for Destination Management Organizations

DMOs and tourism service providers are the most prominent factors in successfully developing a destination and this facilitates destination leadership. The interface between DMOs and industry partnership with organizational strategies can create an organizational setting to obtain more benefits. DMOs have been identified as destination managers; critical for the success of the development of services for tourists and stakeholders. To pursue strategic goals decision makers can guide their organization to develop a unified system for stakeholder collaboration. Several implications from this chapter can be drawn to guide destination manager's decisions.

- **Long-Term Partnerships:** This study identified a long-term perspective on inter-organizational partnership through establishing DMOs to bring in the standard steps for destination development. Facilitating partner involvement can amalgamate setting up benchmarks for destination branding. Thus, while DMOs need to first identify potential partners with long-lasting relationships which are critical for DMOs as it is often one or several partners that ultimately provide competent service.
- **DMOs for Collaboration:** DMOs are at the center of destination development, requiring DMOs to actively search for partners and integrate them into the DMO partner network. In contrast, DMOs often develop new services without any assistance of partners.
- **DMOs, with M for Management:** This chapter indicates the innovative nature of DMOs as they have to constantly develop new ideas and adapt with the environment which is essential for their sustainability. DMOs not only take the role of marketing and managing the destinations but also bring in innovative services to attract the tourists ((Gnoth, 2002); (Wang & Fesenmaier, 2007); (Wang & Xiang, 2007); (Zach, 2008)).

This chapter is focused on the development of a strategic model for destination management which could enhance the tourism prospects for destinations. Determining specialized areas of DMO from the demand and supply perspectives can be thoroughly examined for building competitiveness. In order to achieve prevailing benefits from DMOs and their stakeholders, it becomes important for DMOs to establish unified organizational settings that support the development of inter-organizational relationships to create value for both the DMO and the partnering stakeholders. The DMO model developed through a study "Strategic intervention of destination management organizations to enhance competitiveness of tourism destinations– a model for Karnataka" (Varghese, 2016) can be a most suitable model which can be extended to all popular tourist regions and a macroscopic outlook can be drawn towards building an Inclusive Destination Management system.

Furthermore, an organized approach towards marketing and managing the destinations create value partners. This chapter portrays critical issues, through stakeholder partnerships, as DMOs try to meet the increased challenges of today.

The Unfamiliar Shift

The travel industry, the same as other businesses, is now facing enormous competitive pressures. Travel businesses are extensively vulnerable and the pursuits of opportunities are equally high. The industry is bound to adapt to newer pedagogies of business handling. Tourism of Tomorrow lay down new paradigms in the Travel Trends and creates a compelled scenario which forces to change in the business pattern. Destinations need to capacitate a new dimension to en-cash on tourism. Spontaneous tourism, primarily based on available resources, is definitely a hallmark to create added value generated from tourism, but cannot guarantee that destinations have a firm foothold in the travel market, especially in the dynamic globalization context.

Perhaps, in order to successfully develop tourism, destinations need to establish their strengths on the basis of building sustainable competitive advantages along with continuous thinking and positioning new competitive advantages, and overcoming the inherent disadvantages. Therefore, it is important to accurately assess competitiveness of these destinations. The added value generated from tourism management is insignificant. The hallmark of Branding is considered to be one of the strong strategies to position a market which makes a destination products and services noticeable for the customers among its competitors. Therefore, branding targets customers to understand products and services as expected by the company with real facts and figures.

Figure 2. Destination competency model and managerial paradigm

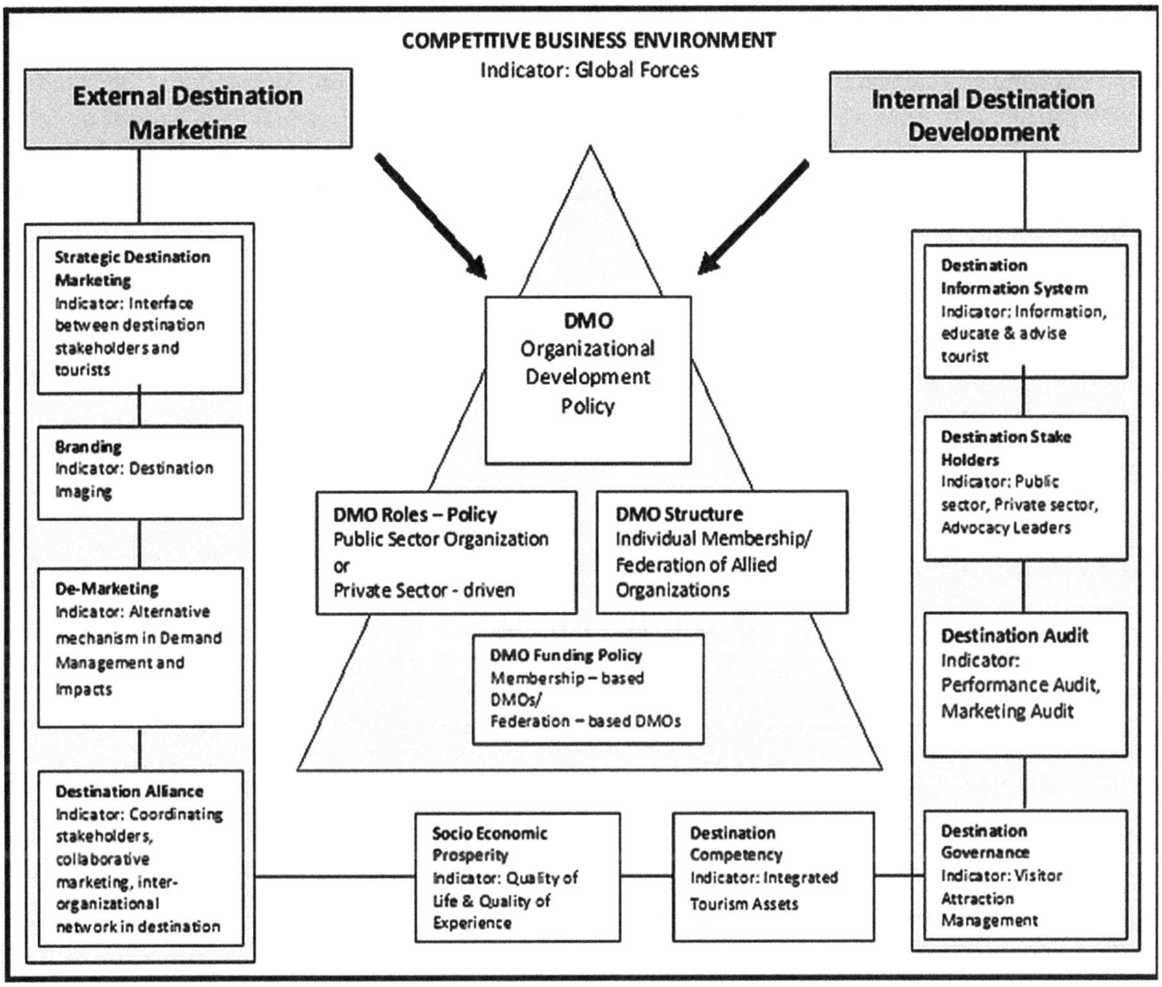

DESTINATION REBRANDING, RESTRUCTURING AND REPOSITIONING (3R'S) WITH DMO'S

DMO and Rebranding

Rebranding is a very important factor in modeling the destination. Rebranding is not merely an explanation to the world about what already exists; rebranding is the process of adding values to the existing product. Comprehensively, it is the process of creating a new name, term, symbol or combination of all these for an already existing and functioning brand with the aim of creating new image in the mind of the stakeholders and competitors. A destination should have clear and positive acceptance in association with trust, quality and integrity which provide the DMOs with a competitive advantage. Tourist's destination choice is influenced by the trust factor, which brings in destination loyalty. In order to model a successful destination, destination brand should go beyond the communication of an image and make a promising brand image. The ultimate aim of rebranding is rejuvenating and revitalizing the product to generate an intense bond between the brand and the consumer(Jesca, Kumbirai, & Brighton, 2014).

Brand image is one of the most significant subjects in destination rebranding literature. The brand image is the mental approach of a brand in consumer mindset. Brand image is a belief developed about the product and service that is derived from different sources of information. A customer understanding about a product is of utmost importance, because they are the ones getting actual feel of the destination influenced by the destination brand image. The brand image is created by the company which shows the customers understanding of brand identity. In any business it is necessary to ensure that customers get maximum satisfaction for products or services with value for money(Jesca, Kumbirai, & Brighton, 2014). A conceptual model is proposed to highlight the significance of stakeholder in destination rebranding. Especially, the destination depended on DMOs and tour operators in the achievement of high brand image, brand loyalty and brand awareness.

Figure 3. Stakeholder involvement in destination rebranding

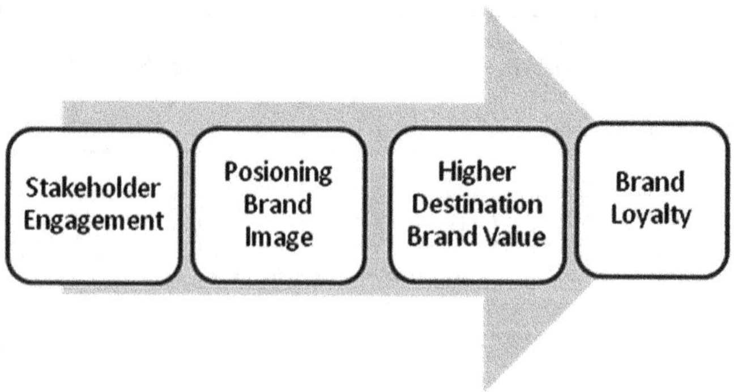

Restructuring Destinations with DMO's

Restructuring the destination management system is considered as the collection and providing details about the destinations to make it easier for the travelers to pre-plan a tour with the abundant information provided by the destination. Information is considered as the lifeblood of tourism industry which is essential for value based model. Destination is perceived as a bunch of services depending on the guests understanding. For example; a tourist from US can perceive Europe as a destination, same way German to Dubai with their own understanding based on the available information online. A traveler's decision making will be based on their anticipations to get themselves satisfied. Thus, DMO research should be to understand the traveller nature and restructure the destination based on the analysis and is considered as the actual motivators for travel(Egger, 2008).

Competitiveness in the field of tourism has made it essential for the destinations to focus on quality and branding as a prominent element that can attract the visitors. While quality adherence and branding are attributes usually associated with products, primarily in the retail segment, these are expanding their applicability and making their presence felt in the field of tourism as well. The concepts of "product concept" and "lifecycle" are becoming increasingly relevant in the management of tourist destinations. Destination Management Organizations play a critical role in the context of facilitating what a new tourist perceives. These concepts and their need in the field of tourism is felt majorly by the people in charge of managing those international tourist destinations that have, over the course of years, lost their attractiveness and have, in turn, faced a depletion in tourist footfall. Apparently, some of the recent scholarly works describe destinations need to be remodeled to restored completely to cater to the ever demanding tourist. It is important to detect the factors that influence the cycle or are consequences of it (Murphy, Pritchard, & Smith, 1999).

As the travel market is totally evolving, it's imperative to take a new way of looking at travel business. Destinations are to create an extra preparedness to equip to newer dynamics of travel technology. For instance, Video KYC (Know your customer) can be great measure to go contactless transactions. Going forward innovative business handling will be the need for travel business. The DMO's can be convergence center to facilitate and be a knowledge center to cater to these measures. When it comes to the refurbishment of a Destination's quality and value propositions, destination as a product must be rejuvenated to cater to the target customer base. Quality and value have been seen as aspects that are critical to revitalize tourism destinations and industries that were on the lower side of the graph. An example of this can be the revitalization of the Northern Ireland tourism industry. They tapped on the quality aspect of their services to their visitors in order to increase the industry performance and the tourist footfall. This was done in a series of steps and initiatives taken by the government. While Northern Ireland focused on the quality aspect to resurrect their dying tourism industry, the Pacific Islands took to value perceptions to increase the attractiveness of these island destinations for tourists (Murphy, Pritchard, & Smith, 1999).

While the importance and applicability of terms like value and quality, used usually in the field of commerce and products, has been established, a structured system that outlines the exact definition of the terms is yet to be established. This can partially be seen as a consequence of the varying priorities of the users of these terms as well as the fact that these users do not hail from a common background. It is therefore that though these terms are self-explanatory in regular usage or pertaining to retail and commerce, a lack of uniformity and standardization has been seen in the usage of these terms when it comes to tourism. Depending on different models of tourism development, it is proposed to understand a destination as combination of products and services available in the location, which can enhance the customer experience (Murphy, Pritchard, & Smith, 1999). A tourist's choice between two destinations may be based largely on this factor, thereby making it important for this aspect to be approached in a more structured and well-defined manner. Value, on the other hand, is seen more as the combined result of the quality that a product is perceived to have and the monetary value or price associated with the product. Both these characteristics will combine to make the value of the product. The aspects of quality and value when used with reference to a product in the retail sector have a more quantified connotation.

However, when the product is tourism or a tourist destination, creating a traveler's perception of value or quality is not an easy task as they are more intangible here. Moreover, the nature of a destination product itself has been a debatable topic, open to interpretations by different individuals. This in turn led to a limited understanding of what role a destination plays in creating these experiences. As a result of these limitations, viz. the intangibility and the nature of a destination product, research in the field of tourism has been limited and restricted. This lack of research, in turn, led to little being known about which are the features that add quality and value to a tourist destination, when seen from the viewpoint of a visitor. In the light of these factors, it becomes important to establish well researched links between the characteristics of a place as a tourist destination and the resulting perceptions of quality and value. This, furthermore, would be fruitful when these links would be able to be used to help the tourist destinations that are struggling with unprecedented competition. For instance, if we juxtapose the destination's product mix with the reaction or impression it creates on visitors, keeping in mind the visitors' core perceptions of quality and value, we may be able to derive the influence of the combination of these elements on future visitation patterns(Murphy, Pritchard, & Smith, 1999).

Figure 4. Restructuring destination branding

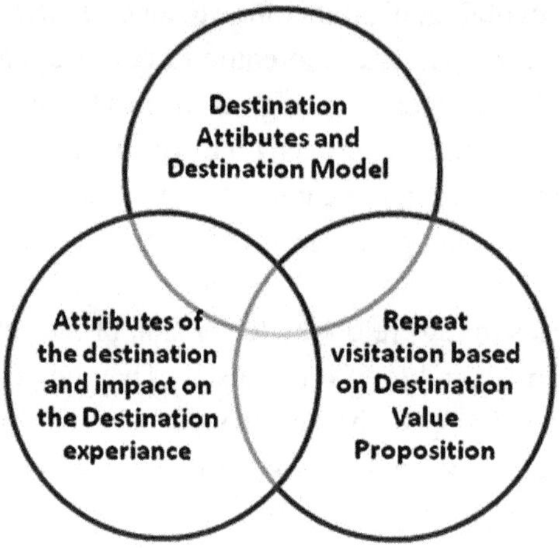

Brand Positioning with DMO

Brand perspective is considered to be one of the most powerful strategies for market positioning which makes the product or services noticeable for the customers among its competitors. Destination branding is a process of formation of a mascot for the purpose of communication to the people along with its features, benefits and values it offers. Also, branding is not merely imposing the created image; influencing the consumers happen through their perceptions. Thus, branding becomes successful when consumer perceptions and company's desire gets a match based on the strategic branding on real facts and figures (Jesca, Kumbirai, & Brighton, 2014). Perhaps positioning is a communications strategy which is described as the way a product is perceived by travellers with the attributes the destination influence the mindset of the people (Chacko & Marcell, 2008).

The main motive of any destination positioning strategy will be to bring in positive images with the existing audience, take corrective measures against negative images or creation of new image. Selection of positioning strategy that forms a distinguished place in customers' mind is essential. With the intense competition in the market for international tourists, most DMOs are constantly developing and refining their market positioning strategies. The basic concept of positioning is not development of something new and unique, but to manipulate what's already existing in the mind, to reinforce the connection that already exist. Tourism destinations worldwide are facing repositioning challenges for modifying the destination image or reclassification of the tourism product in the present positioning strategy (Tkaczynski, Hastings, & Beaumont, 2006). A strategic intelligence body like a DMO can address majority of these challenges. The factors persuade increase in the international tourism and the competition in tourism, changing tourist preferences, attributes of the destinations to rejuvenate them and to create a strong destination image to attract the tourists.

When implemented properly, a DMOs role is capable of bringing about major changes in the economic landscape of a tourist destination. It can act as a trigger to bring in line the entrepreneurial drive of the location to work towards a common goal that includes the collective interests of all the stakeholders instead of focusing on just one. It can work on a shared vision that aims at the holistic development of the destination, while also managing the political processes involved. All these mentioned aspects that come into play in managing a tourist destination are based on a common strategy aimed at a common goal, which requires inputs from all destination managers involved.

A DMO when strategically handled to achieve its predetermined goals is capable of bringing the aims and objectives to life. Strategic positioning can address all the factors that distinguish the structure of a destination or area. These implemented strategies can be seen in the economic as well as competitive performances of not just individual performers but also the performance at a larger level.

BRANDING IN REPOSITIONING

Branding application is a very important process in the destination repositioning. The branding concept existed since a very long time and is given importance in the wide range of academic field. However, branding of a simple product is different from the branding of the destination, due to its fragmented nature. The complexities in reality to apply the concept of practical marketing activities in destination have been investigated in few empirical studies. A destination is a collaborative integration of different stakeholder interest which makes it complex to do branding. DMOs will have to make sure to avoid the conflict of interest among the stakeholders by giving appropriate importance and contributions to everyone in the system. Branding is basically building relationship with the huge customer database. It is very important to maintain the loyal customers for sustainable business model. Branding helps in positioning the product in customers' mindset. The success of the branding is when destination perception among the visitors to be one of a kind. The destinations brand image improves only with the customer's positive experience and feedbacks about the destination.

However, brands have different degrees of functionality and representations for their users. A proper theoretical knowledge and research analysis is required to understand the customer trend and based on the result branding activities can be advanced. The study develops the comprehensive value based and market related interpretation of destination branding which is considered as an important attribute in positioning the brand image. These attributes are utilized in the development of the tourism destination brand locally and globally (Ndlovu, 2009).

Figure 5. 3R's for destination branding

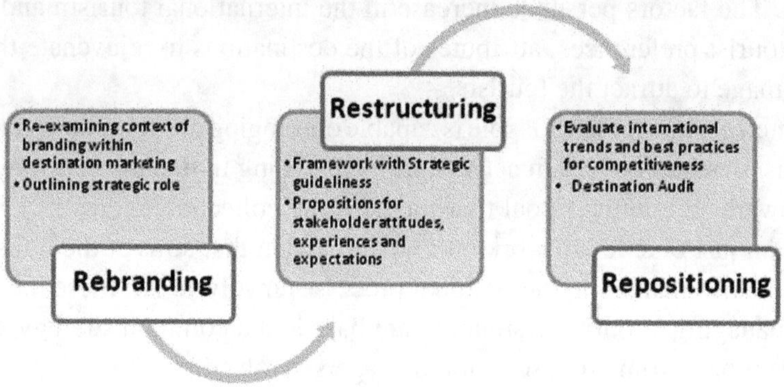

As there is a multiplicity of agencies associated with tourism, there should be an effective mechanism to ensure coordination and collaboration between the various agencies. There should be clarity in roles and responsibilities for the public and private sectors and other stakeholders involved in tourism. The parameters for standardizing and benchmarking destinations and services at the destination have to be strengthened. Quality assurance schemes must be introduced in the state where the quality of operation has to be monitored and ensured.

Destination monitoring systems and destination audit process has to be one of the most important aspects to further improve the current mechanisms at the destinations. "Destination Audit" is an integral component which helps to compare the past performance of the destination with the present and provides an opportunity to develop strategic plans for its future. A performance audit can also be used to compare the performance of one destination with another to assess the competitiveness (p.28). The predominant purpose of conducting a destination audit is to take precautionary measures against crisis situations, ensure good management techniques, identifying competitiveness weaknesses and at the same time, deploying resources and identifying better opportunities, ensuring stakeholder participation and so on.

Destination audit includes components like visitor statistics, development strategies and plans, economic development assessments, annual reports and other documents, reports on competitor destinations which can be availed from DMOs, government departments, research bureau and economic development agencies, tourism enterprises etc. (p.5) The audit process highlights the different stages, primarily, determination of vision, goals and objectives to establish principal competitors and market structure, followed by identifying a destination's competitiveness and examining the destination's environment. Then each dimension of destination competitiveness is evaluated to appraise the present performance of the destination. Lastly, the audit includes, recommending a strategy to enhance the competitiveness and sustainability of the destination and also review the destination's competitive vision and position.

CONCLUSION

The destinations have to be managed well before being marketed to the tourists. High footfall at the destination can be attracted by creating activities and facilities at the destination like adventure tourism, heritage walks, interaction with the local community, relaxation and rejuvenation activities and so on that keeps the tourists engaged during their period of stay at the destination. The local community involved in the tourism activities at the destination ensures better coordination and support extended by the host community in tourism. A country's ability to create added value and consequently upsurge the national wealth by handling its resources and processes defines Destination competitiveness. An appropriate approach in destination management with the help of planning and development will help in effective destination branding.

REFERENCES

Andrews, P. (2009). Identity, rhetoric and culture. *Cambridge Journal of Education*, *39*(2), 159–161. doi:10.1080/03057640902934057

Bornhost, T., Ritchie, J., & Sheehan, L. (2010). Determinants of tourism success for DMOs and destination: An empirical examination of stakeholders' perspectives. *Tourism Management*, 31.

Buhalis, D. (2000). Marketing the competitive destination of the future- Growth strategies for accomodation establishments in alpine regions. *Tourism Management*, *21*(1), 97–116. doi:10.1016/S0261-5177(99)00095-3

Carlsen, J. (1999). A system approach to island tourism destination management. *Systems Research and Behavioral Science*, *16*(4), 321–327. doi:10.1002/(SICI)1099-1743(199907/08)16:4<321::AID-SRES255>3.0.CO;2-5

Castelltort, M., & Mader, G. (2010). A Monetary Public Value (MPV) analysis. *Tourism Management*, 31.

Chacko, H. E., & Marcell, M. H. (2008). Repositioning a Tourism Destination. *Journal of Travel & Tourism Marketing*, *23*(2-4), 223–235. doi:10.1300/J073v23n02_17

Egger, R. (2008). *Restructuring the Destination Management system paradigm*. Unpublished manuscript.

Gnoth, J. (2002). Leveraging export brands through a tourism destination brand.J Brand Management. *Journal of Brand Management*, *9*(4), 262–280. doi:10.1057/palgrave.bm.2540077

Kim, D. Y., Hwang, Y. H., & Fesenmaier, D. (2005). Modeling Tourism Advertising Effectiveness. *Journal of Travel Research*, *44*(1), 42–49. doi:10.1177/0047287505276590

Kim, S., Shaw, T., & Schneider, H. (2003). Website design benchmarking within industry groups. *Internet Research*, *13*(1), 17–26. doi:10.1108/10662240310458341

Lehto, G., Morrison, A., & O'Leary, J. (2006). The economic value portofolio matrix: A target market selection tool for destination marketing organizations. *Tourism Management*, 27.

Lindgaard, G., Fernandes, G., Dudek, C., & Brown, J. (2006). Attention web designers: You have 50 milliseconds to make a good first impression. *Behaviour & Information Technology*, *25*(2), 115–126. doi:10.1080/01449290500330448

Luna-Nevarez, C., & Hyman, M. (2012). Common practices in destination website design. *Journal of Destination Marketing & Management*, *1*(1-2), 94–106. doi:10.1016/j.jdmm.2012.08.002

Maccannell, D. (1999). The tourist and the local. *Tourist Studies*.

Magas, D. (2010). Why the destination management organization? *Tourism and Hospitality Management, 2010*, 1041–1047.

Marzano, G., & Scott, N. (2009). Power of destination branding. *Annals of Tourism Research*, 250.

Min, Martin, & Jung. (2013). *The Sociological Imagination*. Oxford University Press.

Morrison, A., Bruen, S., & Anderson, D. (1998). Convention and Visitors bureaus in the USA. *Journal of Travel and Tourism*.

Murphy, P., Pritchard, M. P., & Smith, B. (1999). The destination product and its impact on traveller perceptions. *Tourism Management*.

Ndlovu, J. (2009). *Branding as a strategic tool of reposition a destination: A survey of key tourism stakeholders in zimbawe*. Unpublished manuscript.

Pestana, C., Laurent, B., Nicolas, P., Elisabeth, R., Bernardin, S., & Assaf, G. (n.d.). Performance of French destinations: Tourism attraction perspectives. *Tourism Management*.

Pike, S., & Ryan, C. (2004, May). Destination positioning analysis through a comparison of cognitive, affective and conative perceptions. *Journal of Travel Research*, *42*(4), 333–342. doi:10.1177/0047287504263029

Prahalad, C., & Ramaswamy, V. (2004). Co- creation experiences: The next practice in value creation. *Journal of Interactive Marketing*, *18*(3), 5–14. doi:10.1002/dir.20015

Presenza, A., Sheehan, L., & Ritchie, J. (2005). *Towards a model of the roles and activities of destination management organizations*. Unpublished manuscript.

Ritchie, J., & Crouch, G. I. (2003). *The competitive destination. A sustainable tourism perspective*. CABI. doi:10.1079/9780851996646.0000

Saftic, D., & Rafajac, O. (2010). DMO model as a step forward in the quality of the management of istrian tourist destination. *Conference. Tourism and Hospitality Management*.

Stepchenkova, S., & Zhan, F. (2013). Visual destination images of Peru: Comparative content analysis of DMO and user generated photography. *Tourism Management*, *36*, 590–601. doi:10.1016/j.tourman.2012.08.006

Stephan, B. (2009). The role of user generated content in Tourists. *Journal of Hospitality Marketing & Management*.

Tkaczynski, A., Hastings, K., & Beaumont, N. (2006). Factors influencing repositioning of a tourism destination. *ANZMAC 2006 Conference Proceedings*.

Vajcnerova, I., & Ryglova, K. (2010). The application of quality management principles into the management of the destination. The Business Review, 306-311.

Varghese, B. (2013). Intervention of destination management organization's in tourist destinations for branding, image building and competitiveness- A conducive model for karnataka. *International Journal of Investment and Management*, *2*(3), 50–56. doi:10.11648/j.jim.20130203.13

Varghese, B. (2016). *Centre for Research-Christ University*. Academic Press.

Wang, W., & Fesenmaier, D. (2007). Collaborative destination marketing: A case study of Elkhart county, Indiana. *Tourism Management, 28*(3), 863–875. doi:10.1016/j.tourman.2006.02.007

Wang, Y., & Xiang, Z. (2007). Towards a Theoretical Framework of Collaborative Destination Marketing. *Journal of Travel Research, 46*(1), 75–85. doi:10.1177/0047287507302384

Zach, C. (2008). Modeling and recognition of landmark image collections using iconic scene graphs. *European conference on computer vision.*

Zupanovic, I. (2010). New postulates in strategic management of integrated tourist destination product in Montenegro. *Tourism and Hospitality Management.*

Zuzic, K. (2012). Destination management systems: Examples of Italian models and the proposal of an Istrian model. *Tourism and Hospitality Management.*

ADDITIONAL READING

Araujo, L. M. D., & Bramwell, B. (1999). Stakeholder assessment and collaborative tourism planning: The case of Brazil's Costa Dourada project. *Journal of Sustainable Tourism, 7*(3-4), 356–378. doi:10.1080/09669589908667344

Beesley, L. (2005). The management of emotion in collaborative tourism research settings. *Tourism Management, 26*(2), 261–275. doi:10.1016/j.tourman.2003.11.019

d'Angella, F., & Go, F. M. (2009). Tale of two cities' collaborative tourism marketing: Towards a theory of destination stakeholder assessment. *Tourism Management, 30*(3), 429–440. doi:10.1016/j.tourman.2008.07.012

Del Chiappa, G., Atzeni, M., & Ghasemi, V. (2018). Community-based collaborative tourism planning in islands: A cluster analysis in the context of Costa Smeralda. *Journal of Destination Marketing & Management, 8*, 41–48. doi:10.1016/j.jdmm.2016.10.005

Jacques, M., Baccarani, C., Baglieri, D., & Consoli, R. (2009). Collaborative innovation in tourism: Managing virtual communities. *The TQM Journal.*

Jamal, T., & Jamrozy, U. (2006). Collaborative networks and partnerships for integrated destination management. *Tourism management dynamics: Trends, management and tools*, 164-172.

Lamsfus, C., Martín, D., Alzua-Sorzabal, A., & Torres-Manzanera, E. (2015). Smart tourism destinations: An extended conception of smart cities focusing on human mobility. In *Information and communication technologies in tourism 2015* (pp. 363–375). Springer. doi:10.1007/978-3-319-14343-9_27

Luque-Martinez, T., Del Barrio-García, S., Ibanez-Zapata, J. A., & Molina, M. Á. R. (2007). Modeling a city's image: The case of Granada. *Cities (London, England), 24*(5), 335–352. doi:10.1016/j.cities.2007.01.010

Mariani, M., Baggio, R., Buhalis, D., & Longhi, C. (Eds.). (2014). Tourism management, marketing, and development: volume I: the importance of networks and ICTs (Vol. 1). New York, Springer.

Prentice, R., & Andersen, V. (2000). Evoking Ireland: Modeling tourism propensity. *Annals of Tourism Research, 27*(2), 490–516. doi:10.1016/S0160-7383(99)00085-7

Ramkissoon, H., & Nunkoo, R. (2011). City image and perceived tourism impact: Evidence from Port Louis, Mauritius. *International Journal of Hospitality & Tourism Administration, 12*(2), 123–143. doi :10.1080/15256480.2011.564493

Selin, S. (1994). Collaborative alliances: New interorganizational forms in tourism. *Journal of Travel & Tourism Marketing*, 2(2-3), 217–227. doi:10.1300/J073v02n02_13

Wang, X., Zhang, J., Gu, C., & Zhen, F. (2009). Examining antecedents and consequences of tourist satisfaction: A structural modeling approach. *Tsinghua Science and Technology, 14*(3), 397–406. doi:10.1016/S1007-0214(09)70057-4

Wang, Y., & Fesenmaier, D. R. (2007). Collaborative destination marketing: A case study of Elkhart county, Indiana. *Tourism Management, 28*(3), 863–875. doi:10.1016/j.tourman.2006.02.007

KEY TERMS AND DEFINITIONS

Destination Audit: Process of a comprehensive, systematic, and periodic examination of a destination to identify the strengths and weaknesses of the destination and its resources.

Destination Competitiveness: Ability of a destination to compete effectively and profitably in the marketplace.

Destination Stakeholders: Destination stakeholders are the key participants or the beneficiaries from the tourism industry.

External Destination Marketing (EDM): All the marketing and promotional campaigns and activities undertaken to attract more tourists to the destination.

Internal Destination Development (IDD): All activities relating to the development of a destination, effective funding, deployment of resources, maintaining sustainability.

Chapter 12
Visitor Perception and Expectation of Heritage Tourism at Mahabalipuram Monuments

Bindu V. T. Nair
https://orcid.org/0000-0002-4966-4962
Avinashilingam Institue for HomeScience and Higher Education for Women, Coimbatore, India

Sathiyabamavathy K.
Avinashilingam Institute of Home Science and Higher Education for Women, Coimbatore, India

ABSTRACT

This chapter explores visitor perception and expectation of heritage tourism at Mahabalipuram group of monuments. Visitor opinions were collected in the form of constructed questionnaire to study the perception and expectation of Mahabalipuram as the visitors can act as ambassadors for heritage tourism development. The research objectives of this study were therefore to assess tourists' motives for visiting Mahabalipuram, their expectation of the heritage destination, and their perception after visiting the monument. The competitiveness as a destination and the gap between tourists' perceptions and expectations were assessed in the current study. The present study discloses the gap of visitor expectations and their perceived feelings of the heritage destination. Visitor perception on the major 5A's were assessed in this study, which pave the way for better positioning and planning of the destination.

INTRODUCTION

India is a land of cultural and natural treasures. Indian cultural and heritage centers are unique and exhaustive. Indian Cultural and heritage tourism plays a pivotal role in the tourism receipts of the country. Heritage tourism acts as a medium to reconstruct and communicate the national solidarity. Mahabalipuram is one of the UNESCO recognized world Heritage Site of seventh and eighth century

DOI: 10.4018/978-1-7998-4330-6.ch012

in India attracting visitors globally. October 11 2019, was one of the most remarkable day of Mamallapuram where Indian Prime Minister Mr. Narendra Modi hosted the meeting with Chinese President Xi Jinping. Mahabalipuram monuments were contributed by the Pallava dynasty and monolithic rock cut architecture is their specialty. Mahabalipuram is situated 50 km south of Chennai in the Coromandel Cost of Bay of Bengal. The historic city had foreign trade even in the seventh and eighth century with China, Sri Lanka and other South-Asian Countries. Understanding the potential and heritage and cultural background of the city, the present study revolves around Mahabalipuram monuments. Mahabalipuram or Mamallapuram is a historic city. During the reign of the Pallava dynasty, between the 3rd century CE and 7th century CE, it became an important centre of art, architecture and literature. Mahabalipuram was already a thriving sea port on the Bay of Bengal before this time. The site has 400 ancient monuments and Hindu temples, including one of the largest open-air rock reliefs in the world: the Descent of the Ganges or Arjuna's Penance. The monuments were built during the Pallava dynasty. Known as the Seven Pagodas in many colonial-era publications, they are also called the Mamallapuram temples or Mahabalipuram temples in contemporary literature. The site, restored after 1960, has been managed by the Archaeological Survey of India. As per the Archeological Survey of India further investigations in Mahabalipuram helps in protecting coastal monuments that are threatened by the incursion of the sea and there are plenty for chances of discovering more manmade structures from underwater and restoring them. Preserving Heritage monuments will helps to build social capital, promotes the preservation of local customs, culture and traditions, improves the community's image and pride and on the whole promotes positive behavior of the community. In this high tourism competitive environment a tourism destination should constantly strengthen its competitive edge by meeting the needs of visitors more than other destinations. Comprehending the visitor´s perception of heritage could be beneficial for segmenting the visitors (Biran, 2011). Tourism is one of the main driving forces for socio-economic growth in developing country like India. Government of India and Ministry of Tourism and Culture encourage heritage tourism development in India by offering several benefits to the Indian states that are predominantly famous for attracting tourists. India's rich heritage is sufficiently reflected in the various temples, palaces, monuments, and forts which is found everywhere in the country. Cultural heritage destination showcases the masterpieces of ancestor's creativity to reflect the traditional values of the society. It enhances the self esteem and pride of local people, offers opportunities to communicate and understand tourists with diverse cultural background from various part of the world. As per the Archaeological Survey of India, among the most popular heritage destinations in India Mahabalipuram attracts thousands of visitors every year globally (MoT). This echoes the significance of this monument in the heritage sites of the country. Mamallapuram also called as Mahabalipuram is a town in southeastern Indian state of Tamilnadu that is best known for UNESCO recognized World Heritage Site of 7th and 8th century Group of Monuments at Mamallapuram. The town was named after Pallava king NarasimhavarmanI, who was also called as Mahamalla (Great Warrior). Due to economic prosperity, Mamallapuram became the site of group of royal monuments, Rathas (temples in the form of chariots), Mandapas (cave sanctuaries), the giant open air rock relief (Descent of the Ganges) and the Shore temple (dedicated to Hindu gods like Shiva, Durga, Vishnu, Krishna and others) in which most of them are carved out of living rock. The contemporary town plan of Mamallapuram was established by the British Raj in 1827 and recent excavation shows that the town served as main port for trade between countries like China, Srilanka and Rome (Sriharsha, 2018). These monuments are inscribed under the UNESCO World Heritage list in 1984 as a cultural heritage site. Tourism has grown consistently in Mahabalipuram and attained the maximum number of tourist arrivals every year. Most of the residents of Mahabalipuram are dependent on various activities of tourism industry directly and indirectly. Hence it is necessary to carry out in-depth study among the various factors influencing tourist activities in Mahabalipuram to promote Sustainable Tourism Development.

BACKGROUND

Park (2010) defined Heritage is a symbolic embodiment of past, reconstructed and reconstituted in the collective memories and traditions of contemporary societies rather than being perceived as a mere apotheosis of by-gone times. Heritage Tourism is a type of Special Interest Tourism which helps the tourists to learn the potentials of past and their lifestyle (Li & Lo, 2004). For past few decades, heritage and tourism industries are complimentary to each other. Heritage Tourism has achieved significant interest economically and politically for regional, state and national benefits (Jamal & Kim, 2005).

Tourist satisfaction is a result of experience to meet with the expected level or more (Engel et al., 1995). Satisfaction is the most important tool for marketing the destination lucratively (Kozak and Rimmington (2000). Many studies have been conducted for identifying customer satisfaction like Expectancy Disconfirmation Model (Oliver, 1980) and Service Quality Model (Parasuraman et al., 1985). Lee et al. (2011) described that the complete understanding of tourist motivations and expectations would enhance the service quality of tourism destination. This would in turn increase the satisfaction levels of tourists by reducing their complaints. Tribe & Snaith (1998) measured the holiday satisfaction of tourists by developing a research instrument called HOLSAT. This model is used to find the key attributes of holiday destination and the attitude of tourists towards it either satisfaction or dissatisfaction. During early times, studies mainly focused on positive economic impacts but later due to the witness of negative impacts coupled with tourism industry enabled a balanced approach for investigating positive and negative impacts of tourism (Latkova, 2008). Having said this, Ko & Stewart (2002) identified the importance of analyzing the relationship between the resident perceptions of tourism impacts and their attitudes for further tourism development. They found that community satisfaction of the residents was mainly related to their perceptions of positive and negative impacts of tourism. It directly influences their attitudes for additional enrichment. Vargas Sanchez et al. (2009) studied the resident's attitude by determining the perceived tourism impacts and community satisfaction for further tourism development. Philips and Jones (2006) presented an assessment of beach erosion and sea level rise due to climatic changes and uncovered a significant threat to the economy of many localities and regions. Carr (1999) studied the behavior of young men and women tourists and found very few differences with respect to the leisure activities of them. Beerli and Martin (2004) determined the associations between characteristics of tourists are given much importance as compared to the demographic variables to enhance delight of tourists and also identified that motivation lead to form perceptions about the destination. In this vein, Prakobsiri (2007) stated that sustainable development is the tool to improve the quality of life not by increasing materialistic behavior rather preserving the environment and resources. He identified six components for Sustainable tourism development such as resources, accessibility, amenities, safety, carrying capacity and community participation of which amenities and community participation need improvements in Kwan Phayao Lake Rim Communities in Phayao Province, Thailand. Poria et al (2003) identified that there is a difference between the perception of heritage tourists and tourists as heritage attractions. They defined heritage tourists as one who perceive the attraction as their own heritage and whose behavior is different from other tourists. Kerstetter et al. (2001) found that heritage tourists are highly knowledgeable, likely to stay for long duration, interested to pay more time during their tour and happy to spend more than the ordinary tourists. In order to satisfy the fondness of tourists with the outcome of technological innovations and increasing diverse needs of tourists, results found changes in heritage tourism both in the production (supply) and consumption (demand) patterns of tourists (Apostolakis, 2003). McDowall (2010) compared the overall satisfaction of local visitors and nonlocal visitors. He found difference in the satisfaction levels between then in the attributes such as duration of the festival and quality of the program. His findings also revealed that the cultural event and family activity were the strong motivating factor behind their trip for domestic and foreign tourists. Valle et al. (2006) found that the tourist satisfaction is a major factor of destination loyalty and identified that the most satisfied tourists were more determined in revising and recommending the destination. Petrick (2004) identified that the first-time visitors and less loyal tourists are less price sensitive than loyal tourists. Chaudhary (2000) conducted a gap analysis between tourist expectations and satisfaction to find out pre and post trip opinion of international tourists about the destinations in India. Her findings clarified strengths and weaknesses of tourism destinations in India. Hui et al. (2007) examined the expectations, perceptions and overall satisfaction of different types of tourist towards the products and services based on the tourist origin. The local people are one of the major resources for the tourism development and they are ultimately affected by the various practices of tourism development (Sunitkul et al, 2010). Particulars of the self reported opinion by the residents on each variable are more informative for guiding the decisions for tourism development (Dyer, 2007). A clear understanding and thorough knowledge of local residents‘ perceptions about tourism is essential for anticipating the chances and conflicts from within tourism activities and for pursuing sustainable tourism (Sunitkul, 2010). Although residents view tourism as a form of socio-economic development positively, they do not appreciate and support the practices of young people and rising prices (Goodwin et al., 1997). People may inspire tourism as a tool to fulfill their desires or to match them with the lifestyle exhibited by tourists. The tourism literature also emphasizes the importance of both push and pulls factors in shaping tourist motivations and in choosing vacation destinations (Crompton, 1979; Guillet et al., 2011; Jensen, 2011; S. S. Kim, Lee, & Klenosky, 2003; Kozak, 2002; Sangpikul, 2008; Yoon & Uysal, 2005). Therefore the destination choice process might be related to tourists' assessments of destination attributes and their perceived utility values in meeting their needs (push and pull factors). Push factors are useful in explaining the desire to travel, while pull factors help explain preferences for travel type or the choice of destination (Crompton, 1979; Jensen, 2011; Kozak, 2002). In this background the present study focuses on the pull factors of the destination basically the 5A's and the other related factors.

MAIN FOCUS OF THE CHAPTER

Mahabalipuram was a seaport even at the beginning of the Christian era. The epigraphical sources confirm that Pallava kings had active contacts with Ceylon, China and the Southeast Asian countries (Vora and Sundaresh, 2003). There were other references to it from the Periplus of the Erythraean Sea in the Greek work of the first century AD and also by Ptolemy, the Greek geographer of the second century. The Chinese Traveller of the seventh century Hiuen Tsang mentioned that this was the seaport of the Pallavas. Many Indian Colonists had traveled to South-East Asia from this port town. It was also a port for the spread of Hindu and Buddhist religion and culture in South-East Asia. Roman coins of Theodosius (4th century AD) found from the region suggest that Mababalipuram also had trade contact with the Roman world around Christian era (Vora and Sundaresh, 2003). Mahabalipuram is an attractive name based on a stirring incident in the puranas and it is represented in a magnificent relief in the Varaha Mandapa. Though it leads to some misconception, Jouveau Dubreuil and other eminent scholars asserted that Mahabalipuram existed long before Mamalla (Ramaswami, 1989). Cultural tourism is gaining more popularity in the contemporary tourism industry & acts as a magnet to attract both domestic and foreign tourists. It permits destinations to enlarge the interest of the tourists, lengthen their stays and reduce seasonality (Patuelli et al., 2013). Unlike other cultural and heritage attractions in Tamilnadu, Mahabalipuram is the most populous tourism destination in the State which attracts maximum tourists every year in the state next to Chennai. Hence, the descriptive study of each indicator in the every factor of the impacts of tourism industry can help the decision-maker to formulate various feasible strategies in Heritage destination. The importance of sustainable tourism development in the destination and more specifically with cultural heritage seems to be well evident in these arguments. All these discussions evoked interest to study the Visitors Perception and Expectation of Heritage Destination Mahabalipuram for sustainable tourism development.

Statement of the Problem

Mahabalipuram as a UNESCO heritage site, offers a fertile ground to the world with several promotional campaigns and publicity. Apart from domestic tourists, foreign tourists from Europe, Australia, America and other parts of Asia visit Mahabalipuram round the year. Although Mahabalipuram provides with economic benefits to the government and local people, it creates several problems such as environmental degradation, increase vehicular traffic and accidents, and lack of co-operation among local people, tourism providers and Archaeological Survey of India (ASI) officials in tourism management plans. Majority of hoteliers are more tourists centric without adhering to the environmental management practices directly leading to depletion of resources. Thus questions for the sustainability of Mahabalipuram as tourism destination need to be answered with proper justification. In order to find practical solutions for sustainable tourism development, perceptions of residents and tourists have been studied to guide planners to implement the principles of sustainable tourism. An in-depth study with the following objectives has revealed several hidden obstructions for the development of tourism in sustainable manner. Tourists have their perceptions regarding a destination's competitiveness and their motivation to travel, and tourism products providers have their own ideas of what tourists want (Headley & Choi, 1992). In this globalized tourism competitiveness, a tourism destination should constantly strengthen its competitive edge by meeting the needs of tourists more than other destinations. (Pansiri Jaloni, 2014). One way to achieve this is through narrowing the gap between tourists' perceptions and expectations. The research objectives of this study were therefore to; assess tourists' motives for visiting Mahabalipuram, their expectation of the heritage destination and their perception after visiting the monument. The concept of competitiveness as a destination as well as the gap between tourists' perceptions and expectation was assessed in the current study.

Methodology

The perception and expectation of visitors were carefully assessed using a questionnaire by conducting visitors survey, as it was a pilot study convenience sampling were used to procure data . All those who were willing and interested to participate were approached for survey. Various factors of visitor perception in Mahabalipuram for Gap analysis has been addressed with the help of visitor perception on 5A's,

Attraction, Accessibility, Accommodation, Activities, Hospitality Services and Cost effectiveness in the study area. The research population consists of all National and International tourists who visited the UNESCO world Heritage destination Mahabalipuram. Hard copies of the constructed questionnaire were circulated in Mahabalipuram to collect the responses of the visitors. Nearly 145 questionnaires were distributed, and 138 questionnaires were collected back due to incompleteness and discrepancies only 100 questionnaires were considered for the final analysis. The analysis was conducted during December 2019, January and February 2020. The data's collected were analyzed based on the research objectives using SPSS software and the analysis and interpretation were discussed below.

DISCUSSION

Demographic Profiles of the Visitors in Mahabalipuram Destination

Different destinations attract visitors having different socio economic backgrounds for various reasons. Demographic profile of the visitors of Mahabalipuram and their choices to visit the particular destination were presented in the Table 1.

The demographic profile of visitors of Mahabalipuram during the study period revealed that the majority of the visitors were Male with 55 percent in comparison to female visitors 45 percent which is understood that more male visitors visited Mahabalipuram than females during the study period. Age distribution of visitors revealed that majority of the respondents were between the age group 21 to 40 years (41 percent) followed by 41 to 60 years (29 percent), below 20 years (24 percent) and 61 years above (6 percent) respectively. When studying the nationality of the interviewees, we find that the majority of visitors were Indians with 90 percent and foreigners with 10 percent respectively. Visitors religion who visited Mahabalipuram during the study were Hindus 57 percent, Christian 22 percent and Muslim 21 percent respectively which implies that majority of Hindus visited Mahabalipuram during the study period. Educational status of the visitors revealed that majority of the visitors were under graduates with 52 per cent followed by school education (20 per cent), post graduates (12 per cent) and professional degree holders (12 per cent) respectively. Majority of the visitors were working in private organizations (43 per cent) followed by government job (17 per cent), business people (12 per cent) Students (8 per cent) respectively. It is also inferred from the study that majority of the visitors were earning one lakh to five lakhs (38 per cent), followed by less than one lakh per annum (33 per cent) followed by visitors earning, five lakhs to ten lakhs per annum (16 per cent) and ten lakhs above (13 per cent) respectively. Majority of the respondents were married with 64 per cent followed by singles 36 per cent respectively.

Distribution of Travel Pattern

Travel pattern includes various important components of tourism like number of visits, number of days spent, destination knowledge, best time to visit, awareness of dance festival, visit to the destination during festival and modes of transportation plays a significant role in pulling the visitors towards the destination. The results of the present study on travel pattern of the visitors were presented in the form of the Table 2.

Table 1. Demographic profiles

Demographic Factors	Particulars	Frequency	Percent
Gender	Male	55	55.00
	Female	45	45.00
Age	Below 20 years	24	24.00
	21 to 40 years	41	41.00
	41 to 60 years	29	29.00
	61 above	6	06.00
Nationality	Indian	90	90.00
	Foreigner	10	10.00
Religion	Hindu	57	57.00
	Muslim	21	21.00
	Christian	22	22.00
Education Level	School education	20	20.00
	UG degree	52	52.00
	PG degree	12	12.00
	Professional	12	12.00
	Others	4	04.00
Occupation	Government job	17	17.0
	Private job	43	43.00
	Business	12	12.00
	Student	08	08.00
Annual Income	Less than 1 lakh	33	33.00
	1 lakh to 5 lakh	38	38.00
	5 lakh to 10 lakh	16	16.00
	10 lakh above	13	13.00
Marital Status	Single	34	34.00
	Married	66	66.00

(Source: Primary data)

The distribution of travel pattern results revealed that majority of the visitors were first time visitor (42 per cent) followed by one to three time visitors (34 per cent) and more than three times visitors (24 per cent). Majority of visitors has spent two days in Mahabalipuram to explore (40 per cent) followed by one day visitors (32 per cent) and more than two days spending visitors (28 per cent) respectively. Visitors also expressed that they came to know about Mahabalipuram and its festivals through newspaper mostly (40 per cent) followed by various other sources like Magazines (18 per cent), Word of Mouth (16 per cent), other sources (16 percent) and Internet (10 per cent) respectively. Visitors felt that Mahabalipuram is suitable to visit during Winter (34 per cent), Throughout the year (34 per cent), during Summer (20 per cent) and during Festival time (12 per cent) respectively. It is also understood from the study that majority of visitors are not aware of the dance festival happening every year (52 per cent) and visitors who has already heard about dance festival (48 per cent). Since majority of the visitors have not heard about Dance festivals happening in Mahabalipuram, they have not visited the destination during the festival (66 per cent) followed by visitors who visited Mahabalipuram during the festival season (34 per cent). It is also understood through the study that majority of the visitors travelled to the destination through Bus (38 per cent), Train (32 per cent) and about 26 per cent of the visitors visited the destination using their own transportation arrangements and very few visited the destination using Taxi (4 per cent).

GAP ANALYSIS

A paired t-test is used to compare visitor expectation and perception on various factors of 5A's of the destinations and other related factors.

Table 2. Distribution of travel pattern

Factors	Constructs	Frequency	Percentage
Number of Visits	First time visitor	21	42.0
	1-3 times	17	34.0
	More than 3 times	12	24.0
Number of days spent	1 day trip	16	32.0
	2 days trip	20	40.0
	3 days and more	14	28.0
Destination Knowledge	Magazine	9	18.0
	Newspaper	20	40.0
	Internet	5	10.0
	Word of Mouth	8	16.0
	Others	8	16.0
Time to Visit	Summer	10	20.0
	Winter	17	34.0
	Festival time	6	12.0
	Throughout the year	17	34.0
Awareness on Dance Festival	Yes	24	48.0
	No	26	52.0
Visited during the Festival	Yes	17	34.0
	No	33	66.0
Mode of Transport	Train	16	32.0
	Bus	19	38.0
	Taxi	2	4.0
	Own arrangement	13	26.0

(Source: Primary data)

H0: There is no significant difference between Expectations and Perceptions of Visitors of on Attraction of Mahabalipuram

Table 3 explains the Expectation and Perception about the factor Attraction by the visitors of Mahabalipuram revealed that the significance value is (0.000) so that null hypothesis is rejected. The mean value of Expectation on Attraction is 20.09 where as perception is 15.16 which shows that there is a negative gap, their expectation is higher where as their perception is lower. We can conclude that there is gap exist between Attraction Expectation and Attraction Perception of the destination Mahabalipuram.

Table 3a. Expectation and perception of the attraction by the visitors of Mahabalipuram

Attraction		Mean	N	Std. Deviation	Std. Error Mean
Pair 1	Attraction Expectation	20.09	100	3.607	.361
	Attraction Perception	15.1600	100	1.73333	.17333

Table 3b. Expectation and perception of the attraction by the visitors of Mahabalipuram

Attraction		Paired Differences					t	df	Sig. (2-tailed)
		Mean	Std. Deviation	Std. Error Mean	95% Confidence Interval of the Difference				
					Lower	Upper			
Pair 1	Attraction Expectation – Attraction Perception	4.93000	3.90093	.39009	4.15597	5.70403	12.638	99	.000

(Source: Primary data)

H0: There is no significant difference between Expectations and Perceptions of Visitors on Accessibility of Mahabalipuram

Table 5 explains the Expectation and Perception on the Accessibility by the visitors of Mahabalipuram revealed that the significance value is (0.001) which is less than 0.05 at 95% level of significance so that null hypothesis is rejected. The mean value of expectation on Accessibility is 20.15 where as perception is 18.93 which shows that there is a narrow gap, their expectation is higher where as their perception is lower. We can conclude that there is gap exist between the Accessibility Expectation and Accessibility Perception of the destination Mahabalipuram.

Table 4. Expectation and perception of the accessibility by the visitors of Mahabalipuram

Accessibility		Mean	N	Std. Deviation	Std. Error Mean
Pair 1	Accessibility Expectation	20.15	99	2.862	.288
	Accessibility Perception	18.9394	99	2.17508	.21860

Table 5. Paired samples test

Accessibility		Paired Differences					t	df	Sig. (2-tailed)
		Mean	Std. Deviation	Std. Error Mean	95% Confidence Interval of the Difference				
					Lower	Upper			
Pair 1	Accessibility Expectation – Accessibility Perception	1.21212	3.46196	.34794	.52165	1.90260	3.484	98	.001

(Source: Primary data)

H0: There is no significant difference between Expectations and Perceptions of Visitors on Amenities of Mahabalipuram

Table 6 explains the Expectation and Perception on the Amenities by the visitors of Mahabalipuram revealed that the significance value is (0.027) which is less than 0.05 at 95% level of significance so that null hypothesis is rejected. The mean value of expectation on Amenities is 19.68 where as perception is 18.95 which shows that there is a narrow gap, their expectation is higher where as their perception is lower. We can conclude that there is gap exist between the Amenities Expectation and Amenities Perception of the destination Mahabalipuram.

Table 6. Expectation and perception of the amenities by the visitors of Mahabalipuram

Amenities		Mean	N	Std. Deviation	Std. Error Mean
Pair 1	Amenities Expectation	19.68	100	2.506	.251
	Amenities Perception	18.9500	100	2.16667	.21667

Table 7. Paired samples test

Amenities		Paired Differences					t	Df	Sig. (2-tailed)
		Mean	Std. Deviation	Std. Error Mean	95% Confidence Interval of the Difference				
					Lower	Upper			
Pair 1	Amenities Expectation – Amenities Perception	.73000	3.26244	.32624	.08266	1.37734	2.238	99	.027

(Source: Primary data)

H0: There is no significant difference between Expectations and Perceptions of Visitors on Accommodation of Mahabalipuram

Table 8 explains the Expectation and Perception on the Accommodation by the visitors of Mahabalipuram revealed that the significance value is (0.000) which is less than 0.05 at 95% level of significance so that null hypothesis is rejected. The mean value of expectation on Accommodation is 20.56 where as perception is 18.90 which shows that there is a negative narrow gap, their expectation is higher where as their perception is lower. We can conclude that there is gap exist between the Accommodation Expectation and Accommodation Perception of the destination Mahabalipuram.

Table 8. Expectation and perception of the accommodation by the visitors of Mahabalipuram

Accommodation		Mean	N	Std. Deviation	Std. Error Mean
Pair 1	Accommodation Expectation	20.5600	100	2.74256	.27426
	Accommodation Perception	18.9000	100	2.15322	.21532

Table 9. Paired samples test

Accommodation		Paired Differences					t	df	Sig. (2-tailed)
		Mean	Std. Deviation	Std. Error Mean	95% Confidence Interval of the Difference				
					Lower	Upper			
Pair 1	Accomodation – Accomodation Perception	1.66000	3.53116	.35312	.95934	2.36066	4.701	99	.000

(Source: Primary data)

H0: There is no significant difference between Expectations and Perceptions of Visitors on Activities of Mahabalipuram

Table 10 explains the Expectation and Perception on the Activities by the visitors of Mahabalipuram revealed that the significance value is (0.000) which is less than 0.05 at 95% level of significance so that null hypothesis is rejected. The mean value of expectation on Amenities is 20.94 where as perception is 18.95 which shows that there is a narrow gap, their expectation is higher where as their perception is lower. We can conclude that there is gap exist between Amenities Expectation and Amenities Perception of the destination Mahabalipuram.

Table 10. Expectation and perception of the activities by the visitors of Mahabalipuram

Activities		Mean	N	Std. Deviation	Std. Error Mean
Pair 1	Activities Expectation	20.94	100	2.766	.277
	Activities Perception	18.9500	100	2.16667	.21667

Table 11. Paired samples test

Activities		Paired Differences					t	df	Sig. (2-tailed)
		Mean	Std. Deviation	Std. Error Mean	95% Confidence Interval of the Difference				
					Lower	Upper			
Pair 1	Activities Expectation – Activities Perception	1.99000	3.45971	.34597	1.30352	2.67648	5.752	99	.000

(Source: Primary data)

H0: There is no significant difference between Expectations and Perceptions of Visitors on Hospitality Services of Mahabalipuram

Table 12 explains the Expectation and Perception on the Hospitality Services by the visitors of Mahabalipuram revealed that the significance value is (0.000) which is less than 0.05 at 95% level of significance so that null hypothesis is rejected. The mean value of expectation on Hospitality Services is 21.00 where as perception is 18.95 which shows that there is a narrow gap, their expectation is higher where as their perception is lower. We can conclude that there is gap exist between the Hospitality Services Expectation and Hospitality Services Perception of the destination Mahabalipuram.

Table 12. Expectation and perception of the hospitality services by the visitors of Mahabalipuram

Hospitality Services		Mean	N	Std. Deviation	Std. Error Mean
Pair 1	Hospitality Expectation	21.00	100	2.274	.227
	Hospitality Perception	18.9500	100	2.16667	.21667

Table 13. Paired samples test

Hospitality Services		Paired Differences					T	Df	Sig. (2-tailed)
		Mean	Std. Deviation	Std. Error Mean	95% Confidence Interval of the Difference				
					Lower	Upper			
Pair 1	Hospitality Expectation – Hospitality Perception	2.05000	3.04304	.30430	1.44619	2.65381	6.737	99	.000

(Source: Primary data)

H0: There is no significant difference between Expectations and Perceptions of Visitors of on Cost Effectiveness of Mahabalipuram

Table 14 explains the Expectation and Perception on the Cost Effectiveness by the visitors of Mahabalipuram revealed that the significance value is (0.000) which is less than 0.05 at 95% level of significance so that null hypothesis is rejected. The mean value of Expectation on Cost Effectiveness is 20.52 where as perception is 18.95 which shows that there is a narrow gap, their expectation is higher where as their perception is lower. We can conclude that there is gap exist between Cost-Effectiveness Expectation and Cost Effectiveness Perception of the destination Mahabalipuram.

Table 14. Expectation and perception of the cost effectiveness by the visitors of Mahabalipuram

Cost-effectiveness		Mean	N	Std. Deviation	Std. Error Mean
Pair 1	Cost-effectiveness Expectation	20.52	100	2.809	.281
	Cost-effectiveness Perception	18.9500	100	2.16667	.21667

Table 15. Paired samples test

Cost-effectiveness		Paired Differences					T	df	Sig. (2-tailed)
		Mean	Std. Deviation	Std. Error Mean	95% Confidence Interval of the Difference				
					Lower	Upper			
Pair 1	Cost-effectiveness Expectation – Cost-effectiveness Perception	1.57000	3.54838	.35484	.86592	2.27408	4.425	99	.000

(Source: Primary data)

CONCLUSION

Mahabalipuram is a unique heritage destination in various aspects. The demographic profile of visitors of Mahabalipuram revealed that more male visitors (55 percent) visited Mahabalipuram than female visitors (45 percent) during the study period. Age distribution of visitors revealed that majority of the respondents were between the age group 21 to 40 years (41 percent), and 90 percent of Indians and 10 percent of Foreigners visited the destination during the study period. Visitor's religion who visited Mahabalipuram during the study period were mostly Hindus (57 per cent) followed by considerable amount of other religious visitors visited during the study period. Educational status of the visitors revealed that majority of the visitors were under graduates with 52 per cent, most of the visitors were working in private organizations (43 per cent). It is also inferred from the study that majority of the visitors were earning 5 to 10 lakhs per annum (38 per cent) and Majority of the respondents were married with 64 per cent followed by singles 36 per cent respectively.

The distribution of travel pattern results revealed that majority of the visitors were first time visitors (42 per cent), most of the visitors has spent two days in Mahabalipuram to explore (40 per cent) followed by one day visitors (32 per cent) and Visitors also expressed that they came to know about Mahabalipuram and its festivals through newspaper mostly (40 per cent) than other sources of information. Visitors felt that Mahabalipuram is suitable to visit during winter (34 per cent), it is also understood from the study that majority of visitors are not aware of the dance festival happening every year (52 per cent). Since most of the visitors have not heard about Dance festivals happening in Mahabalipuram, they have not visited the destination during the festival (66 per cent), it is also understood through the study that majority of the visitors travelled to the destination through Bus (38 per cent) and also by Train (32 per cent). Measures to create awareness on dance festival needs to be done to pull special interest visitors to the destination. This will also entangle the cultural ethos of the heritage centre.

Paired Sample T-test is carried out to find out the perception of the major stake holder of the destination "tourists" for destination competitiveness in Mahabalipuram by analyzing their expectation and perception on 5A's factors of tourism in this heritage destination. The results of the present study revealed that there is narrow difference between Visitors expectation and perception on the factors Attraction, Accessibility, Amenities, Accommodation, Activities, Hospitality Services and Cost effectiveness of the destination Mahabalipuram, hence that visitors´ expectation has not been met by the destination. Studies on push factors are many (Bansal & Eiselt, 2004; Crompton, 1979; Jensen, 2011; Kozak, 2002; Prayag & Ryan, 2010; Yoon & Uysal, 2005). On the other hand, Kozak (2002) used culture, pleasure-seeking/fantasy, relaxation, and being physical as pull factors. The Gap analysis of Pull and push factors is also studied widely (Pansiri & Mmereki, 2011; Pearce, 2001; Pike & Mason, 2011). Taking cues from this the present study reveals basically the pull factors which encompasses the 5A's of a destination. Hence the existing monuments in Mahabalipuram destination should be well conserved and positioned correctly to portray the significant heritage architecture. Recently the central and state have identified the potential of the destination by hosting the remarkable meet of Indian Prime Minster with Chinese President. This may provide a face lift to the destination in the global heritage map.

REFERENCES

Anand, A. (2008). *Tourism Perspective in the 21st Century*. Shubhi Publications.

Andriotis, K. (2005). Community Groups' Perceptions of and Preferences for Tourism Development: Evidence from Crete. *Journal of Hospitality & Tourism Research (Washington, D.C.)*, *29*(1), 67–90. doi:10.1177/1096348004268196

Apostolakis, A. (2003). The Convergence Process In Heritage Tourism. *Annals of Tourism Research*, *30*(4), 795–812. doi:10.1016/S0160-7383(03)00057-4

Apostolakis, A. (2003). The Convergence Process In Heritage Tourism. *Annals of Tourism Research*, *30*(4), 795–812. doi:10.1016/S0160-7383(03)00057-4

Aref, F., Ma'rof, R., & Sarjit, S. G. (2009). Community Perceptions toward Economic and Environmental Impacts of Tourism on Local Communities. *Asian Social Science*, *5*(7), 130–137. doi:10.5539/ass.v5n7p130

Beerli, A., & Martín, D. J. (2004). Tourists' Characteristics and the Perceived Image of Tourist Destinations: A Quantitative Analysis – A Case Study of Lanzarote, Spain. *Tourism Management*, *25*(5), 623–636. doi:10.1016/j.tourman.2003.06.004

Carr, N. (1999). A Study of Gender Differences: Young Tourist Behaviour in a UK Coastal Resort. *Tourism Management*, *20*(2), 223–228. doi:10.1016/S0261-5177(98)00087-9

Chambers, W. (1788). Some accounts of the Sculptures and ruins at Mavalipuram. In 2000 years of Mamallapuram. New Delhi Publication.

Chaudhary, M. (2000). India's Image as a Tourist Destination a Perspective of Foreign Tourist. *Tourism Management*, *21*(3), 293–297. doi:10.1016/S0261-5177(99)00053-9

Choi, H. C., & Murray, I. (2010). Resident Attitudes toward Sustainable Community Tourism. *Journal of Sustainable Tourism*, *18*(4), 575–594. doi:10.1080/09669580903524852

Choi, H. S., & Sirakaya, E. (2005). Measuring Residents' Attitude toward Sustainable Tourism: Development of Sustainable Tourism Attitude Scale. *Journal of Travel Research*, *43*(4), 380–394. doi:10.1177/0047287505274651

Christou, E. (2005). Heritage and cultural tourism: A marketing-focused approach. In M. Sigala & D. Leslie (Eds.), *International Cultural Tourism: Management, Implications and Cases*. Elsevier Butterworth-Heinemann. doi:10.1016/B978-0-7506-6312-0.50004-2

Crompton, J. L. (1979). Motivations for pleasure vacation. *Annals of Tourism Research*, *6*(4), 408–424. doi:10.1016/0160-7383(79)90004-5

Davis, R., Misra, S., & van Auken, S. (2002). A gap analysis approach to marketing curriculum assessment: A study of skills and knowledge. *Journal of Marketing Education*, *24*(3), 218–224. doi:10.1177/0273475302238044

Dredge, D. (2010). Place Change and Tourism Development Conflict: Evaluating Public Interest. *Tourism Management*, *31*(1), 104–112. doi:10.1016/j.tourman.2009.01.004

Drost, A. (1996). Developing Sustainable Tourism for World Heritage Sites. *Annals of Tourism Research*, *23*(2), 479–492. doi:10.1016/0160-7383(96)83345-7

Dwyer, L., & Kim, C. (2003). Destination competitiveness: Determinants and indicators. *Current Issues in Tourism*, *6*(5), 369–414. doi:10.1080/13683500308667962

Dwyer, L., Livaic, Z., & Mellor, R. (2003). Competitiveness of Australia as a tourist destination. *Journal of Hospitality and Tourism Management*, *10*(1), 60–78.

Fazili, A. I., & Ashraf, S. H. (2006). Tourism. In *India Planning & Development*. Sarup and Sons.

Guillet, B. D., Lee, A., Law, R., & Leung, R. (2011). Factors affecting outbound tourists' destination choice: The case of Hong Kong. *Journal of Travel & Tourism Marketing*, *28*(5), 556–566. doi:10.1080/10548408.2011.588120

Hall, C.M., & Piggin, R. (2001). *Tourism and World Heritage in OECD Countries*. Academic Press.

Hallowell, R. (1996). The Relationships of Customer satisfaction, Customer Loyalty, and Profitability: An Empirical Study. *International Journal of Service Industry Management, 7*(4), 27–42. doi:10.1108/09564239610129931

Headley, D. E., & Choi, B. (1992). Achieving service quality through gap analysis and a basic statistical approach. *Journal of Services Marketing, 6*(1), 5–14. doi:10.1108/08876049210035683

http://www.tourismconcern.org.uk/media/2002/WSSD%20sustainable%20tourism%20bri efing.htm

Huang, C.-H., Tsaur, J.-R., & Yang, C.-H. (2012). Does World Heritage List Really Induce More Tourists? Evidence from Macau. *Tourism Management, 33*(6), 1450–1457. doi:10.1016/j.tourman.2012.01.014 PMID:32287738

Hui, T. K., Wan, D., & Ho, A. (2007). Tourists' Satisfaction, Recommendation and Revisiting Singapore. *Tourism Management, 28*(4), 965–975. doi:10.1016/j.tourman.2006.08.008

Jensen, J. M. (2011). The relationships between socio-demographic variables, travel motivations and subsequent choice of vacation. In *2nd International Conference on Economics, Business and Management* (*Vol. 22*, pp. 37–44). Academic Press.

Kaul, H., & Gupta, S. (2009). Sustainable Tourism in India. *Worldwide Hospitality and Tourism Themes, 1*(1), 12–18. doi:10.1108/17554210910949841

Kerstetter, D. L., Confer, J. J., & Graefe, A. R. (2001). An Exploration of the Specialization Concept within the Context of Heritage Tourism. *Journal of Travel Research, 39*(3), 267–274. doi:10.1177/004728750103900304

Kozak, M. (2002). Comparative analysis of tourist motivations by nationality and destinations. *Tourism Management, 23*(3), 221–232. doi:10.1016/S0261-5177(01)00090-5

Kozak, M., Baloglu, S., & Bahar, O. (2010). Measuring destination competitiveness: Multiple destinations versus multiple nationalities. *Journal of Hospitality Marketing & Management, 19*(1), 56–71. doi:10.1080/19368620903327733

Kozak, M., & Rimmington, M. (1999). Measuring tourist destination competitiveness: Conceptual considerations and empirical findings. *International Journal of Hospitality Management, 18*(3), 273–283. doi:10.1016/S0278-4319(99)00034-1

Lee, S., Jeon, S., & Kim, D. (2011). The Impact of Tour Quality and Tourist Satisfaction on Tourist Loyalty: The Case of Chinese Tourists in Korea. *Tourism Management, 32*(5), 1115–1124. doi:10.1016/j.tourman.2010.09.016

Longhurst, A. H. (1930). *Pallava Architecture volume 3. Archaeological Survey of India in Indian History and Architecture website*. Retrieved from http://puratattva.in/2010/09/14/mahabalipuram-the-workshop-of-pallavas-part-v- 41.html

Mahabalipuram – The Workshop of Pallavas – Part I. (2010a, August 3). *Indian History and Architecture A passage to the Indian History and Monuments*. Retrieved from http://puratattva.in/2010/08/03/mahabalipuram-the-workshop-of-pallavas-part-i- 45.html

Mahabalipuram – The Workshop of Pallavas – Part II. (2010b, August 4). *Indian History and Architecture A passage to the Indian History and Monuments.* Retrieved from http://puratattva.in/2010/08/04/mahabalipuram-the-workshop-of-pallavas-part-ii- 44.html

Mahabalipuram – The Workshop of Pallavas – Part IV. (2010c, August 23). *Indian History and Architecture A passage to the Indian History and Monuments.* Retrieved from http://puratattva.in/2010/09/03/mahabalipuram-the-workshop-of-pallavas-part-iv- 42.html

Mahabalipuram – The Workshop of Pallavas – Part V. (2010d, September 14). *Indian History and Architecture A passage to the Indian History and Monuments.* Retrieved from http://puratattva.in/2010/09/14/mahabalipuram-the-workshop-of- pallavas-part-v-41.html

McDowall, S. (2010). A Comparison between Thai Residents and Non-Residents in their Motivations, Performance Evaluations, and Overall Satisfaction with a Domestic Festival. *Journal of Vacation Marketing*, *16*(3), 217–233. doi:10.1177/1356766709356139

Min, H., & Min, H. (1996). Competitive benchmarking of Korean luxury hotels using the analytic hierarchy process and competitive gap analysis. *Journal of Services Marketing*, *10*(3), 58–72. doi:10.1108/08876049610119794

Mok, C., & Armstrong, R. W. (1994). Perception of Australia as a Holiday Destination: A Case of Hong Kong and Taiwanese Tourists. *Australian Journal of Hospitality Management*, *1*(2), 25–28.

Murphy, A. E. (2003). Illustrating the utility of a modified gap analysis as a regional tourism planning tool: Case study of potential Japanese and German travelers to the Cowichan Region. *Journal of Travel Research*, *41*(4), 400–409. doi:10.1177/0047287503041004009

Murphy, P. E. (1983). Perceptions and attitudes of decision-making groups in tourism centers. *Journal of Travel Research*, *21*(2), 8–12. doi:10.1177/004728758302100302

Nicholas, L. N., Thapa, B., & Ko, Y. J. (2009). Residents' Perspectives of a World Heritage Site The Pitons Management Area, St. Lucia. *Annals of Tourism Research*, *36*(3), 390–412. doi:10.1016/j.annals.2009.03.005

Omerzel, D. G. (2006). Competitiveness of Slovenia as a tourist destination. *Managing Global Transitions International Research Journal*, *4*(2), 167–189.

Palmer, P., & O'Neill, M. A. (2004). Wine production and tourism: Adding service to a perfect partnership. *The Cornell Hotel and Restaurant Administration Quarterly*, *45*(3), 269–284. doi:10.1177/0010880404263075

Pansiri, J. (2014). Tourist Motives and Destination Competitiveness:A Gap Analysis Perspective. *International Journal of Hospitality & Tourism Administration*, *15*(3), 217–247. doi:10.1080/15256480.2014.925718

Pansiri, J., & Courvisanos, J. (2010). Attitude to risk in technology-based strategic alliances for tourism. *International Journal of Hospitality & Tourism Administration*, *11*(3), 275–302. doi:10.1080/15256480.2010.498283

Pansiri, J., & Mmereki, R. N. (2010). Using the SERVQUAL model to evaluate the impact of public service reforms in the provision of primary health care in Botswana. *Journal of African Business*, *11*(2), 219–234. doi:10.1080/15228916.2010.509005

Pansiri, J., & Mmereki, R. N. (2011). An evaluation of international tourist motivation and tourism destination competitiveness. In M. J. Gross (Ed.), *Tourism: Creating a brilliant blend: Proceedings of the 21st Annual Conference* (pp. 1–16). Adelaide, Australia: CAUTHE 2011.

Parasuraman, A., Berry, L. L., & Zeithaml, V. A. (1991). Refinement and reassessment of the SERVQUAL Scale. *Journal of Retailing*, *67*(4), 420–450.

Parasuraman, A., & Zeithaml, V. A. (1994). Alternative scales for measuring service quality: A comparative assessment based on psychometric and diagnostic criteria. *Journal of Retailing*, *70*(3), 201–230. doi:10.1016/0022-4359(94)90033-7

Parasuraman, A., Zeithaml, V. A., & Berry, L. (1985). A conceptual model of service quality and its implications for future research. *Journal of Marketing*, *49*(4), 41–50. doi:10.1177/002224298504900403

Parasuraman, A., Zeithaml, V. A., & Berry, L. L. (1985). A conceptual model of service quality and its implications for future research. *Journal of Marketing*, *49*(4), 41–50. doi:10.1177/002224298504900403

Parasuraman, A., Zeithaml, V. A., & Berry, L. L. (1988). SERVQUAL: A multiple item scale for measuring consumer perceptions of service quality. *Journal of Retailing*, *64*(1), 5–6.

Parasuraman, A., Zeithaml, V. A., & Berry, L. L. (1994). Reassessment of expectations as a comparison standard in measuring service quality: Implications for further research. *Journal of Marketing*, *58*(1), 111–124. doi:10.1177/002224299405800109

Park, H. y. (2010). Heritage Tourism Emotional Journeys into Nationhood. *Annals of Tourism Research*, *37*(1), 116–135. doi:10.1016/j.annals.2009.08.001

Peterson, K. (1994). The heritage resource as seen by the tourist: The heritage connection. Journal of Tourism: An Exploration.

Petrick, J. F. (2004). Are Loyal Visitors Desired Visitors? *Tourism Management*, *25*(4), 463–470. doi:10.1016/S0261-5177(03)00116-X

Philips, M., & Jones, A. (2006). Erosion and Tourism Infrastructure in the Coastal Zone: Problems, Consequences and Management. *Tourism Management*, *27*(3), 517–524. doi:10.1016/j.tourman.2005.10.019

Poria, Y., Butler, R., & Airey, D. (2003). The Core of Heritage Tourism. *Annals of Tourism Research*, *30*(1), 238–254. doi:10.1016/S0160-7383(02)00064-6

Rajani, M. B., & Kasturirangan, K. (2013). Sea-Level Changes and its Impact on Coastal Archaeological Monuments: Seven Pagodas of Mahabalipuram, a Case Study. *Photonirvachak (Dehra Dun)*, *41*(2), 461–468. doi:10.100712524-012-0210-y

Ramsey, D., & Everitt, J. (2008). If you dig it, they will come! Archaeology heritage sites and tourism development in Belize, Central America. *Tourism Management*, *29*(5), 909–916. doi:10.1016/j.tourman.2007.11.002

Schouten, F. (1995). Improving visitor care in heritage attractions. *Tourism Management, 16*(4), 259–261. doi:10.1016/0261-5177(95)00014-F

Stankey, G. (1973). Visitor Perception Wilderness Recreation Carrying Capacity. Utah: USDA Forest Service. *Tourism Recreation Research, 26*(1), 103–105.

World Travel and Tourism Council. (1998). *Agenda 21 for the travel and tourism industry—towards environmentally sustainable development.* WTTC.

Tribe, J., & Snaith, T. (1998). From SERVQUAL to HOLSAT: Holiday satisfaction in Varadero, Cuba. *Tourism Management, 19*(1), 25–34. doi:10.1016/S0261-5177(97)00094-0

Truong, T.-H., & Foster, D. (2006). Using HOLSAT to evaluate tourist satisfaction at destinations:The case of Australian holidaymakers in Vietnam. *Tourism Management, 27*(5), 842–855. doi:10.1016/j.tourman.2005.05.008

Valle, P. O., & Silva, J. A., Mendes, J., & Guerreiro, M. (2006). Tourist Satisfaction and Destination Loyalty intention: A Structural and Categorical Analysis. Int. *Journal of Business Science and Applied Management, 1*(1), 26–44.

Vora, K. H., & Sundaresh. (2003). *Mahabalipuram: a saga of glory to tribulations.* Academic Press.

World Commission on Environmental and Development (WCED). (1987). *Our Common Future.* Oxford University Press.

World Summit on Sustainable Development. (2002). *Sustainable Tourism and the Earth Summit. Briefing Paper.* Author.

World Tourism Organisation. (2008). *UNWTO World Tourism Barometer.* Madrid: Author.

World Tourism Organization. (1980). *Manila Declaration on World Tourism.* www.univeur.org/CMS/UserFiles/65.%20Manila.PDF

WWF. (2002). World-Wide Fund for Nature International. UNEP World Conservation Monitoring Center.

Yale, P. (1991). *From Tourist Attractions to Heritage Tourism. Huntingdon.* ELM Publications.

Zouni, G., & Kouremenos, A. (2008). Do tourism providers know their visitors? An investigation of tourism experience at a destination. *Tourism and Hospitality Research, 8*(4), 282–297. doi:10.1057/thr.2008.30

ADDITIONAL READING

Alegre, J., & Garau, J. (2010). Tourist satisfaction and dissatisfaction. *Annals of Tourism Research, 37*(1), 52–73. doi:10.1016/j.annals.2009.07.001

Anwar, S. A., & Sohail, M. S. (2004). Festival tourism in the United Arab Emirates: First-time versus repeat visitor perceptions. *Journal of Vacation Marketing, 10*(2), 161–170. doi:10.1177/135676670401000206

Barker, M., Page, S. J., & Meyer, D. (2003). Urban visitor perceptions of safety during a special event. *Journal of Travel Research, 41*(4), 355–361. doi:10.1177/0047287503041004004

George, R. (2010). Visitor perceptions of crime-safety and attitudes towards risk: The case of Table Mountain National Park, Cape Town. *Tourism Management, 31*(6), 806–815. doi:10.1016/j.tourman.2009.08.011

Hall, C. M., Sharples, L., Mitchell, R., Macionis, N., & Cambourne, B. (Eds.). (2004). *Food tourism around the world.* Routledge. doi:10.4324/9780080477862

Kozak, M. (2001). Comparative assessment of tourist satisfaction with destinations across two nationalities. *Tourism Management, 22*(4), 391–401. doi:10.1016/S0261-5177(00)00064-9

Kozak, M., & Rimmington, M. (2000). Tourist satisfaction with Mallorca, Spain, as an off-season holiday destination. *Journal of Travel Research, 38*(3), 260–269. doi:10.1177/004728750003800308

Murphy, L., Benckendorff, P., & Moscardo, G. (2007). Destination brand personality: Visitor perceptions of a regional tourism destination. *Tourism Analysis, 12*(5-6), 419–432. doi:10.3727/108354207783227948

Nield, K., Kozak, M., & LeGrys, G. (2000). The role of food service in tourist satisfaction. *International Journal of Hospitality Management, 19*(4), 375–384. doi:10.1016/S0278-4319(00)00037-2

Oppermann, M. (2000). Tourism destination loyalty. *Journal of Travel Research, 39*(1), 78–84. doi:10.1177/004728750003900110

Pizam, A., Neumann, Y., & Reichel, A. (1978). Dimentions of tourist satisfaction with a destination area. *Annals of Tourism Research, 5*(3), 314–322. doi:10.1016/0160-7383(78)90115-9

Rand, G. E. D., Heath, E., & Alberts, N. (2003). The role of local and regional food in destination marketing: A South African situation analysis. *Journal of Travel & Tourism Marketing, 14*(3-4), 97–112. doi:10.1300/J073v14n03_06

Yüksel, A., & Yüksel, F. (2003). Measurement of tourist satisfaction with restaurant services: A segment-based approach. *Journal of Vacation Marketing, 9*(1), 52–68. doi:10.1177/135676670200900104

KEY TERMS AND DEFINITIONS

Loyalty Intention: The intention to repeat a purchase after having a good experience with the product.

Sustainable Tourism: Net of practices and theories aimed at protecting the non-renewable resources that make tourism possible.

Tourism Management: A discipline emerged to enhance the performance of tourism industry and tourist destinations

Tourist Destination: It is the place designed to receive and recreate tourists.

Tourist Satisfaction: The concept refers to the levels of conformity expressed by consumers after having a tourist experience.

Compilation of References

Abdollahfam, R., & Keshavarz, P. (2015). Feasibility Study of Tourism Development Impacts on Stable Regional Development Dimensions: The Case Study. *Life Science Journal*, *12*(7), 88–96.

Abubakar, A. M., Behravesh, E., Rezapouraghdam, H., & Yildiz, S. B. (2019). Applying artificial intelligence technique to predict knowledge hiding behavior. *International Journal of Information Management*, *49*, 45–57. doi:10.1016/j.ijinfomgt.2019.02.006

Agarwal, S., & Singh, R. (2016). 21 Corporate social responsibility practices in the Indian food industry. *A Stakeholder Approach to Managing Food: Local, National, and Global Issues*, 297-305.

Aggarwal, V. S., Handa, M., & Singh, K. A. (2010). Expectation and Satisfaction of Adventure Tourists for Land, Air and Water based Sports in India. *DIAS Technology Review*, *7*(2), 9–19.

Aien, S., Kordjamshidi, M., & Pourfallah, H. (2012). *An investigation on thermal performance of rocky architecture approaching thermal comfort with less energy load.* 28th international plea conference opportunities. Limits & needs towards an environmentally responsible architecture, Lima, Peru.

Airbnb. (2016). *Airbnb and the Rise of the Millennial Travel.* Retrieved from: https://www.airbnb.com/

Airbnb. (2017). *Across the BRICS: How Airbnb connects the emerging economies.* Retrieved from: https://www.airbnb.com/BRICS

Airbnb. (2019). *Disaster Response Program.* Retrieved from: https://www.airbnb.com/disastor-response

Akama, J. S., & Kieti, D. M. (2003). Measuring tourist satisfaction with Kenya's Wildlife safari: A case study of Tsavo West National Park. *International Journal of Tourism Management*, *24*(1), 73–81. doi:10.1016/S0261-5177(02)00044-4

Akbarpour, M., Mosaviaraghi, M., Rahimi, E., Rahimi, M., & Yaqobi, R. (2013). The Approach of the Host Community to the Expansion of Tourism in the Rural Areas (Case Study: Kandovan village, Oskoo City, Iran). *Global Journal of Science. Engineering and Technology*, *13*(1), 15–29.

Alfreds, D. (2016). *Here's how much money SA Airbnb hosts earn, Fin24.* Retrieved from: https://www.fin24.com/Tech/News

Alipour, H., Malazizi, N., & Rezapouraghdam, H. (2018, June). Complementing Sustainability through Green Marketing: From Tourism Operator's Perspective. In *8th Advances in Hospitality and Tourism Marketing And Management (Ahtmm) Conference* (p. 636). Academic Press.

Alipour, H., Olya, H. G., Hassanzadeh, B., & Rezapouraghdam, H. (2017). Second home tourism impact and governance: Evidence from the Caspian Sea region of Iran. *Ocean and Coastal Management*, *136*, 165–176. doi:10.1016/j.ocecoaman.2016.12.006

Alipour, H., Rezapouraghdam, H., & Esmaeili, B. (2020). Public Health and Tourism; A Personalist Approach to Community Well-Being: A Narrative Review. *Iranian Journal of Public Health*, *49*(3), 447–453.

Alipour, H., Rezapouraghdam, H., & Hasanzade, B. (2019). A Holistic Analysis of the Second-home Tourism Impacts in the Caspian Sea Region of Iran. *Experiencing Persian Heritage (Bridging Tourism. Theory into Practice*, *10*, 275–293.

Almeida-Alonso, M. M., & Millán-Borrajo, F. (2019). Are Social Media Data Pushing Overtourism? The Case of Barcelona and Chinese Tourists. *Sustainability*, *11*(12), 3356. doi:10.3390u11123356

Alonso Ferreras, V. H. (2002). Quality, competition and client Satisfaction in Organisation Offering tourism Sciences. *Studies and Perspective in Tourism.*, *11*(3-4), 253–268.

Altin, M., Schwartz, Z., & Uysal, M. (2017). Where you do it matters: The impact of hotels' revenue-management implementation strategies on performance. *International Journal of Hospitality Management*, *67*, 46–52. doi:10.1016/j.ijhm.2017.08.001

Anand, A. (2008). *Tourism Perspective in the 21st Century*. Shubhi Publications.

Andrews, P. (2009). Identity, rhetoric and culture. *Cambridge Journal of Education*, *39*(2), 159–161. doi:10.1080/03057640902934057

Andriotis, K. (2005). Community Groups' Perceptions of and Preferences for Tourism Development: Evidence from Crete. *Journal of Hospitality & Tourism Research (Washington, D.C.)*, *29*(1), 67–90. doi:10.1177/1096348004268196

Apleni, L. (2017). 'Residents' perceptions on urban tourism as a catalyst for economic development: A case study of Buffalo City, South Africa. *African Journal of Hospitality, Tourism and Leisure*, *6*(3), 1–12.

Apleni, L., Vallabh, D., & Henama, U. S. (2017). Motivation for tourist's participation in religious tourism in Eastern Cape: A case study of Buffalo City, South Africa. *African Journal of Hospitality, Tourism and Leisure*, *6*(2), 1–14.

Apostolakis, A. (2003). The Convergence Process In Heritage Tourism. *Annals of Tourism Research*, *30*(4), 795–812. doi:10.1016/S0160-7383(03)00057-4

Aref, F., Ma'rof, R., & Sarjit, S. G. (2009). Community Perceptions toward Economic and Environmental Impacts of Tourism on Local Communities. *Asian Social Science*, *5*(7), 130–137. doi:10.5539/ass.v5n7p130

Arival Travel. (2018). *Airbnb*. www.arival.travel

Arksus, A., & Tarkan, E. (2002). Leadership on hospitality industry: An study of case in Turkey. *Studies and Perspective in Tourism.*, *11*(3-4), 253–268.

Arowosafe, F. C., & Emmanuel, A. A. (2014). Investigating indicators for tourist satisfaction at Mole National Park, Ghana. *American Journal of Tourism Management*, *4*(1), 1–6.

Asghari Kaljahi, E., & Amini Birami, F. (2015). Engineering geological properties of the pyroclastic cone-shaped rocky houses of Kandovan, Iran. *Bulletin of Engineering Geology and the Environment*, *74*(3), 959–969. doi:10.100710064-014-0679-4

Ashrafi, N. (2013). Preservation of Kandovan, Based on Pathology of Population Decline. *International Journal of Architecture and Urban Development*, *3*(4), 71–80.

Assefa, B. (2011). *Assessing Tourist Satisfaction in Ethiopia* (Master Thesis). Addis Ababa University.

Aubke, F., Wöber, K., Scott, N., & Baggio, R. (2014). Knowledge sharing in Revenue management teams: Antecedents and consequences of group cohesion. *International Journal of Hospitality Management*, *41*, 149–157. doi:10.1016/j.ijhm.2014.05.010

Avraham, E. (2015). Destination image repair during crisis: Attracting tourism during the Arab Spring uprisings. *Tourism Management*, *47*, 224–232. doi:10.1016/j.tourman.2014.10.003

Ayikoru, M., Tribe, J., & Airey, D. (2009). Reading tourism education: Neoliberalism unveiled. *Annals of Tourism Research*, *36*(2), 191–221. doi:10.1016/j.annals.2008.11.001

Badenhorst-Weiss, J., & Nel, J.D. (2011). A conceptual framework to analyse supply-chain designs. *Acta Commercii*, 1-16.

Ballart Hernandez, J., & Tresseras, J. (2001). *Gestión del patrimonio cultural*. Ariel.

Ball, R. (1988). *Seasonality in the UK labour market*. Averbury.

Balmford, A., Beresford, J., Green, J., Naidoo, R., Walpole, M., & Manica, A. (2009). A global perspective on trends in nature-based tourism. *PLoS Biology*, *7*(6), e1000144. doi:10.1371/journal.pbio.1000144 PMID:19564896

Battour, M., Ismail, N. M., Battour, M., & Awais, M. (2014). Islamic tourism: An empirical examination of travel motivation and satisfaction in Malaysia. *Current Issues in Tourism*. Advance online publication. doi:10.1080/13683500.2014.965665

Bauman, Z. (2007). *Consuming Life*. Polity Press.

Beck, J., Knutson, B., Cha, J., & Kim, S. (2011). Developing Revenue managers for the Lodging Industry. *Journal of Human Resources in Hospitality & Tourism*, *10*(2), 182–194. doi:10.1080/15332845.2011.536941

Beerli, A., & Martín, D. J. (2004). Tourists' Characteristics and the Perceived Image of Tourist Destinations: A Quantitative Analysis – A Case Study of Lanzarote, Spain. *Tourism Management*, *25*(5), 623–636. doi:10.1016/j.tourman.2003.06.004

Beerli, A., & Martin, J. D. (2004). Factors influencing destination image. *Annals of Tourism Research*, *31*(3), 657–681. doi:10.1016/j.annals.2004.01.010

Benko, G. (2000). Estrategias de comunicación y marketing urbano. EURE, 26(79), 67-76. doi:10.4067/S0250-71612000007900004

Benner, M. (2019). From overtourism to sustainability: A research agenda for qualitative tourism development in the Adriatic. *Munich Personal RePEc Archive*, 1–18. Retrieved from https://mpra.ub.uni-muenchen.de/92213/1/MPRA_paper_92213.pdf

Bessiere, J. (1998). Local development and heritage: Traditional food and cuisine as tourist attractions in rural areas. European Society for rural. *Sociology*, *38*(1).

Bigne, E., & Andreu, L. (2004). Emotion in segmentation: An empirical study. *Annals of Tourism Research*, *31*(3), 517–539. doi:10.1016/j.annals.2003.12.018

Bizcommunity. (2017). Airbnb Experiences one year later: Foodies and millennials lead the way. *Tours News*. https://www.bizcommunity.com/Article/1/

Bochner, S., & Coulon, L. (1997). A culture assimilator to Train Australian Hospitality Industry Workers serving Japanese Tourists. *Journal of Tourism Studies*, *8*(2).

Boorstin, D. (1964). *The Image: A Guide to Pseudo-Events in America*. Atheneum.

Bornhost, T., Ritchie, J., & Sheehan, L. (2010). Determinants of tourism success for DMOs and destination: An empirical examination of stakeholders' perspectives. *Tourism Management*, 31.

Boston Consulting Group. (2013). *Traveling with Millennials*. Retrieved from: https://www.bcg.com

Bouchair, A. (2015). Vernacular architecture: Towards a sustainable future. CRC Press, Taylor and Francis Group.

Boyer, M. (1982). *Le tourisme*. Editions du Seuil.

Boyer, M. (2003). *Le tourisme en France*. Editions EMS.

Brenner, L., & Aguilar-Garcı'a, A. (2002). Luxury tourism and regional economic development in Mexico. *The Professional Geographer*, *54*(4), 500–520. doi:10.1111/0033-0124.00346

Bruner, E. M. (1994). Abraham Lincoln as authentic reproduction: A critique of postmodernism. *American Anthropologist*, *96*(2), 397–415. doi:10.1525/aa.1994.96.2.02a00070

Brymer, R. A. (1991). Employee empowerment: A guest driven leadership strategy. *The Cornell Hotel and Restaurant Administration Quarterly*, *32*(1), 58–68.

Buhalis, D. (2000). Marketing the competitive destination of the future- Growth strategies for accomodation establishments in alpine regions. *Tourism Management*, *21*(1), 97–116. doi:10.1016/S0261-5177(99)00095-3

Burawoy, M. (1979). *Manufacturing Consent. Changes in the Labor Process under monopoly Capitalism*. University of Chicago Press.

Business Partners. (2018). *Fear of failure hinders South Arica's aspiring enterpreneurs*. Retrieved from: https://www.businesspartners./en-za/media-centre/media-release/

Cameron, L., & Potvin, C. (2016). Characterizing desired futures of Canadian communities. *Futures*, *82*, 37–51. doi:10.1016/j.futures.2016.05.003

Capanegra, A. (2007). *La Política Turística Argentina del Siglo XX; emergencia e institucionalización del turismo como Política y Planificación pública 1930-2001*. Jornadas Nacionales de Sociología 2007, Mesa 37: Sociología del Turismo, Facultad de Ciencias Sociales, Universidad de Buenos Aires.

Capocchi, A., Vallone, C., Pierotti, M., & Amaduzzi, A. (2019). Overtourism: A Literature Review to Assess Implications and Future Perspectives. *Sustainability*, *11*(12), 3303. doi:10.3390u11123303

Carlsen, J. (1999). A system approach to island tourism destination management. *Systems Research and Behavioral Science*, *16*(4), 321–327. doi:10.1002/(SICI)1099-1743(199907/08)16:4<321::AID-SRES255>3.0.CO;2-5

Carman, M. (2006). *Las trampas de la cultura: los intrusos y los nuevos usos de la cultura en el barrio del Abasto*. Editorial Paidós.

Carrion, F. (2000). *Lugares o flujos centrales: los centros históricos urbanos, Santiago de Chile, CEPAL, Serie Medio Ambiente y Desarrollo. CEDEM*. Anuarios Estadísticos. Centro de Estudios para el Desarrollo Económico Metropolitano. Secretaría de Desarrollo Económico. Gobierno de la Ciudad de Buenos Aires.

Carr, N. (1999). A Study of Gender Differences: Young Tourist Behaviour in a UK Coastal Resort. *Tourism Management*, *20*(2), 223–228. doi:10.1016/S0261-5177(98)00087-9

Castelltort, M., & Mader, G. (2010). A Monetary Public Value (MPV) analysis. *Tourism Management*, 31.

Cetin, G., Demirciftci, T., & Bilgihan, A. (2016). Meeting Revenue management challenges: Knowledge, skills and abilities. *International Journal of Hospitality Management*, 57.

Chacko, H. E., & Marcell, M. H. (2008). Repositioning a Tourism Destination. *Journal of Travel & Tourism Marketing, 23*(2-4), 223–235. doi:10.1300/J073v23n02_17

Chambers, R., & Conway, G. (1992). *Sustainable rural livelihoods: Practical concepts for the 21st century*. IDS discussion paper 296. Brighton, UK: Institute of Development Studies, University of Sussex.

Chambers, W. (1788). Some accounts of the Sculptures and ruins at Mavalipuram. In 2000 years of Mamallapuram. New Delhi Publication.

Chaney, S., & Ryan, C. (2012). Analyzing the Evolution of Singapore's World Gourmet Summit: An Example of Gastronomic Tourism. *International Journal of Hospitality Management, 31*(2), 309–318. doi:10.1016/j.ijhm.2011.04.002

Charles, K., & Marshall, L. (1992). Motivational preferences of Caribbean Hotel workers: an explanatory study. *International Journal of Contemporary Hospitality Management, 4*(3), 25-29.

Chaudhary, M. (2000). India's Image as a Tourist Destination a Perspective of Foreign Tourist. *Tourism Management, 21*(3), 293–297. doi:10.1016/S0261-5177(99)00053-9

Cheer, J. M., Milano, C., & Novelli, M. (2019). Tourism and community resilience in the Anthropocene: Accentuating temporal overtourism. *Journal of Sustainable Tourism, 27*(4), 554–572. doi:10.1080/09669582.2019.1578363

Cherif, K. (2017). *Economie du tourisme - un investissement d'avenir pour l'Algérie*. Casbah Editions, Alger.

Chesney-Lind, M., & Lind, I. Y. (1986). Visitors as Victims: Crimes against Tourists in Hawaii. *Annals of Tourism Research, 13*(2), 167–191. doi:10.1016/0160-7383(86)90036-8

Choi, H. C., & Murray, I. (2010). Resident Attitudes toward Sustainable Community Tourism. *Journal of Sustainable Tourism, 18*(4), 575–594. doi:10.1080/09669580903524852

Choi, H. S., & Sirakaya, E. (2005). Measuring Residents' Attitude toward Sustainable Tourism: Development of Sustainable Tourism Attitude Scale. *Journal of Travel Research, 43*(4), 380–394. doi:10.1177/0047287505274651

Chon, K. S., & Olsen, M. D. (1991). Functional and Symbolic Approaches to Consumer Satisfaction/Dissatisfaction. *Journal of the International Academy of Hospitality Research, 28*, 1–20.

Christou, E. (2005). Heritage and cultural tourism: A marketing-focused approach. In M. Sigala & D. Leslie (Eds.), *International Cultural Tourism: Management, Implications and Cases*. Elsevier Butterworth-Heinemann. doi:10.1016/B978-0-7506-6312-0.50004-2

Chua, E. L., Chiu, J. L., & Bool, N. C. (2019). Sharing Economy: An Analysis of Airbnb Business Model and the Factors that Influence Consumer Adoption. *Review of Integrative Business and Economic Research, 8*(2), 19–37.

Churchil, G., & Surprenant, C. (1982). An investigation into the determinants of customer satisfaction. *JMR, Journal of Marketing Research, 19*(4), 491–504. doi:10.1177/002224378201900410

Ciccollela, P. (1999). Globalización y dualización en la Región Metropolitana de Buenos Aires. Grandes inversiones y reestructuración socio-territorial en los años noventa. *Revista Eure, 25*(26), 5–27.

Claudio, M., Joseph, M. C., & Marina, N. (2018). Overtourism: a growing global problem. *The Conservation*. Retrieved from https://theconversation.com/overtourism-a-growing-global-problem-100029

Clemons, S.D. & Woodruff, R.B. (1992). Broadening the view of Consumer (Dis) satisfaction: A proposed Means – end Disconfirmation model of CS/D. *American Marketing Association, 13*(5), 413-421.

Cooper, C., & Hall, C. M. (2008). *Contemporary tourism: an international approach*. Elsevier.

Corigliano, M. A. (2002). The route to quality: Italian gastronomy networks in operation. In A.-M. Hjalager & G. Richards (Eds.), *Tourism and gastronomy* (pp. 166–185). Routledge.

Cortez Puya, T. (2005). *Recuperación de patrimonio cultural urbano como recurso turístico*. Universidad Complutense de Madrid.

Costa, M. J. (2017). *Total Hotel Revenue Management em Portugal* (Master Thesis). Escola Superior de Hotelaria e Turismo, IPP, Porto, Portugal.

Costa, V. F., Monte, A. P., & Fernandes, P. O. (2013). Tourism revenue for the north region of Portugal: an econometric analysis. *III Congresso International de Turismo, ESG/IPCA*, 241-259.

Crompton, J. L. (1979). Motivations for pleasure vacation. *Annals of Tourism Research*, *6*(4), 408–424. doi:10.1016/0160-7383(79)90004-5

Cross, R. G., Higbie, J. A., & Cross, Z. N. (2011). Milestones in the application of analytical pricing and revenue management. *Journal of Revenue and Pricing Management*, *10*(1), 8–18. doi:10.1057/rpm.2010.39

Cruz, M. S., & Zaragoza, M.P.P. (2019). Analysis of the accommodation density in coastal tourism areas of insular destinations from the perspective of overtourism. *Sustainability, 11*(11). doi:10.3390u11113031

Curtis, M., & Zahrn, F. (2014). Forecasting for Revenue Management: An Introduction. *The International Journal of Applied Forecasting*, *34*, 32–38.

Dabour, N. (2003). Problems and prospects of sustainable tourism development in the OIC countries: Ecotourism. *Journal of Economic Cooperation, 24*, 25-62.

Daft, R. L. (2011). *Leadership* (5th ed.). Cengage.

Damayanti, V., Ekasari, A. M., & Syaodih, E. (2019). The Determinants of Cultural Tourism Attractions Based on Tourist Satisfaction in the Sundapolis Area, Bandung City. *Advances in Social Science. Education and Humanities Research*, *409*(1), 538–541.

Dar, H. (2018). Satisfaction of Domestic Tourists Visiting Gulmarg in Jammu and Kashmir, India. *International Journal on Recent Trends in Business and Tourism*, *2*(1), 16–22.

Dasgupta, D. (2011). *Tourism Marketing*. Pearson Education.

Dávid, L. (2011). Tourism ecology: Towards the responsible, sustainable tourism future. *Worldwide Hospitality and Tourism Themes*, *3*(3), 210–216. doi:10.1108/17554211111142176

Davidson, R., & Cope, B. (2003). Business Travel. Prentice Hall.

Davis, P. (2006). Beyond Human Resource management in co-operatives. *Cross Cultural Management*, *13*(1), 69–95. doi:10.1108/13527600610643493

Davis, R., Misra, S., & van Auken, S. (2002). A gap analysis approach to marketing curriculum assessment: A study of skills and knowledge. *Journal of Marketing Education*, *24*(3), 218–224. doi:10.1177/0273475302238044

De Kadt, E. (1979). *Tourism: Passport to Development. In Perspectives on thesocial andcultural effects of tourism in developing countries*. World Bank Press.

De Villiers, J. (2018). There are now 40% more Airbnb hosts over the age of 60 in SA than a year ago. *Business Insider*. Retrieved from: https:www.businessinsider.co.za/-south-african

Deery, M., & Jago, L. (2009). A framework for work-life balance practices: Addressing the needs of the tourism industry. *Tourism and Hospitality Research*, *9*(2), 97–108. doi:10.1057/thr.2009.4

Demen-Meyer, C. (2005). Le tourisme: Essai de definition. *Management & Avenir*, *3*(1), 7–25.

Department of Tourism. Rajasthan. (2019). *Tourism Department Annual Progress Report 2018-19*. Published By Govt. of Rajasthan. Retrieved from http://www.tourism.rajasthan.gov.in/annual-progress-report.html

Destiny. (2018). Airbnb hosts across SA made R3.8bn in the past year. *Destiny Connect*. Retrieved from: https://www.destinyconect.com/

Dewailly, J. M. (2006). *Tourism & Geography: between peregrinity and chaos*. Harmattan.

DigitalTrends. (2018). *Airbnb targets crious travelers as 'experiences' expands to 200 U.S. cities*. Retrieved from: https://www.digitaltrends.co.za/web

Dincera, F. I., Yimaza, S., & Dincera, M. Z. (2015). The Economic Contribution of Turkish Tourism Entrepreneurship on the Development of Tourism Movements In Islamic Countries. *Procedia: Social and Behavioral Sciences*, *195*, 413–422. doi:10.1016/j.sbspro.2015.06.483

Dordevic, M. (2010). Supply Chain Management in tourism. In *4th International Quality Conference*. University of Kragujevac.

Dredge, D. (2010). Place Change and Tourism Development Conflict: Evaluating Public Interest. *Tourism Management*, *31*(1), 104–112. doi:10.1016/j.tourman.2009.01.004

Drost, A. (1996). Developing Sustainable Tourism for World Heritage Sites. *Annals of Tourism Research*, *23*(2), 479–492. doi:10.1016/0160-7383(96)83345-7

Duan, W., Gu, B., & Whinston, A. B. (2008). Do online reviews matter? An empirical investigation of panel data. *Decision Support Systems*, *45*(4), 1007–1016. doi:10.1016/j.dss.2008.04.001

Dwyer, L. (2008). Destination compititiveness and policy making for poverty reduction. *Sustaining Quality of Life through Tourism*, 24–28.

Dwyer, L., & Kim, C. (2003). Destination competitiveness: Determinants and indicators. *Current Issues in Tourism*, *6*(5), 369–414. doi:10.1080/13683500308667962

Dwyer, L., Livaic, Z., & Mellor, R. (2003). Competitiveness of Australia as a tourist destination. *Journal of Hospitality and Tourism Management*, *10*(1), 60–78.

Early, P. C. (1989). Social Loafing and collectivism: A comparison of the United States and the People's in the Republic of China. *Administrative Science Quarterly*, *34*(4), 565–581. doi:10.2307/2393567

Echtner, C. M., & Ritchie, J. B. (1991). The meaning and measurement of destination image. *Journal of Tourism Studies*, *2*(2), 2–12.

Eftekhari, A. R., Mahdavi, D., & Pourtaheri, M. (2011). *Evaluation of tourism sustainability in cultural and historical villages of Iran: The emphasis on sustainable tourism development paradigm*. Scientific Information Database. Retrieved from http://www.sid.ir

Egger, R. (2008). *Restructuring the Destination Management system paradigm*. Unpublished manuscript.

Elhannani, F. E., Bousalem, A. B., & Benbouziane, M. (2018). Economic Diversification and Trade Openness in Algeria Empirical Investigation. *Topics in Middle Eastern and African Economies Proceedings of Middle East Economic Association, 20*(1), 1–09.

Expedia. (2018). *Millennial Traveller Report: Why millennials will shape the next 20 years of travel.* Future Foundation.

Fainstein, S., & Judd, D. (1999). *The Tourist city.* Yale University Press.

Faria, A. G. (2015). *O Departamento de Revenue Management de um grupo hoteleiro: Análise dos Tivoli Hotels & Resorts.* Escola Superior de Turismo e Tecnologia do Mar, Instituto Politécnico de Leiria.

Fazili, A. I., & Ashraf, S. H. (2006). Tourism. In *India Planning & Development.* Sarup and Sons.

Fernandes, C. F., George, B., & Mishra, A. (2016). Delivering Hospitality through Franchising: The Case of Monginis, India. In Strategic Tools and Methods for Promoting Hospitality and Tourism Services (pp. 139-148). IGI Global.

Fields, K. (2002). Demand for the gastronomy tourism product: motivational factors. In A.-M. G. R. Hjalager (Ed.), *Tourism and gastronomy* (pp. 36–50). Routledge.

Fitriana, R., Saragih, J., & Luthfiana, N. (2017, December). Model business intelligence system design of quality products by using data mining in R Bakery Company. [). IOP Publishing.]. *IOP Conference Series. Materials Science and Engineering, 277*(1), 012005.

Font, X., Tapper, R., Schwartz, K., & Kornilaki, M. (2008). Sustainable Supply Chain Management in Tourism. *Business Strategy and the Environment, 17*(4), 260–271. doi:10.1002/bse.527

Forno, F., & Garibaldi, R. (2015). Sharing Economy in Travel and Tourism: The Case of Home-Swapping in Italy. *Journal of Quality Assurance in Hospitality & Tourism, 16*(2), 202–220. doi:10.1080/1528008X.2015.1013409

Franch, M., Martini, U., Buffa, F., & Parisi, G. (2008). 4l tourism (landscape, leisure, learning and limit) responding to new motivations and expectations of tourist to improbé the competitiveness of alpine destination in sustainable way. *Tourism Review, 63*(1), 4–14. doi:10.1108/16605370810861008

Fuller, J.. Matzler, K., & Faullant, R. (2006). *Asymetric effects in customer satisfaction.* Academic Press.

Gabel, D. (2016). Uber and the persistence of market power. *Journal of Economic Issues, 50*(2), 527–534. doi:10.1080/00213624.2016.1179060

Garreton, M. (2001). *El espacio cultural latinoamericano. Bases para una política cultural de integración.* Santiago de Chile, Fondo de Cultura Económica.

Gartell, R. B. (1988). *Destination Marketing.* Kent Hunt Publisher.

Ghaderi, Z., Hatamifar, P., & Khalilzadeh, J. (2018). *Analysis of tourist satisfaction in tourism supply chain management.* Anatolia. doi:10.1080/13032917.2018.1439074

Ghasemzadeh, B. (2013). An opportunity for tourism development with troglodytic architecture. *Research Journal of Applied Sciences, Engineering and Technology, 5*(12), 3294–3297. doi:10.19026/rjaset.5.4570

GhouchiA. (2014). http://press.jamejamonline.ir/newspreview/1532554389176220663

Giddens, A. (1991). *Modernity and self-identity.* Polity Press.

Gilliland, M. (2010). Defining "demand" for demand forecasting. *Foresight: The International Journal of Applied Forecasting.*

Giri, A. V. (2018). *Hurdles and Barriers in the Development of Bakery Business*. Idea Publishing.

Gnoth, J. (2002). Leveraging export brands through a tourism destination brand.J Brand Management. *Journal of Brand Management*, *9*(4), 262–280. doi:10.1057/palgrave.bm.2540077

Gomes, J. (2004). *Como aumentar as receitas de um negócio: o conceito de gestão de receitas*. Revista Hotéis de Portugal.

Goodwin, H. (2017). *The Challenge of Overtourism*. Responsible Tourism Partnership. doi:10.1016/S0140-6736(84)91114-0

Gorelik, A. (2006). Buenos Aires, de la crisis al boom. *Revista Punto de Vista*, *84*, 1–8.

Gotham, K. F. (2005). Tourism gentrification: The case of new Orleans' vieux carre (French Quarter). *Urban Studies (Edinburgh, Scotland)*, *42*(7), 1099–1121. doi:10.1080/00420980500120881

Gottdiener, M. (1994). *The New Urban Sociology*. McGraw-Hill.

Grimes, S., Bouchair, A., & Tebouche, H. (2017). Sustainability of the Expansion Areas for Coastal Touristic Sites "E.A.C.T.S" Such as the case of El-Aouana in Algeria: Indicators for considering biodiversity. *Energy Procedia*.

Guattari, F., & Rolnik, S. (2005). *Micropolíticas*. Editorial Tinta Limón.

Guillet, B. D., Lee, A., Law, R., & Leung, R. (2011). Factors affecting outbound tourists' destination choice: The case of Hong Kong. *Journal of Travel & Tourism Marketing*, *28*(5), 556–566. doi:10.1080/10548408.2011.588120

Gursoy, D., & Rutherford, D. (2004). Host attitudes toward tourism: An improved structural model. *Annals of Tourism Research*, *31*(3), 495–516. doi:10.1016/j.annals.2003.08.008

Guttentag, D. (2006). *Airbnb: Why tourists choose it and how they get it*. University of Waterloo.

Hair, J., Anderson, R. E., Tatham, R. L., & Black, W. C. (1995). *Multivariate data analysis* (4th ed.). Prentice-Hall Inc.

Haji Ebrahim Zargar, A. (1994). *Assessing the rocky phase of Kandovan Village*. Shahid Beheshti University.

Halewood, C., & Hannam, K. (2001). Viking heritage tourism: Authenticity and commodification. *Annals of Tourism Research*, *28*(3), 565–580. doi:10.1016/S0160-7383(00)00076-1

Hall, C.M., & Piggin, R. (2001). *Tourism and World Heritage in OECD Countries*. Academic Press.

Hall, C. M., & Sharples, E. (2003). *The consumption of experiences or the experience of consumption? An introduction to the tourism of taste*. In C. M. Hall, E. Sharples, R. Mitchell, N. Macionis, & B. Cambourne (Eds.), *Food tourism around the world. Development, management and markets* (pp. 1–24). Butterworth Heinemann.

Hall, C. M., & Sharples, E. (Eds.). (2008). *Food and wine festivals and events around the world: Development, management and markets*. Butterworth Heinemann. doi:10.4324/9780080887951

Hallowell, R. (1996). The Relationships of Customer satisfaction, Customer Loyalty, and Profitability: An Empirical Study. *International Journal of Service Industry Management*, *7*(4), 27–42. doi:10.1108/09564239610129931

Hall, S. (1995). Hotel's success breeds happy employees. *Hotel and Hotel Management.*, *210*(18), 3–4.

Harvey, D. (1990). *The Condition of Posmodernity*. Blackwell.

Harvey, D. C. (2001). Heritage pasts and heritage presents: Temporality, meaning and the scope of heritage studies. *International Journal of Heritage Studies*, *7*(4), 319–338. doi:10.1080/13581650120105534

Hassan, M. M., & Shahnewaz, M. (2014). Measuring tourist service satisfaction at destination: A case study of Cox's Bazaar sea beach, Bangladesh. *American Journal of Tourism Management*, *3*(1), 32–43.

Hayes, D. K., & Miller, A. A. (2011). *Revenue management for the Hospitality Industry*. John Wiley & Sons, Inc.

Haywood, J., Mayock, P., Freitag, J., Owoo, K. A., & Fiorilla, B. (2017). *Airbnb & Hotel Performance: An analysis of properietary data in 13 global markets*. Retrieved from: https://wwww.str.com

Headley, D. E., & Choi, B. (1992). Achieving service quality through gap analysis and a basic statistical approach. *Journal of Services Marketing*, *6*(1), 5–14. doi:10.1108/08876049210035683

Heijden Van der, K., Bradfield, R., Burt, G., Cairns, G., & Wright, G. (2002). The sixth sense: accelerating organizational learning with scenarios. Chichester, UK: Wiley.

Helmore, K., & Singh, N. (2001). Sustainable livelihoods – Building on the wealth of the poor. Bloomfield: Kumarian Press.

Henama, U. S. (2017a). Airbnb a solution for lack of tourism stock in Knysna. *Cape Argus*. Retrieved from://www.pressreader.com/south-africa/cape-argus

Henama, U. S. (2013). Attracting Indian Outbound Tourists to South Africa: A BRICS Perspective. *India Quarterly*, *69*(3), 229–247. doi:10.1177/0974928413489466

Henama, U. S. (2014a). Health tourism in South Africa: Opportunities and possibilities. *African Journal of Hospitality, Tourism and Leisure*, *3*(1), 1–14.

Henama, U. S. (2014b). The demise of 1Time Airline and the reaction of various interest groups. *African Journal of Hospitality, Tourism and Leisure*, *3*(20), 1–11.

Henama, U. S. (2017b). International Migration and National Tourism Industry Development: The Case of South Africa. The EUrASEANSs. *Journal of Global Socio-Economic Dynamics*, *2*(3), 72–82.

Henama, U. S. (2018a). A literature review into South Africa's attempt to create a nudist beach at Ray Nkonyeni Municipality, KwaZulu-Natal. *African Journal of Hospitality, Tourism and Leisure*, *7*(3), 1–14.

Henama, U. S. (2018b). Disruptive entrepreneurship using Airbnb: The South African experience. *African Journal of Hospitality, Tourism and Leisure*, *7*(1), 1–13.

Henama, U. S. (2019a). The Sharing Economy in South Africa's Tourism Industry: The Case of Uber E-Hailing Taxi Services. In O. Oncioiu (Ed.), *Improving Business Performance Through Innovation in the Digital Economy*. IGI Publishing.

Henama, U. S. (2019b). *Tourism as a Neoliberal Economic Messiah: The Case of South Africa. Neoliberalism in the Tourism and Hospitality Sector*. IGI. doi:10.4018/978-1-5225-6983-1.ch003

Henama, U. S., & Sifolo, P. P. S. (2017). Tourism Migration in South Africa: Current Dynamics, Immediate Challenges and Future Prospects. *International Journal of Innovation in the Digital Economy*, *8*(1), 47–60. doi:10.4018/IJIDE.2017010103

Henderson, L. (2016). Innovation or Rule Breakers? Regulating Uber, Airbnb & The Sharing Economy (Bachelor's Dissertation). Otago: University of Otago.

Hinton, P. R., Brownlow, C., McMurray, I., & Cozens, B. (2004). *SPSS Explained*. Routledge. doi:10.4324/9780203642597

Hofstadter, R. (1992). *Social Darwinism in American Thoughts*. Beacon Press.

Hofstede, G. (1991). *Cultures and Organization: Software of the mind*. Mcgraw-Hill.

Holmes, T. (2015). Airbnb in SA selling trust to the wary. *Mail and Guardian*. Retrieved from: https://www.mg.co.za

Hong, J. (2018). Rise of the sharing economy and the future of travel and tourism industry. *Journal of Hotel & Business Management*, *7*(2), 5–11. doi:10.4172/2169-0286.1000180

http://www.tourismconcern.org.uk/media/2002/WSSD%20sustainable%20tourism%20bri efing.htm

Huang, C.-H., Tsaur, J.-R., & Yang, C.-H. (2012). Does World Heritage List Really Induce More Tourists? Evidence from Macau. *Tourism Management*, *33*(6), 1450–1457. doi:10.1016/j.tourman.2012.01.014 PMID:32287738

Hudson, S. (1999). Consumer behaviour related to tourism. In A. P. Y. Mansfeld (Ed.), *Consumer behaviour in travel and tourism* (pp. 7–32). Haworth Press.

Huefner, R. J. (2014). An introduction to revenue management: Exploring common techniques and the CPA's role. *The CPA Journal*, *84*(6), 17–21.

Huh, J. (2002). *Tourist Satisfaction with Cultural/ Heritage Sites: The Virginia Historic Triangle* (Thesis). Virginia Polytechnic Institute and State University, Blacksburg, VA.

Hui, T. K., Wan, D., & Ho, A. (2007). Tourists' Satisfaction, Recommendation and Revisiting Singapore. *Tourism Management*, *28*(4), 965–975. doi:10.1016/j.tourman.2006.08.008

Ignatov, E., & Smith, S. (2006). Segmenting Canadian Culinary Tourists. *Current Issues in Tourism*, *9*(3), 235–255. doi:10.2167/cit/229.0

Ilgren, E. B., Kazemain, H., & Hoskins, J. A. (2014). Kandovan the next 'Capadoccia'?–A potential public health issue for erionite related mesothelioma risk. *Epidemiology, Biostatistics, and Public Health*, *12*(1).

Imani, B., Khalifeh, E., & Ebrahimi, K. (2017). Assessing the Impacts of Implementing a Guidance Plan in Rural Sustainable Development (Case Study: Kandovan Village, East Azerbaijan Province). *Journal of Research and Rural Planning*, *6*(2), 57–71.

Inskeep, E. (1994). National and regional tourism planning. Routledge.

IOL. (2018). *Airbnb opens experiences platform for submission from hosts across SA*. Retrieved from: https://www.iol.com.za/news/south-africa/

Irna. (2017). Retrieved from http://www.irna.ir/ eazarbaijan/fa/ News/ 82010759.

Issacs, L. (2017). Fedhasa demands Airbnb be regulated. *IOL*. Retrieved from: https://www.iol.co.za/capetimes/news/

Ivanov, S. (2014). *Hotel Revenue Management: From Theory to Practice*. Zangador. Zangador Ltd.

Jamal, S. (2017). *Airbnb offers free accommodation for people displaced by Western Cape disasters*. Retrieved from: https://www.timeslives.co.za/news/south-africa/

Jameson, F. (1991). *Postmodernism, or, the cultural logic of late capitalism*. Duke University Press.

Jaremen, D. E., & Nawrocka, E. (2015). Relations in the tourism supply chain. *Seria. Administracja i Zarzadzanie*, *104*, 98–110.

Jen, C. T. (2013). The influence of conflict centrality and task interdependency on individual performance and job satisfaction. *International Journal of Conflict Management*, *24*(2), 126–147. doi:10.1108/10444061311316762

Jensen, J. M. (2011). The relationships between socio-demographic variables, travel motivations and subsequent choice of vacation. In *2nd International Conference on Economics, Business and Management* (*Vol. 22*, pp. 37–44). Academic Press.

Jeong, J. S., Jeong, H. N., Ko, K. H., Moon, J. H., & Kang, H. G. (2013). Success Factors of Paris Baguettes Bakery Franchise Business: Industry Competition and Core Competence Analysis. *Asia Marketing Journal*, *15*(2), 1–27.

Jones, A., & Jenkins, I. (2002). 'A taste of Wales - Blas Ar Gymru': Institutional malaise in promoting Welsh food tourism products. In A.-M. Hjalagar & G. Richards (Eds.), *Tourism and Gastronomy* (pp. 114–131). Routledge.

Judd, D. (1999). *Constructing the Tourist Bubble. In The Tourist city.* Yale University Press.

Judd, D. (2003). Urban tourism and the geography of the city. *Eure magazine. Santiago de Chile, 29*(87), 51–62.

Jungleworks. (2019). *How Airbnb works – insights into business and revenue model.* USA Clicklabs.

Kadri, B. (2007). La ville et le tourisme: Relation ancienne, complexité nouvelle et défi conceptuel. *Téoros, 12*(1), 76–79.

Kar, A., & Kundu, S. (2014). The changing perspective of food habits in Kolkata: A case on Monginis Kolkata. *International Journal of Business Performance Management, 15*(1), 64–71.

Karami, E., & Sharifi, E. S. (2017). Rocky Habitats Architecture (Kandowan Village as study sample). *Archipolis, 1*(1), 1–10.

Karatepe, O. M., Rezapouraghdam, H., & Hassannia, R. (2020). Job insecurity, work engagement and their effects on hotel employees' non-green and nonattendance behaviors. *International Journal of Hospitality Management, 87*, 102472. doi:10.1016/j.ijhm.2020.102472

Kasavana, M., & Brooks, R. (1998). *Front Office Procedures* (5th ed.). Educational Institute of the American Hotel and Motel Association.

Kaswengi, J. (2011). Revenue management: fermeture d'allotements et coût d'opportunité dans le tourisme B to B. *Revue Management et Avenir*, (43), 57-76.

Kaul, H., & Gupta, S. (2009). Sustainable Tourism in India. *Worldwide Hospitality and Tourism Themes, 1*(1), 12–18. doi:10.1108/17554210910949841

Kerstetter, D. L., Confer, J. J., & Graefe, A. R. (2001). An Exploration of the Specialization Concept within the Context of Heritage Tourism. *Journal of Travel Research, 39*(3), 267–274. doi:10.1177/004728750103900304

Khan, S., Sara, L. M., Sydenstricker-Neto, J., Sutherland, C., & Hordijk, M. (2015). Scenario building as a process and tool in urban governance. In *Geographies of Urban Governance* (pp. 193–214). Springer. doi:10.1007/978-3-319-21272-2_10

Khorasani, M., Hatami, L., & Kiakojoori, D. (2017). Strategic Planning of Rural Tourism Development Using SOAR Model: A Case Study of Kandovan Village. *Journal of Sustainable Rural Development., 1*(2), 171–188. doi:10.29252/jsrd.01.02.171

Kim, D. Y., Hwang, Y. H., & Fesenmaier, D. (2005). Modeling Tourism Advertising Effectiveness. *Journal of Travel Research, 44*(1), 42–49. doi:10.1177/0047287505276590

Kimes, S. E. (2008). Hotel Revenue Management: Today and Tomorrow. Cornell University School of Hotel Administration, 8(14).

Kim, S., Shaw, T., & Schneider, H. (2003). Website design benchmarking within industry groups. *Internet Research, 13*(1), 17–26. doi:10.1108/10662240310458341

Kiumarsi, S., Jayaraman, K., Isa, S. M., & Varastegani, A. (2014). Marketing strategies to improve the sales of bakery products of small-medium enterprise (SMEs) in Malaysia. *International Food Research Journal, 21*(6), 18–24.

Klempner, G. (2004). Ethics and Advertising. *Philosophy for Business*, (9).

Koens, K., Postma, A., & Papp, B. (2018). Is overtourism overused? Understanding the impact of tourism in a city context. *Sustainability, 10*(12), 1–15. doi:10.3390u10124384

Kokaz Pucciani, K., & Murphy, H. C. (2011). An investigation of data management and property management systems in hotels. *Tourism and Hospitality Management, 17*(1), 101–104.

Korstanje, M. (2012). Reconsidering cultural tourism: An anthropologist's perspective. *Journal of Heritage Tourism, 7*(2), 179–184. doi:10.1080/1743873X.2011.639883

Korstanje, M. E., & Ivanov, S. H. (2012). Tourism as a form of new psychological resilience: The inception of dark tourism. *CULTUr-revista de Cultura e Turismo, 6*(4), 56–71.

Korstanje, M., & Seraphin, H. (2017). Revisiting the sociology of consumption in tourism. In R. K. Dixit (Ed.), *The Routledge Handbook of Consumer Behaviour in Hospitality and Tourism* (pp. 16–25). Routledge.

Kosow, H., & Gaßner, R. (2008). *Methods of future and scenario analysis: Overview, assessment, and selection criteria. DIE Research Project Development Policy: Questions for the Future.* Deutsches.

Kotler, P. (2000). *Marketing Management, Millenium* (10th ed.). Prentice Hall.

Kotler, P., Armstrong, G., Brown, L., & Chandler, S. A. (1998). *Marketing* (4th ed.). Prentice Hall.

Kozak, M. (2002). Comparative analysis of tourist motivations by nationality and destinations. *Tourism Management, 23*(3), 221–232. doi:10.1016/S0261-5177(01)00090-5

Kozak, M., Baloglu, S., & Bahar, O. (2010). Measuring destination competitiveness: Multiple destinations versus multiple nationalities. *Journal of Hospitality Marketing & Management, 19*(1), 56–71. doi:10.1080/19368620903327733

Kozak, M., & Rimmington, M. (1999). Measuring tourist destination competitiveness: Conceptual considerations and empirical findings. *International Journal of Hospitality Management, 18*(3), 273–283. doi:10.1016/S0278-4319(99)00034-1

KPMG. (2018). *Tomorrow's experience, today Harnessing a customer first approach in a changing world. KPMG Global Customer Experience Excellence report.* KPMG International Cooperative.

Kumaran, D. S., Kannan, D. R., & Milton, T. (2013). Tourist Satisfaction with Cultural/Heritage Sites at Madurai, *Asia Pacific. Journal of Marketing Management, 2*(3), 62–68.

Kuščer, K., & Mihalič, T. (2019). Residents' Attitudes towards Overtourism from the Perspective of Tourism Impacts and Cooperation–The Case of Ljubljana. *Sustainability, 11*(6), 1823. doi:10.3390u11061823

Laffont, J. J., & Tirole, J. (1991). Privatization and incentives. *JL Econ. & Org., 7*(special), 84–105. doi:10.1093/jleo/7.special_issue.84

Lai, K., & Li, X. (2016). Tourism destination image: Conceptual problems and definitional solutions. *Journal of Travel Research, 55*(8), 1065–1080. doi:10.1177/0047287515619693

Lalicic, L., & Weismayer, C. (2017). The role of aunthenticity in Airbnb Experiences. In R. Schegg & B. Stangl (Eds.), *Edited. Information and Communication Technologies in Tourism.* Springer Publishing.

Lane, L. (2016). Are millennial travel trends shifting in 2016? *Forbes.* Retrieved from: https://www.forbes.com/sites/lealane

Larry, Y., & Munhtuya, G. (2005). A Comparative Analysis of International Tourist Satisfaction in Mongolia. *Tourism Management Journal, 27*(6), 1331–1342.

Lee, K. S. (2012). Development of baking technology and analysis of domestic and international bakery market and trends. *Food Science and Industry, 45*(4), 16–20.

Lee, S., Jeon, S., & Kim, D. (2011). The Impact of Tour Quality and Tourist Satisfaction on Tourist Loyalty: The Case of Chinese Tourists in Korea. *Tourism Management, 32*(5), 1115–1124. doi:10.1016/j.tourman.2010.09.016

Legohérel, P., Poutier, E., & Fyall, A. (2013). Revenue Management for Hospitality & Tourism. Goodfellow Publishers Limited.

Lehto, G., Morrison, A., & O'Leary, J. (2006). The economic value portofolio matrix: A target market selection tool for destination marketing organizations. *Tourism Management*, 27.

Leiper, N. (1979). The Framework of Tourism: Towards a Definition of Tourism, Tourist, and the Tourist Industry. *Annals of Tourism Research*, *6*(4), 390–407. doi:10.1016/0160-7383(79)90003-3

Leiper, N. (1993). Defining tourism and related concepts: tourist, market, industry and tourism system. In *VNR'S Encyclopedia of hospitality and Tourism*. Springer.

Levinson, H. (1976). *Psychological Man*. The Levinson Institute.

Lewis, R. (1988). *Chambers Richard*. Marketing Leadership in Hospitality.

Lillo Bañuls, A. (2007). El capital humano como factor estrategico para la competitividad del sector turístico. *Cuadernos de Turismo*, *19*, 47–69.

Lindgaard, G., Fernandes, G., Dudek, C., & Brown, J. (2006). Attention web designers: You have 50 milliseconds to make a good first impression. *Behaviour & Information Technology*, *25*(2), 115–126. doi:10.1080/01449290500330448

Lindroth, K., Ritalahti, J., & Soisalon-Soininen, T. (2007). Creative Tourism in Destination Development. *Tourism Review*, *62*(3-4), 53–57. doi:10.1108/16605370780000322

Lin, L. F., & Cai, X. W. (2016). Research on influencing factors of tourist satisfaction in Hengshan Mountain. *Journal of Hunan Finance and Economics University*, *1*(1), 130–134.

Littlewood, K. (2005). Special Issue Papers: Forecasting and control of passenger bookings. *Journal of Revenue and Pricing Management*, *4*(2), 111–123. doi:10.1057/palgrave.rpm.5170134

Lominé, L., & Edmunds, J. (2007). *Key concepts in tourism*. Palgrave Macmillan. doi:10.1007/978-1-349-92040-2

Longhurst, A. H. (1930). *Pallava Architecture volume 3. Archaeological Survey of India in Indian History and Architecture website*. Retrieved from http://puratattva.in/2010/09/14/mahabalipuram-the-workshop-of-pallavas-part-v- 41.html

Long, L. M. (2004). *Culinary tourism*. The University Press of Kentucky.

Lozato-Giotart, J.-P. (1993). *Géographie du tourisme: de l'espace regardé à l'espace consommé*. Masson.

Luhmann, N. (1979). *Trust and power*. John Willey & Sons.

Luna-Nevarez, C., & Hyman, M. (2012). Common practices in destination website design. *Journal of Destination Marketing & Management*, *1*(1-2), 94–106. doi:10.1016/j.jdmm.2012.08.002

MacCannell, D. (1976). *The Tourist*. Schocken Books.

Maccannell, D. (1999). The tourist and the local. *Tourist Studies*.

Magas, D. (2010). Why the destination management organization? *Tourism and Hospitality Management*, *2010*, 1041–1047.

Mahabalipuram – The Workshop of Pallavas – Part I. (2010a, August 3). *Indian History and Architecture A passage to the Indian History and Monuments*. Retrieved from http://puratattva.in/2010/08/03/mahabalipuram-the-workshop-of-pallavas-part-i- 45.html

Mahabalipuram – The Workshop of Pallavas – Part II. (2010b, August 4). *Indian History and Architecture A passage to the Indian History and Monuments.* Retrieved from http://puratattva.in/2010/08/04/mahabalipuram-the-workshop-of-pallavas-part-ii- 44.html

Mahabalipuram – The Workshop of Pallavas – Part IV. (2010c, August 23). *Indian History and Architecture A passage to the Indian History and Monuments.* Retrieved from http://puratattva.in/2010/09/03/mahabalipuram-the-workshop-of-pallavas-part-iv- 42.html

Mahabalipuram – The Workshop of Pallavas – Part V. (2010d, September 14). *Indian History and Architecture A passage to the Indian History and Monuments.* Retrieved from http://puratattva.in/2010/09/14/mahabalipuram-the-workshop-of-pallavas-part-v-41.html

Maitland, R. (2010). Everyday life as a creative experience in cities. *International Journal of Culture, Tourism and Hospitality Research, 4*(3), 176–185. doi:10.1108/17506181011067574

Maleki, S., & Bigdeli Rad, V. (2017). Social Sustainability Measures for Rural Areas in Iran. *Space Ontology International Journal, 6*(1), 79–84.

Marais, M., Du Plessis, E., & Saayman, M. (2017). Critical success factors of a business tourism destination: Supply side analysis. *Acta Commercii, 17*(1), 1–12. doi:10.4102/ac.v17i1.423

Marcuse, P. (1986). Abandoment, gentrification and Displacement: Linkage in New York City. In N. Smith & P. Wiliams (Eds.), *Gentrification of the city.* Allen & Unwin.

Martín, J. M. M., Martínez, J. M. G., & Fernández, J. A. S. (2018). An analysis of the factors behind the citizen's attitude of rejection towards tourism in a context of overtourism and economic dependence on this activity. *Sustainability, 10*(8), 2851. Advance online publication. doi:10.3390u10082851

Marzano, G., & Scott, N. (2009). Power of destination branding. *Annals of Tourism Research*, 250.

McDowall, S. (2010). A Comparison between Thai Residents and Non-Residents in their Motivations, Performance Evaluations, and Overall Satisfaction with a Domestic Festival. *Journal of Vacation Marketing, 16*(3), 217–233. doi:10.1177/1356766709356139

McIntosh, A. (2004). The family meal and its significance in global times. In R. Grew (Ed.), *Food in global history* (pp. 239–263). New Global History Press.

Mckercher, B., & Lau, G. (2007). Understanding the movements of tourists in a destination: Testing the importance of markets in the tourist attraction system. *Asian Journal of Tourism and Hospitality Research., 1*(1), 39–53.

McMichael, P. (2011). *Development and Social Change.* Sage.

Mcnaughton, D. (2006). The host as uninvited guest; hospitality, violence and tourism. *Annals of Tourism Research, 33*(3), 645–665. doi:10.1016/j.annals.2006.03.015

Mgwadlamba, N. (2017). Letter: Iconic Kynsna forest still standing. *Cape Time.* Retrieved from: https://www.iol.co.za

Min, Martin, & Jung. (2013). *The Sociological Imagination.* Oxford University Press.

Miner, J. B. (1992). *Industrial-organizational psychology.* McGraw-Hill.

Min, H., & Min, H. (1996). Competitive benchmarking of Korean luxury hotels using the analytic hierarchy process and competitive gap analysis. *Journal of Services Marketing, 10*(3), 58–72. doi:10.1108/08876049610119794

Ministry of Regional Planning. (2008). *Environment and Tourism, "Schéma Directeurd' Aménagement Touristique 2025*. Official Document.

Ministry of Tourism and Handicrafts. (n.d.). *Synthesis of tourist flows in Algeria in 2012*. Author.

Minuciu, R. (2003). *Tourism economy*. Ed. Uranus.

Mitchell, T. R., & Daniels, D. (2003). Motivation. In Handbook of Psychology (Vol. 12, pp. 225–54). New York: Wiley.

Mohamadi, T. (2012). *Narrated by the Actuary of Eastern Azerbaijan Travel Facilitating Organization*. Retrieved from https://www.yjc.ir/fa/print/3670234

Mok, C., & Armstrong, R. W. (1994). Perception of Australia as a Holiday Destination: A Case of Hong Kong and Taiwanese Tourists. *Australian Journal of Hospitality Management*, *1*(2), 25–28.

Molefe, L. P., Tauoatsoala, P., Sifolo, P. P. S., Manavhela, P., & Henama, U. S. (2018). 'The effects of tourism supply chain management practices on tourism operations in Pretoria, South Africa'. *African Journal of Hospitality, Tourism and Leisure*, *7*(2), 1–15.

Moller, C., Weiermair, K., & Wintersberger, E. (2007). The changing Travel Behaviour of Austria's Ageing Population and its impact on Tourism. *Tourism Review*, *62*(3-4), 15–19. doi:10.1108/16605370780000316

Morgan, N., Pritchard, A., & Pride, R. (2011). *Destination brands: Managing place reputation*. Routledge.

Morrison, A., Bruen, S., & Anderson, D. (1998). Convention and Visitors bureaus in the USA. *Journal of Travel and Tourism*.

Mosedale, J. (2016). Neoliberalism and the political economy of tourism. In J. Modesale (Ed.), *Neoliberalsm and the political economy of tourism* (pp. 1–19). Routledge. doi:10.4324/9781315597782

Moslehpour, M., Aulia, C. K., & Masarie, C. E. (2015). Bakery product perception and purchase intention of Indonesian consumers in Taiwan. *International Journal of Business and Information*, *10*(1), 8–20.

Moss, L. A. (Ed.). (2006). *The amenity migrants: Seeking and sustaining mountains and their cultures*. Cabi. doi:10.1079/9780851990842.0000

Muhana, E. (2006). Sustainable Tourism Development and Environmental Management for Developing Countries. *Problems and Perspectives in Management*, *4*(2), 14–34.

Muller, H., & Wyss, U. (2007). Aypical Working Hours and their impacts on Leisure – an unofficial Swiss Time Use Study. *Tourism Review*, *62*(1), 14–24. doi:10.1108/16605370780000158

Murphy, A. E. (2003). Illustrating the utility of a modified gap analysis as a regional tourism planning tool: Case study of potential Japanese and German travelers to the Cowichan Region. *Journal of Travel Research*, *41*(4), 400–409. doi:10.1177/0047287503041004009

Murphy, P. E. (1983). Perceptions and attitudes of decision-making groups in tourism centers. *Journal of Travel Research*, *21*(2), 8–12. doi:10.1177/004728758302100302

Murphy, P., Pritchard, M. P., & Smith, B. (1999). The destination product and its impact on traveller perceptions. *Tourism Management*.

Nasirahmadi, A., Madarsara, F. A., & Aghdam, H. R. (2014). Cultural issues and teaching literature for language learning. *Procedia: Social and Behavioral Sciences*, *98*, 1325–1330. doi:10.1016/j.sbspro.2014.03.549

Ndlovu, J. (2009). *Branding as a strategic tool of reposition a destination: A survey of key tourism stakeholders in zimbawe.* Unpublished manuscript.

Nedelea, A. (2010). *Comparative Tourism Marketing Case Studies, Dehli.* Abhijeet Publications.

Needle, D. (2004). *Business in Context: An Introduction to Business and Its Environment.* Cengage Learning Business Press.

Nicholas, L. N., Thapa, B., & Ko, Y. J. (2009). Residents' Perspectives of a World Heritage Site The Pitons Management Area, St. Lucia. *Annals of Tourism Research, 36*(3), 390–412. doi:10.1016/j.annals.2009.03.005

Nielsen. (2017). *Young and Ready to Travel (And Shop): A look at Millennial Travellers.* Retrieved from: https://www.nielsen.com

Nkanyeni, U. (2017). Airbnb signs Africa's first city agreement with Cape Town. *Traveller24.* Retrieved from: https://www.traveller24.com/FindYourEscape/

Ocampo, A., & Barrientos, A. (2008). From the bakery to the brain business: Developing inducible yeast models of human neurodegenerative disorders. *BioTechniques, 45*(4), vii–xiv.

Okumus, B., & McKercher, B. (2007). Incorporating local and international cuisines in the marketing of tourism destinations: The cases of Hong Kong and Turkey. *Tourism Management, 28*(1), 253–261. doi:10.1016/j.tourman.2005.12.020

Oliveira, W. (2000). *CRM e e-business.* Matosinhos, Edições Centro Atlântico.

Oliver, R. L., & Bearden, W. O. (1985). Disconfirmation Processes and Consume Evaluations in Product Usage. *Journal of Business Research, 13*(3), 235–246. doi:10.1016/0148-2963(85)90029-3

Omerzel, D. G. (2006). Competitiveness of Slovenia as a tourist destination. *Managing Global Transitions International Research Journal, 4*(2), 167–189.

Oskam, J., & Boswijk, A. (2016). Airbnb: The future of networked hospitality business. *Journal of Tourism Futures, 2*(1), 22–42. doi:10.1108/JTF-11-2015-0048

Palmer, P., & O'Neill, M. A. (2004). Wine production and tourism: Adding service to a perfect partnership. *The Cornell Hotel and Restaurant Administration Quarterly, 45*(3), 269–284. doi:10.1177/0010880404263075

Pansiri, J., & Mmereki, R. N. (2011). An evaluation of international tourist motivation and tourism destination competitiveness. In M. J. Gross (Ed.), *Tourism: Creating a brilliant blend: Proceedings of the 21st Annual Conference* (pp. 1–16). Adelaide, Australia: CAUTHE 2011.

Pansiri, J. (2014). Tourist Motives and Destination Competitiveness:A Gap Analysis Perspective. *International Journal of Hospitality & Tourism Administration, 15*(3), 217–247. doi:10.1080/15256480.2014.925718

Pansiri, J., & Courvisanos, J. (2010). Attitude to risk in technology-based strategic alliances for tourism. *International Journal of Hospitality & Tourism Administration, 11*(3), 275–302. doi:10.1080/15256480.2010.498283

Pansiri, J., & Mmereki, R. N. (2010). Using the SERVQUAL model to evaluate the impact of public service reforms in the provision of primary health care in Botswana. *Journal of African Business, 11*(2), 219–234. doi:10.1080/15228916.2010.509005

Pantazatou, K. (2018). Taxation of the Sharing Economy in the European Union. In Cambridge Handbook of Law and Regulation of the Sharing Economy. CUP. doi:10.1017/9781108255882.028

Parasuraman, A., Berry, L. L., & Zeithaml, V. A. (1991). Refinement and reassessment of the SERVQUAL Scale. *Journal of Retailing, 67*(4), 420–450.

Parasuraman, A., & Zeithaml, V. A. (1994). Alternative scales for measuring service quality: A comparative assessment based on psychometric and diagnostic criteria. *Journal of Retailing*, *70*(3), 201–230. doi:10.1016/0022-4359(94)90033-7

Parasuraman, A., Zeithaml, V. A., & Berry, L. (1985). A conceptual model of service quality and its implications for future research. *Journal of Marketing*, *49*(4), 41–50. doi:10.1177/002224298504900403

Parasuraman, A., Zeithaml, V. A., & Berry, L. L. (1988). SERVQUAL: A multiple item scale for measuring consumer perceptions of service quality. *Journal of Retailing*, *64*(1), 5–6.

Parasuraman, A., Zeithaml, V. A., & Berry, L. L. (1994). Reassessment of expectations as a comparison standard in measuring service quality: Implications for further research. *Journal of Marketing*, *58*(1), 111–124. doi:10.1177/002224299405800109

Park, H. y. (2010). Heritage Tourism Emotional Journeys into Nationhood. *Annals of Tourism Research*, *37*(1), 116–135. doi:10.1016/j.annals.2009.08.001

Park, H. Y. (2013). *Heritage tourism*. Routledge. doi:10.4324/9781315882093

Parliamentary Monitoring Group. (2019). *Report of the Portfolio Committee on Tourism for Budget Vote No. 33*. Retrieved from: https://pmg.org.za/tabled-committee-report/3843/

Patterson, P. G. (1993). Expectations and product performance as determinants of satisfaction for a high-involvement purchase. *Psychology and Marketing*, *10*(5), 449–465. doi:10.1002/mar.4220100507

Pearce, D. (2008). A needs-functions model of tourism distribution. *Annals of Tourism Research*, *35*(1), 148–168. doi:10.1016/j.annals.2007.06.011

People Travel. (2015). Retrieved from https://financialtribune.com/articles/people-travel/16570/kandovan-village-may-be-nominated-for-unesco-list

Perkumienė, D., & Pranskūnienė, R. (2019). Overtourism: Between the Right to Travel and Residents' Rights. *Sustainability*, *11*(7), 2138. doi:10.3390u11072138

Perlik, M. (2006). 15 The Specifics of Amenity Migration in the European Alps. *The amenity migrants: Seeking and sustaining mountains and their cultures*, 215.

Pestana, C., Laurent, B., Nicolas, P., Elisabeth, R., Bernardin, S., & Assaf, G. (n.d.). Performance of French destinations: Tourism attraction perspectives. *Tourism Management*.

Peterson, K. (1994). The heritage resource as seen by the tourist: The heritage connection. Journal of Tourism: An Exploration.

Petrick, J. F. (2004). Are Loyal Visitors Desired Visitors? *Tourism Management*, *25*(4), 463–470. doi:10.1016/S0261-5177(03)00116-X

Philips, M., & Jones, A. (2006). Erosion and Tourism Infrastructure in the Coastal Zone: Problems, Consequences and Management. *Tourism Management*, *27*(3), 517–524. doi:10.1016/j.tourman.2005.10.019

Pigram, J. J., & Wahab, S. (Eds.). (2005). *Tourism, development and growth: the challenge of sustainability*. Routledge. doi:10.4324/9780203975138

Pike, S., & Ryan, C. (2004, May). Destination positioning analysis through a comparison of cognitive, affective and conative perceptions. *Journal of Travel Research*, *42*(4), 333–342. doi:10.1177/0047287504263029

Pilcher, J. M. (2006). *Food in world history*. Routledge.

Pimonratanakan, S., & Ayasanond, C. (2018). The Supply Chain Management of Pomleo Production, Process and Tourism Sustainability in Nakhon Chaisri, Nakhon Pathom Province, Thailand. *African Journal of Hospitality, Tourism and Leisure*, *7*(5), 1–14.

Pizam, A. (1997). Article. *Journal of Travel Research*, *36*(1), 23–28. doi:10.1177/004728759703600104

Pizam, A., Neumann, Y., & Reichel, A. (1978). Dimensions of tourist satisfaction with a destination. *Annals of Tourism Research*, *5*(3), 314–322. doi:10.1016/0160-7383(78)90115-9

Polzin, S. E., Chu, X., & Godfrey, J. (2014). The impact of millennials' travel behaviour on future personal vehicle travel. *Energy Strategy Reviews*, *5*, 59065. doi:10.1016/j.esr.2014.10.003

Poria, Y., Butler, R., & Airey, D. (2003). The Core of Heritage Tourism. *Annals of Tourism Research*, *30*(1), 238–254. doi:10.1016/S0160-7383(02)00064-6

Prahalad, C., & Ramaswamy, V. (2004). Co- creation experiences: The next practice in value creation. *Journal of Interactive Marketing*, *18*(3), 5–14. doi:10.1002/dir.20015

Prats, L. (1997). *Antropología y Patrimonio*. Ariel Antropología.

Prayag, G., Hosany, S., Muskat, B., & Del, C. G. (2017). Understanding the relationships between tourists' emotional experiences, perceived overall image, satisfaction, and intention to recommend. *Journal of Travel Research*, *56*(1), 41–54. doi:10.1177/0047287515620567

Prayag, G., Hosany, S., & Odeh, K. (2013). The role of tourists' emotional experiences and satisfaction in understanding behavioral intentions. *Journal of Destination Marketing & Management*, *2*(2), 118–127. doi:10.1016/j.jdmm.2013.05.001

Presenza, A., Sheehan, L., & Ritchie, J. (2005). *Towards a model of the roles and activities of destination management organizations.* Unpublished manuscript.

Putri, M. (2017). Tourist Satisfaction at Cultural Destination-A Case Study of Saung Angklung Udjo Bandung, West Java (Indonesia). *International Journal of Tourism & Hospitality Reviews*, *4*(1), 35–43. doi:10.18510/ijthr.2017.415

PWC. (2015). *The Sharing Economy. Consumer Intelligence Series.* Retrieved from: https://pwc.com/CISsharing

Queensland Tourism Industry Council. (2014). *The Sharing Economy: How it will impact the tourism landscape and what businesses can do.* Queensland Tourism Industry Council.

Rajani, M. B., & Kasturirangan, K. (2013). Sea-Level Changes and its Impact on Coastal Archaeological Monuments: Seven Pagodas of Mahabalipuram, a Case Study. *Photonirvachak (Dehra Dun)*, *41*(2), 461–468. doi:10.100712524-012-0210-y

Ramsey, D., & Everitt, J. (2008). If you dig it, they will come! Archaeology heritage sites and tourism development in Belize, Central America. *Tourism Management*, *29*(5), 909–916. doi:10.1016/j.tourman.2007.11.002

Rawlings, L. K. (2016). Airbnb takes on more travel pie. *IT Web*. Retrieved from: https://www.itweb.co.za/content

Razani, M., Emami, S. M. A., & Baghbanan, A. (2013). *The Necessity of Applying Decay Mechanisms Investigation in the Rock Hewn Architecture of Kandovan; a Historical Village in Azerbaijan, Iran.* Iranian National Symposium on the Application of Scientific Analyses in Archaeometry & the Conservation of Cultural Heritage. Tabriz, Iran.

Razani, M., Emami, S. M. A., & Baghbanan, A. (n.d.). *National conference on the application of scientific analysis in the anthropology and restoration of cultural heritage*. Academic Press.

Reisinger, Y. (2009). *International Tourism: Cultures and Behavior*. Butterworth - Heinemann Publications.

Renko, N., Sustic, I., & Butigan, R. (2011). Designing marketing strategy using the five competitive forces model by michael e. Porter-Case of Small Bakery in Croatia. *International Journal of Management Cases, 13*(3), 376–385.

Renko, S., Renko, N., & Polonijo, T. (2010). Understanding the role of food in rural tourism development in a recovering economy'. *Journal of Food Products Marketing, 16*(3), 309–324. doi:10.1080/10454446.2010.485096

Report of the National Economic and Social Council (CNES). (2000). *Contribution to the redefinition of national tourism policy*. Author.

Rezapouraghdam, H., Behravesh, E., Ari, E., & Doh, W. L. (2018). Cost-benefit analysis of educational tourism in North Cyprus: a qualitative study of the socio-economic impacts. *e-Review of Tourism Research, 15*(6), 457-479. https://www.researchgate.net/profile/Elaheh_Behravesh/publication/324601003_Cost-Benefit_Analysis_of_Educational_Tourism_in_North_Cyprus_A_Qualitative_Study_of_the_Socio-economic_Impacts/links/5b8cddf6299bf1d5a73a0de4/Cost-Benefit-Analysis-of-Educational-Tourism-in-North-Cyprus-A-Qualitative-Study-of-the-Socio-economic-Impacts.pdf

Rezapouraghdam, H., Alipour, H., & Arasli, H. (2018). Workplace spirituality and organization sustainability: A theoretical perspective on hospitality employees' sustainable behavior. *Environment, Development and Sustainability*, 1–19.

Rezapouraghdam, H., Alipour, H., & Darvishmotevali, M. (2018). Employee workplace spirituality and pro-environmental behavior in the hotel industry. *Journal of Sustainable Tourism, 26*(5), 740–758. doi:10.1080/09669582.2017.1409229

Rezapouraghdam, H., Behravesh, E., Ari, E., & Doh, W. L. (2018). Cost-Benefit Analysis of Educational Tourism in North Cyprus: A Qualitative Study of the Socio-economic Impacts. *Ereview of Tourism Research, 15*(6).

Richards, G., & Hall, D. (Eds.). (2003). *Tourism and sustainable community development* (Vol. 7). Psychology Press.

Rifkin, J. (2000). *La era del acceso*. Editorial Paidós.

Ritchie, J., & Crouch, G. I. (2003). *The competitive destination. A sustainable tourism perspective*. CABI. doi:10.1079/9780851996646.0000

Roblek, V., Stok, Z. M., & Mesko, M. (2016). Complexity of a Sharing Economy for Tourism and Hospitality. *Tourism & Hospitality Industry 2016, Congress Proceedings*, 378-387.

Rodriguez, d. B. I., & San Martin, H. (2008). Tourist satisfaction a cognitive-affective model. *Annals of Tourism Research, 35*(2), 551-573.

Rogelberg, S. G., & Brooks-Laber, M. E. (2002). Securing our collective duture: Challenges facing those designing and doing research in industrial and organization psychology. In S. G. Rogelberg (Ed.), *Handbook of research methods in industrial and organizational psychology* (pp. 479–485). Blackwell.

Ross, G. F. (2006). Tourism industry employee workstress—A present and future crisis. *Journal of Travel & Tourism Marketing, 19*(2-3), 133–147. doi:10.1300/J073v19n02_11

Rotmans, J., van Asselt, M., Anastasi, C., Greeuw, S., Mellors, J., Peters, S., Rothman, D., & Rijkens, N. (2000). Visions for a sustainable Europe. *Futures, 32*(9-10), 809–831. doi:10.1016/S0016-3287(00)00033-1

Rust, R. T., Zahorik, A. J., & Keininghan, T. L. (1993). *Return on Quality*. Probus Publishing.

Saftic, D., & Rafajac, O. (2010). DMO model as a step forward in the quality of the management of istrian tourist destination. *Conference. Tourism and Hospitality Management.*

Samuel, R. (2012). *Theatres of memory: Past and present in contemporary culture*. Verso Books.

Sanchez Canizares, S. M. (2007). La satisfaccion laboral en los establecimientos hoteleros. Analisis empirico en la provincia de Cordoba. *Cuadernos de Turismo*, *20*(1), 223–249.

Santos, N., & Fernandes, P. (2010). Análise e Previsão da Procura Turística na Região Norte de Portugal. *Proceedings of APDR.*

Scarpato, R. (2002a). Gastronomy as a tourist product: the perspectives of gastronomy studies. In A.-M. Hjalager & G. Richards (Eds.), *Tourism and gastronomy* (pp. 51–70). Routledge.

Schofield, J. (2017). How the sharing economy has changed the supply chain. *System ID Barcode Solution.* Retrieved from: https://www.systemid.com/learn/hhh/

Schouten, F. (1995). Improving visitor care in heritage attractions. *Tourism Management*, *16*(4), 259–261. doi:10.1016/0261-5177(95)00014-F

Scoones, I. (1998). *Sustainable rural livelihoods: A framework for analysis.* Institute of Development Studies, University of Sussex.

Sekaran, U. (2000). *Research method for business: A skill building approach.* John Wiley and Sons, Inc.

Sellal, M. (1980). *Production touristique.* Société Nationale d'Edition et de Diffusion.

Séraphin, H., Zaman, M., Olver, S., Bourliataux-Lajoinie, S., & Dosquet, F. (2019). Destination branding and overtourism. *Journal of Hospitality and Tourism Management, 38*, 1–4. doi:10.1016/j.jhtm.2018.11.003

Sharpley, R. (1999). Tourism, tourists and society (2nd ed.). Cambridgeshire, UK: ELM Publications.

Sharpley, H. (2002). The challenges of economic diversification through tourism: The case of Abu Dhabi. *International Journal of Tourism Research*, *4*(2), 221–235. doi:10.1002/jtr.378

Shevel, A. (2016). The fast facts for those considering Airbnb. *Business Day Live.* Retrieved from: http://www.bdlive.co.za

Shrikant, A. (2018). *Can guided tours save Airbnb? How Airbnb is ducking governent regulations with pottery classes and nature walks.* Retrieved from: https://www.vox.com

Sifolo, P. P. S. (2015). A tourism theoretical gap: The case of the Northern Cape Province in South Africa. *African Journal of Hospitality, Tourism and Leisure*, *4*(1), 1–12.

Sifolo, P. P. S., & Henama, U. S. (2019). Implications of Climate Change for Tourism in Africa. *Geo Journal of Tourism and Geosites*, *2*(20), 191–198.

Sifolo, P. P. S., Maimela, M. J., & Tladi, M. P. (2014). Exploitation of Technology by Tshwane Residents for Tourism Development Purposes. World Academy of Sciences, Engineering and Technology. *International Scholarly and Scientific Research & Innovation*, *8*(2), 386–389.

Singh, S. (2002). Tourism in India: Policy Pitfalls. *Asia Pacific Journal of Tourism Research*, *7*(1), 45–59. doi:10.1080/10941660208722109

Skoll, G. R. (2014). *Dialectics in Social Thought: The Present Crisis.* Macmillan. doi:10.1057/9781137387066

Skoll, G. R., & Korstanje, M. (2014). Urban heritage, gentrification, and tourism in Riverwest and El Abasto. *Journal of Heritage Tourism*, *9*(4), 349–359. doi:10.1080/1743873X.2014.890624

Sofield, T. H. B., & Li, F. M. S. (1998). Tourism development and cultural policies in China. *Annals of Tourism Research*, *25*(2), 362–392. doi:10.1016/S0160-7383(97)00092-3

Son, H. (2013). Alternative future scenarios for South Korea in 2030. *Futures, 52*, 27–41. doi:10.1016/j.futures.2013.06.005

Sorsa, P. (1999). *Algeria- the real exchange rate, export diversification and trade protection.* International Monetary Fund, Working paper N° 99/49.

Souissi, M. (2011). Le tourisme de circuit dans le Sahara tunisien: réalités et perspectives. *Insaniyat. Revue algérienne d'anthropologie et de sciences sociales*, (51-52), 271-287.

Stankey, G. (1973). Visitor Perception Wilderness Recreation Carrying Capacity. Utah: USDA Forest Service. *Tourism Recreation Research, 26*(1), 103–105.

Steinmetz, R. (2010). *Food, Tourism and Destination Differentiation: The Case of Rotorua, NewZealand.* http://aut.researchgateway.ac.nz/bitstream/handle/10292/1090/SteinmetzR.pdf?sequence=3

Stepchenkova, S., & Zhan, F. (2013). Visual destination images of Peru: Comparative content analysis of DMO and user generated photography. *Tourism Management, 36*, 590–601. doi:10.1016/j.tourman.2012.08.006

Stephan, B. (2009). The role of user generated content in Tourists. *Journal of Hospitality Marketing & Management.*

Stnciulescu, G., & Tigu, G. (2005). Business strategies for tourism development in Romania. *Journal of Tourism, 2.*

Strydom, A. J., Mangope, D., & Henama, U. S. (2019). Making Community-Based Tourism Sustainable: Evidence from the Free State Province, South Africa. *Geo Journal of Tourism and Geosites, 24*(1), 7–18.

Swarbrooke, J. (2000). Sustainable tourism management. CAB international.

Swart, R. J., Raskin, P., & Robinson, J. (2004). The problem of the future: Sustainability science and scenario analysis. *Global Environmental Change, 14*(2), 137–146. doi:10.1016/j.gloenvcha.2003.10.002

Szivas, E., Riley, M., & Airey, D. (2003). Labor mobility into tourism: Attraction and satisfaction. *Annals of Tourism Research, 30*(1), 64–76. doi:10.1016/S0160-7383(02)00036-1

Taliani, E. C. (2015). Impacto del Revenue management en los Sistemas de Control de Gestión. *Revista Facultad de Ciencias Económicas, 24*(1).

Talluri, K. T., & Van Ryzin, G. (2004). *The theory and practice of revenue management.* Kluwer Academic Publishers. doi:10.1007/b139000

Tannahill, R. (2002). *Food in history.* Headline Book Publishing.

Tao, T. C., & Wall, G. (2009). Tourism as a sustainable livelihood strategy. *Tourism Management, 30*(1), 90–98. doi:10.1016/j.tourman.2008.03.009

Tardieu, J. M. (2004). *Marketing et gestion des services.* Chiron.

Tarlow, P. (2014). *Tourism Security.* Elsevier.

Tauoatsoala, P., Henama, U. S., & Monyane, M. (2015). Service quality as a fundamental cornerstone for visitor satisfaction: Lessons learned from Pretoria, South Africa. *African Journal of Hospitality, Tourism and Leisure, 4*(1), 1–8.

Taylor, J. P. (2001). Authenticity and sincerity in tourism. *Annals of Tourism Research, 28*(1), 7–26. doi:10.1016/S0160-7383(00)00004-9

Temmar, H. (2015). The Algerian economy: 1970-2014. In The public economic governance, University Publications Office.

Temmar. (2015). The Algerian economy: 1970-2014. In *The development strategies*. University Publications Office.

Tkaczynski, A., Hastings, K., & Beaumont, N. (2006). Factors influencing repositioning of a tourism destination. *ANZMAC 2006 Conference Proceedings*.

Toker, B., & Rezapouraghdam, H. (n.d.). *Educational Tourists' Familiarity With the Traditional Foods in Their Host Community: The Case of a Mediterranean Island*. Academic Press.

Tosun, C. (2000). Limits to Community participation in the tourism development process in developing countries. *Tourism Management*, *21*(6), 613–633. doi:10.1016/S0261-5177(00)00009-1

Tosun, C., Temizkan, S. P., Dallen, J. T., & Fyall, A. (2007). Tourist Shopping Experiences and Satisfaction. *International Journal of Tourism Research*, *9*(2), 87–102. doi:10.1002/jtr.595

Tourism Safety and Security. (2009). In *The Sage Handbook of Tourism Studies*. SAGE.

Tourism Update. (2018). *First Airbnb city collaborative agreement in Africa signed*. Retrieved from: https://www.tourismupdate.co.za/article

Tranter, K. A., Stuart-Hill, T., & Parker, J. (2014). *An Introduction to Revenue management for the Hospitality Industry: Principales and Pratices for the Real World*. Pearson.

Tran, X., & Phillip, S. (2010). Financial Performance and Vision Statements of National Tourism Organizations in Northeast Asia. *E-Ertr Review of Tourism Research.*, *8*(3), 57–70.

Tribe, J., & Snaith, T. (1998). From SERVQUAL to HOLSAT: Holiday satisfaction in Varadero, Cuba. *Tourism Management*, *19*(1), 25–34. doi:10.1016/S0261-5177(97)00094-0

Truong, T.-H., & Foster, D. (2006). Using HOLSAT to evaluate tourist satisfaction at destinations:The case of Australian holidaymakers in Vietnam. *Tourism Management*, *27*(5), 842–855. doi:10.1016/j.tourman.2005.05.008

Turner, L. W., Reisinger, Y., & McQuilken, L. (2001). How cultural differences cause dimensions of tourism satisfaction. *Journal of Travel & Tourism Marketing*, *11*(1), 79–101. doi:10.1300/J073v11n01_05

Um, S., Chon, K., & Ro, Y. (2006). Antecedent of revisit intention. *Annals of Tourism Research*, *33*(4), 1141–1158. doi:10.1016/j.annals.2006.06.003

United Nations World Tourism Organisation. (2018). *'Overtourism'? Understanding and Managing Urban Tourism Growth beyond Perceptions*. Author.

UNWTO. (2013). *World Tourism Barometer* (vol. 11). UNWTO.

Urry, J. (1990). *The Tourist Gaze: Leisure and Travel in Contemporary Societies*. Sage Publications.

Urry, J., & Lash, S. (1994). *Economies of signs and space*. Sage.

Useem, M. (1984). *The Inner Circle: large corporations and the rise of business political activity in US and UK*. Oxford University Press.

Vajcnerova, I., & Ryglova, K. (2010). The application of quality management principles into the management of the destination. The Business Review, 306-311.

Valle, P. O., & Silva, J. A., Mendes, J., & Guerreiro, M. (2006). Tourist Satisfaction and Destination Loyalty intention: A Structural and Categorical Analysis. Int. *Journal of Business Science and Applied Management*, *1*(1), 26–44.

Van Hull, P. (2016). Supply Chain-The Next Big Winner from the Sharing Economy. *SDC Executive*. Retrieved from: https://www.sdcexecutive.com/sourcing-procurement/

Vanhove, N. (2005). *The Economics of Tourism Destinations*. Elsevier.

Varghese, B. (2016). *Centre for Research-Christ University*. Academic Press.

Varghese, B. (2013). Intervention of destination management organization's in tourist destinations for branding, image building and competitiveness- A conducive model for karnataka. *International Journal of Investment and Management*, *2*(3), 50–56. doi:10.11648/j.jim.20130203.13

Veblen, T. (1899). *The Theory of the Leisure Class*. Mentor Boos.

Vellinga, R., & Abdelgalil, L. (2007). Trade liberalization and economic diversification: A dynamic CGE Model for the economy of Qatar. *Studies in Business and Economics.*, *13*(02), 1–17. doi:10.29117be.2007.0034

Venktess, K. (2017). South Africans rake in R24.37m from Airbnb. *Fin24*. Retrieved from: https://www.fin24.com/Tech/News/

Virgil, N., & Popsa, R. E. (2014). Business Tourism Market Developments. *Procedia Economics and Finance*, *16*(1), 703–712.

Visit Utah. (2019). *Airbnb Signup*. Retrieved from: https://www.visitutah.com/airbnb-signup/

Vitic, A., & Ringer, G. (2008). Branding post-conflict destinations: Recreating Montenegro after the disintegration of Yugoslavia. *Journal of Travel & Tourism Marketing*, *23*(2-4), 127–137. doi:10.1300/J073v23n02_10

Vora, K. H., & Sundaresh. (2003). *Mahabalipuram: a saga of glory to tribulations*. Academic Press.

Walker, A. (2019). *Airbnb believes it has a R10bn impact on South Africa's economy*. Retrieved from: https://memeburn.com/2019/07/

Walmsley, A. (2017). Overtourism and underemployment: a modern labour market dilemma. In *Responsible Tourism in Destinations 13*. Tackling Overtourism.

Walz, A., Lardelli, C., Behrendt, H., Grêt-Regamey, A., Lundström, C., Kytzia, S., & Bebi, P. (2007). Participatory scenario analysis for integrated regional modelling. *Landscape and Urban Planning*, *81*(1-2), 114–131. doi:10.1016/j.landurbplan.2006.11.001

Wang, W., & Fesenmaier, D. (2007). Collaborative destination marketing: A case study of Elkhart county, Indiana. *Tourism Management*, *28*(3), 863–875. doi:10.1016/j.tourman.2006.02.007

Wang, Y., & Xiang, Z. (2007). Towards a Theoretical Framework of Collaborative Destination Marketing. *Journal of Travel Research*, *46*(1), 75–85. doi:10.1177/0047287507302384

Wee Tracker. (2018). *Can Airbnb master the ABCs of Africa?* Wee Media.

WHC. (2015). Retrieved from: https://whc.unesco.org/ en/ sustainable development

White Paper. (1996). *The Development and Promotion of tourism in South Africa*. Department of Environmental Affairs & Tourism.

Winter, T. (2007). *Post-conflict heritage, postcolonial tourism: tourism, politics and development at Angkor*. Routledge. doi:10.4324/9780203946381

Woodside, A., & Mazanec, J. (2004). *Consumer Psychology of Tourism. Hospitality and Leisure*. CAB International.

World Commission on Environment and Development. (1987). Our common future (Brundtland report). Oxford, UK: Oxford University Press.

World Commission on Environmental and Development (WCED). (1987). *Our Common Future*. Oxford University Press.

World Heritage and Sustainable Development. (n.d.). Retrieved from UNESCO Centre: http://whc.unesco.org/en/sustainabledevelopment

World Summit on Sustainable Development. (2002). *Sustainable Tourism and the Earth Summit. Briefing Paper*. Author.

World Tourism Organisation. (2008). *UNWTO World Tourism Barometer*. Madrid: Author.

World Tourism Organization Guidebook. (n.d.). *Sustainable tourism for development*. Author.

World Tourism Organization. (1980). *Manila Declaration on World Tourism*. www.univeur.org/CMS/UserFiles/65.%20 Manila.PDF

World Tourism Organization. (2000). Highlights 2000 (2nd ed.). Author.

World Travel and Tourism Council. (1998). *Agenda 21 for the travel and tourism industry—towards environmentally sustainable development*. WTTC.

Wortman, A. (1996). Repensando las políticas culturales de la transición. In *Revista Sociedad*. Buenos Aires: Facultad de Ciencias Sociales, UBA.

WWF. (2002). World-Wide Fund for Nature International. UNEP World Conservation Monitoring Center.

Yahyavi, S., & Shaghaghi, S. G. (2012). Investigation of the Structural Characteristics of Kandovan Village from Sustainability Perspective. *Journal of Basic and Applied Scientific Research*, *2*(9), 9200–9204.

Yale, P. (1991). *From Tourist Attractions to Heritage Tourism. Huntingdon*. ELM Publications.

Yamaguishi, T. (1988). Exit from the Group as an individualistic solution to free rider problema in United States and Japan. *Journal of Experimental Social Psychology*, *24*(6), 530–542. doi:10.1016/0022-1031(88)90051-0

Yeoman, I. (2012b). Tourism2050: Scenarios for New Zealand. Victoria University of Wellington.

Yeoman, I., Robertson, M., & Smith, K. (2012). A futurist's view on the future of events. The Routledge handbook of events, 507-525.

Yeoman, I. (2011). Ten years of revenue management. *Journal of Revenue and Pricing Management*, *10*(1), 1–7. doi:10.1057/rpm.2010.53

Yeoman, I. (2012a). *2050: Tomorrow's Tourism*. Channelview. doi:10.21832/9781845413033

Yeoman, I. S., & McMahon-Beattie, U. (2017). The turning points of revenue management: A brief history of future evolution. *Journal of Tourism Futures*, *3*(1), 66–72. doi:10.1108/JTF-11-2016-0040

Yeoman, I., Palomino-Schalscha, M., & McMahon-Beattie, U. (2015). Keeping it pure: Could New Zealand be an eco paradise? *Journal of Tourism Futures*, *1*(1), 19–35. doi:10.1108/JTF-12-2014-0017

Yudice, G. (2002). *El recurso de la cultura*. Gedisa.

Yudice, G., & Miller, T. (2004). *Política cultural*. Editorial Gedisa.

Yuksel, A. (2001). Managing Customer Satisfaction and Retention: A Case of Tourist Destinations. Turkey. *Journal of Vacation Marketing*, *7*(2), 153–168. doi:10.1177/135676670100700205

Zach, C. (2008). Modeling and recognition of landmark image collections using iconic scene graphs. *European conference on computer vision.*

Zamora Gonzalez, J. (2004). Proceso de formación de la lealtad del huesped: Teoría y prueba empírica. *Studies and Perspective in Tourism., 13*(3-4), 197–221.

Zehrer, A., Pechlaner, H., & Raich, F. (2007). Satisfaction Profiles and Tourism Curricula - Tourism Organization under Study. *Tourism Review, 62*(1), 25–33. doi:10.1108/16605370780000159

Zhang, J. J., Joglekar, N., & Rohit, V. (2010). Developing measures for environmental sustainability in hotels: An exploratory study. *Cornell Center for Hospitality Research, 8*(1), 1–17.

Zhang, Z., Chen, R. J. C., Han, L. D., & Yang, L. (2017). Key factors affecting the price of Airbnb listings: A geographically weighted approach. *Sustainability, 9*(9), 3–13. doi:10.3390u9091635

Zouni, G., & Kouremenos, A. (2008). Do tourism providers know their visitors? An investigation of tourism experience at a destination. *Tourism and Hospitality Research, 8*(4), 282–297. doi:10.1057/thr.2008.30

Zupanovic, I. (2010). New postulates in strategic management of integrated tourist destination product in Montenegro. *Tourism and Hospitality Management.*

Zuzic, K. (2012). Destination management systems: Examples of Italian models and the proposal of an Istrian model. *Tourism and Hospitality Management.*

About the Contributors

Maximiliano Korstanje is editor in chief of International Journal of Safety and Security in Tourism (UP Argentina) and Editor in Chief Emeritus of International Journal of Cyber Warfare and Terrorism (IGI-Global US). Korstanje is Senior Researchers in the Department of Economics at University of Palermo, Argentina. In 2015 he was awarded as Visiting Research Fellow at School of Sociology and Social Policy, University of Leeds, UK and the University of La Habana Cuba. In 2017 is elected as Foreign Faculty Member of AMIT, Mexican Academy in the study of Tourism, which is the most prominent institutions dedicated to tourism research in Mexico. He had a vast experience in editorial projects working as advisory member of Elsevier, Routledge, Springer, IGI global and Cambridge Scholar publishing. Korstanje had visited and given seminars in many important universities worldwide. He has also recently been selected to take part of the 2018 Albert Nelson Marquis Lifetime Achievement Award. a great distinction given by Marquis Who´s Who in the world.

Babu George, holds a Ph.D. degree in Management Studies with a focus on Marketing (2004), a Doctor of Business Administration degree with a focus on International Business (2016), and an Ed.S. (Higher Education Innovation and Leadership) degree (2019). He is a graduate of Harvard University's Institute for Educational Management (IEM-2018 batch). For more than seventeen years, he has been serving the international higher education sector as a professor and senior academic administrator. Currently, he manages the international programs of Fort Hays State University's College of Business and also serves as Associate Professor of International Business. In the recent past, he served Swiss Management Center University as Professor of International Marketing Management, University of Nevada Las Vegas as Associate Graduate Professor of Hotel Administration, Alaska Pacific University as Associate Professor of Business Administration and the University of Southern Mississippi as Assistant Professor of Service Industry Management and Tourism. He is visiting professor to several other universities around the world.

Alexandru Nedelea, Ph.D., is an Associate Professor of Marketing at the University Stefan cel Mare of Suceava, Romania. He has written extensively for a wide range of academic and professional audience: Small and Medium Enterprises in Contemporary Society (KMV, Germany), International Tourism: World Geography and Developmental Perspectives (Abhijeet Publications, New Delhi, India), Tourism Marketing (Derc Publishing House, Tewksbury, USA), Knowledge Management Practices, Comparative Tourism Marketing Case Studies (Abhijeet Publications, New Delhi, India). He is editor-in-chief of the journals Ecoforum and Revista de turism (Journal of Tourism: Studies and Research in Tourism) and member of the editorial board of the journals Amfiteatru Economic (indexed in Thomson Reuters – ISI

Web of Knowledge), International Journal of Leisure and Tourism Marketing, Journal of Tourism: An International Research Journal, Tourism Today, Asia-Pacific Journal of Innovation in Hospitality and Tourism, Journal of Marketing and HR, Enlightening Tourism. A Pathmaking Journal, International Journal of Qualitative Research in Services, Journal of Tourism Security, International Journal of Safety and Security in Tourism, Ekonomika, Ekonomika, Journal for Economic Theory and Practice and Social Issues, Journal of Tourism Research, Service Management, The Journal Anuario de Turismo y Sociedad, Tourism Issues, European Journal of Tourism Research, Environmental Economics, The USV Annals of Economics and Public Administration,. He has been accorded membership in scientific associations like International Association of Scientific Experts in Tourism, International Association on Public and Non Profit Marketing, Romanian Marketing Association, International Network for Research in Consumer Behaviour Studies, Regional Science Association International, World Research Club (WRC), European Association of Research on Services (RESER), Marketing and Tourism Research Interest Committees – EuroMed Research Business Institute (EMRBI), Centre International de Recherche et d`Etudes Touristiques (CIRET). He is coordinator of Asian Studies Club Romania-China (www.romania-china.ro, www.romaniajaponia.ro, www.romania-korea.ro, www.romaniaindia.ro), Voyage Club (www.club-voiaj.ro), Bucovina Tourism Marketing Center (www.tourisminbucovina.ro), Volunteer Order (www.ordinulvoluntarilor.ro) and Marketing Club (www.marketingclub.ro).

* * *

Mohammed El Amine Abdelii is researcher in the University of Zaragoza and Salamanca as well as a recognized reviewer of 5 journals indexed in (Scopus, Web of Science) and reviewer books in (IGI Global, Taylor & Francis). He works as Principal Researcher at the University of South-Eastern, located in Norway.

Shazin Aboobacker is a Research Scholar at Christ University and also serves as a Research Associate at the University. Author´ works and publications verse on the role of DMO in tourist destinations and organic destination image.

Habib Alipour holds a PhD in Tourism. He is professor of tourism policy and planning at the Eastern Mediterranean University, Faculty of Tourism, Turkey. His research focuses on environmental impact of tourism, land use planning and tourism, institutional approach to tourism planning and development in developing countries with emphasis on Small Island States. His recent focus has been on co-management as pathway towards sustainable common pool resources and tourism.

Akhshik Arash is a graduate researcher at Eastern Mediterranean University. His research focuses on sustainable tourism, destination management and positive psychology of visitors. He is interested in visitor experience design specifically in protected areas.

Teresa Dieguez is Doctor on Economics, MBA on Prospective and Organizational Strategy, Master on Innovation and Technological Transfer, Specialist on Strategy and Entrepreneurship. Now she is Phd Student on Social Development and Sustainability.

Christo Fernandes was professor of tourism and hospitality at GOA Institute in India. His scope of research verses between branding image and sustainable consumption. Professor Fernandes has a vast experience in conference and events related to sustainable consumption and organic image of tourist destinations.

Marine Ferreira currently works in the commercial and administrative department at AHRESP. Her function is to provide support to member companies, as well as attract new members and also perform billing tasks. She started her professional career with an internship at Europe Hotel Villa Cortés in Tenerife, Spain where she worked in all of the departments of the hotel. Subsequently, and to improve her English skills, she signed up for an extracurricular activity as an intern at Blue-Arrow Company, in London, United Kingdom, in an F&B aspect. Once graduated, she started to work as a storekeeper at Stock Out Lda, at Armani Outlet, where she was quickly promoted to store manager. She performed several management tasks such as stock control & inventory, as well as marketing tasks such as visual merchandising and personalized customer advisement. As store manager, Marine also carried human resources tasks as she was involved in the interviewing and selection process for the store. With a bachelor's degree in Hotel Management and Administration from Institute of Industrial and Management Studies from Porto, Portugal, Marine decided to complete her curricular path by attending the Master's program in Hotel Management - Marketing and Commercial Direction from the Institute of Hotel and Tourism. She concluded the program with the master's thesis entitled: "The Impact of Revenue Management on the Growth of Hospitality in Portugal". In accordance to her master's degree, she was a paid intern at the Revenue Management Academy, where she performed administrative and management tasks, as well as supporting revenue management consultancy at the respective hotels. It was at this company that she developed a deeper understanding about Revenue Management, which was one of the main courses of her master's degree. As it was important to understand how revenue management worked in the hotel industry in Portugal, the internship was a complement to the thesis that she intended to defend in order to gain more concrete and realistic results.

Ashaq Hussain Najar is Assistant Professor at Central University Kashmir, India. He serves as researcher of the hospitality and tourism domain and presently working as assistant professor in Central University Kashmir.

Apleni Lwazi is a Tourism Lecturer at the University of Zululand (UniZulu) in the Republic of South Africa He is responsible for both undergraduate and postgraduate students. He holds an MA in Management in Management Sciences from University of Science & Technology Beijing, China and a PhD candidate at the University of Johannesburg. He has contributed 11 articles in approved scholarly journals and 1 book chapter. These are backed up by 1 local conference (Polokwane, 2016) & 3 international conferences (Atlanta, Casablanca, 2017 & Boston, 2019). His research work has been cited more than 21 times. Lastly, his research interest includes carbon tax, sustainable tourism development, community participation in tourism, 4th industrial revolution (more especially its implications in SA's tourism industry).

Nadia Mansour has been working as visiting researcher at the University of Salamanca, Spain. Now she is editorial board member in leading journals such as Journal of Entrepreneurship Education and Finance and management researcher University.

Bindu Nair is lecturer at Avinashilingam University for Women, Coimbatore, Tamil, India. She extensively has published papers in leading journals regarding cultural tourism and consumer´s loyalty.

Hamed Rezapouraghdam is associated professor at Bahçeşehir Cyprus University (School of Tourism and Hotel Management). He obtained a Master and Ph. D degree in Tourism Management at Isfahan University. Now he works themes related to corporate sustainability, sustainable tourism, ecology and serves as keynote speakers in many worldwide events and conference.

K. Sathiyabamavathy serves as lecturer ta Avinashilingam Institute of Home Science and Higher Education for women, Coimbatore, India. She is expert in themes related to ecology and sustainable tourism.

Priya Singh is doctor of hospitality and a budding researcher and currently working as an assistant professor in Jamia Millia Islamia, New Delhi, India.

Lakhvinder Singh is working as Assistant Professor of Tourism in Government College, Kaithal affiliated to Kurukshetra University, Kurukshetra, Haryana, India. He holds Ph.D. Degree in with UGC-NET qualification in Tourism Administration & Management. He has more than three years of teaching experience and has more than 30 research publications in his credit. He is in the Editorial/Advisory Board of different Journals of International and National repute. His areas of interest include tourist behaviour, tourism marketing, tourism impact, culture & religious tourism, e-tourism and ICT in tourism.

Nabila Smaili is Doctor in Business Administration and serves as associate Professor at the University of Tizi-Ouzou, Algeria. She has published a lot of contribution in leading journals and performed as speaker in countless conferences in her fields of expertise.

Unathi Henama Sonwabile is known on Twitter as @AcademicSangoma and is the leading tourism commentator in South Africa, on radio, television and print media. He is a tourism consultant in addition to being a regular feature in newspaper articles on tourism issues. He teaches tourism at the Tshwane University of University and has published extensively in local and international journal articles. He has attended local and international conferences and is a tourism all-rounder having published in almost all tourism avenues. He studied for his undergraduate studies in tourism at the Central University of Technology located in Bloemfontein. He finalised his Masters in Africa Studies at the University of the Free State.

Peter Tarlow was the director and rabbi of Texas A&M Hillel for thirty years. Tarlow retired on July 1, 2013, becoming its Rabbi Emeritus. Tarlow came to Texas A&M Hillel from Santiago, Chile and upon retirement he assumed the leadership of the Center for Latino – Jewish Relations. As distinguished from Rabbi Peter Tarlow, Dr. Peter E. Tarlow is a world-renowned speaker and author specializing in issues of tourism security, safety, and health. Tarlow is a founder and president of Tourism & More Inc. (T&M). Tarlow is a member of International Editorial Boards for academic tourism around the world. He trained police departments for three international Olympic Games and the 2014 World Cup. Tarlow lectures on security issues, life safety issues, and event risk management issues around the world. He also is a member of the faculty of Texas A&M 's Medical School and a police chaplain for the College Station, Texas police department.

Bindi Varghese is a Doctorate in Commerce, specializing in tourism. As an academician and tourism professional, she has over 17 years of academic and one year of Industrial experience. Currently, she is affiliated with Christ University, as an Associate Professor and is the research coordinator at School of Business Studies and Social Sciences. She has served many educational institutions in South India and has served as a national and international expert, for a decade among the educational institutions of India. Currently, she is actively associated with Indian Tourism Congress (ITC) and Kerala Development Society (KDS), New Delhi. With a passion for research, she has guided MPhil and Ph.D. scholars. The active researches undertaken include Impact Assessment Studies, Medical Tourism, Destination Management Organization and Ecological Studies. Dr. Bindi completed a major research project on the title "Strategic Intervention of Destination Management Organizations to Enhance Competitiveness of Tourism Destinations– A Model for Karnataka" funded by Christ University. The study has been presented to the Karnataka Government as an annexure. This action research proposes a strategic model to the Government of Karnataka to improve the quality of tourism in the State. Along with her academic expertise, she is also an issue editor for 'ATNA- Journal of Tourism Studies', published by Christ University, Bengaluru. She has extensively written for national and international peer-reviewed journals and very well recognized in the academic circle in India. She has authored one book; on Medical Tourism in India: by an international publisher in Germany, and has also contributed chapters to edited books and has published several articles in areas of destination Management, governance, Medical Tourism, E-Tourism etc. To her credit, she has edited a book on "Evolving Paradigms in Tourism and Hospitality in Developing Countries: A Case Study of India". The book is published by CRC Press, Taylor and Francis group -International publisher in US and released in 2018. As an educationist, with her experience and expertise, she is a frequent invited to address students from other institutions and at various industry summits. Apart from the academic journal, Dr. Bindi also writes for leading newspapers and travel magazines in tourism and hospitality.

Index

M

P

R

S

T

U

www.ingramcontent.com/pod-product-compliance
Lightning Source LLC
LaVergne TN
LVHW062156300926
849161LV00010B/2677